Hybrid Membrane Systems for Water Purification

Hybrid Membrane Systems for Water Purification

Technology, Systems Design and Operation

Rajindar Singh
USFilter PWS, Inc.,
Colorado Springs, USA

ELSEVIER

Amsterdam • Boston • Heidelberg • London • New York • Oxford
Paris • San Diego • San Francisco • Singapore • Sydney • Tokyo

ELSEVIER B.V.
Radarweg 29, PO Box 211
1000 AE Amsterdam
The Netherlands

ELSEVIER Inc.
525 B Street, Suite 1900
San Diego, CA 92101-4495
USA

ELSEVIER Ltd
The Boulevard, Langford Lane
Kidlington, Oxford OX5 1GB
UK

ELSEVIER Ltd
84 Theobalds Road
London WC1X 8RR
UK

First edition 2006

ISBN-13: 978-1-85617-442-8
ISBN-10: 1-856-17442-5

06 07 08 09 10 10 9 8 7 6 5 4 3 2 1

Printed in Great Britain

'Water is the source of life, sap for all things'

— Guru Nanak

'Air is our guru, water our father,
And the great earth our mother . . .'

— Guru Nanak

To My Parents,
S. Niranjan Singh and Kuldip Kaur.
In Memoriam

Contents

Preface

According to a recent UN report (The New York Times, July 26, 2005), more than a billion people lack access to safe water; polluted water contributes to the death of about 15 million children each year. The United Nations Educational Scientific and Cultural Organisation (UNESCO) estimates that by the year 2050, between two and seven billion people will face water shortages. Further, it is estimated that the amount of water available per person will shrink by a third during the next two decades. For example, according to a recent World Bank study, three out of five aquifers in India will be in critical condition by the year 2025, and, by 2050, demand will exceed all available supplies (Bloomberg News, October 28, 2005). These Malthusian predictions are slowly being recognised. Anyone who is from a developing country knows only too well about the chronic shortage of water and energy, especially in urban areas, even in countries that are not in drought affected regions of the world. No doubt membrane technology will play an important role in mitigating this acute problem around the world.

Membrane engineering is a 'Green Technology'; membrane processes are both energy efficient and environment friendly. In today's world with the emerging economies of India and China combined with those of the advanced economies, energy consumption and environmental degradation is a prime worry. According to the US Department of Energy, the United States with 4.8% of world's population consumed 25% of world's energy in 2003. Membrane technology, unlike traditional separation processes such as distillation and evaporation is independent of thermodynamically imposed efficiency limitations on heat utilisation. These energy limitations of traditional separation processes, along with water shortage predictions, provided the impetus for the development and the commercialisation of membrane processes during the last 40 years. For example, the reverse osmosis (RO) membrane process is ten times more efficient than the thermal processes for seawater desalination.[1]

Until the late 1970s, RO separation was mainly a demonstration technology aimed at proving itself a viable industrial process. Since then, and especially in the 1990s, membrane-based separation processes integrated with traditional separation systems have been successfully deployed in the desalination, wastewater and municipal water treatment, semiconductor, pharmaceutical, food, beverage, and gas separation industries. Membrane separation processes used in water treatment minimise waste and pollution unlike coagulation-clarification and ion exchange. It is becoming the technology of choice for making potable water around the world – desalination, reuse of municipal water, wastewater reclaim and recycle – as well as for pollution control treatment and zero discharge, thanks to innovations such as submerged membrane bioreactors and membrane degasifiers.

Developments in membrane technology such as high flux and high rejection RO membranes have reduced capital and operating costs drastically in the last 20 years. For example, the cost of seawater RO desalination has dropped from $1.75/m^3 ($6.50/kgallons) in 1990 to approximately $0.75/m^3 ($2.80/kgallons) today.[2] Standard seawater RO membrane elements have achieved flow rates of 28 m^3/day (7500 gallons/day) at 99.75% salt rejection compared to 15 m^3/day (4000 gallons/day) at 99.4% rejection 20 years ago. Further, efforts are underway to reduce energy consumption of seawater RO desalination systems to 1.7 kW-h/m^3 of RO permeate produced, more than an order of magnitude reduction from the earliest days of seawater RO desalting. Recent examples of large-scale membrane desalination plants include: (a) The largest seawater RO desalination plant in the world with a capacity of 165,000 m^3/day (300,000 m^3/day by end of 2006) containing 40,000 spiral-wound, thin-film composite membrane elements was commissioned in Ashkelon, Israel in 2005[2,3]; (b) A 50,000 m^3/day (13.2 million gallons/day) seawater RO desalination plant containing 1,000 hollow fibre membrane elements operating at 60% product water recovery and employing ultrafiltration for pretreatment was commissioned in Osaka, Japan in June 2005[4]; (c) A 45,000 m^3/day seawater RO desalination plant in under construction in northern Chile[4]; and (d) A 57,000 m^3/day RO desalination plant for treating groundwater is under construction in El Paso, Texas.[4]

Renewable energy sources such as wind power and solar energy have been successfully integrated with seawater RO desalination plants of low to medium capacity.[5] Such alternate sources can help mitigate water shortages in remote areas. Similarly, one can envisage fuel cells 'powered' by digester gas[6] supplying power to membrane systems for treating wastewater, purifying well water, or processing farm products, for example, processing maple juice by reverse osmosis.

The purpose of this book is to introduce engineers, consultants, designers, industrial managers and students to the broad field of membrane technology, including hybrid membrane systems, with a special focus on water purification, reclamation, and reuse.

References

1 W.J. Koros, Evolving beyond the thermal age of separation processes: membranes can lead the way, AIChE J., 50 (10) (2004), 2326–2334.

2 I. Barbour, Innovations in membrane technology are meeting challenges in water reuse and desalination, Ultrapure Water, 22 (6) (2005), 16–20.

3 Chemical Engineering Progress, November (2005), 13.

4 Membrane Quarterly, NAMS, October (2005), 14–15.

5 RE powers desalination, Filt. & Sep. May (2001), 30–31.

6 R. Singh, Will developing countries spur fuel cell surge? Chem. Eng. Progress, March (1999), 59–66.

Acknowledgements

This work consumed me for nearly two years, working weeknights and weekends when all other activities such as tennis, cricket, and general reading were brought to a screeching halt. I would like to thank my forbearing and loving family: my wife, Dr. Rashna Batliwala Singh, son, Samir Indar Singh, and daughter, Namrita Shirin Singh. I would also like to thank my editor, Dr. Geoff Smaldon, who tracked me down after I 'fled,' with no return address, to the mountains of Colorado from the hills of western Massachusetts. I am grateful for his encouragement and patience as I went through many self-imposed revisions and delays. I would like to acknowledge Mitch Summerfield, Vice-President and General Manager, USFilter PWS, Inc., Colorado Springs, for his support, enthusiasm, and official approval.

'Membrane technology' first inspired me when I read articles by A.S. Michaels, M.C. Porter, and S. Sourirajan as a graduate student at the University of Massachusetts at Amherst. This interest was sustained over the years while working at UMass, Burns & Roe, Engelhard, and USFilter. I would like to thank all the mentors and peers with whom I worked, especially Tan Ammerlaan, Sal Cabibbo, Don Guy, Bob Farrauto, Art Kaufman, Al Ko, Bob Laurence, Mike Malone, Mohammad Rahman, John Tembrock, Wolf Vieth and Chiu Wang.

Chapter 1

Introduction to membrane technology

'No power on earth can stop an idea whose time has come'.

Victor Hugo

1.1 Technology overview

Synthetic membrane processes perform versatile functions. In liquid separations, for example, they are used to separate particles that span four orders of magnitude from dissolved ions to bacteria (Figure 1.1). More than 95% of sales are for liquid separations. Since 1965, membrane technology has transformed from laboratory development to proven industrial applications. Membranes are used for desalination of seawater and brackish water, potable water production, and for treating industrial effluents. Membrane processes are used for the concentration of and purification of food and pharmaceutical products, in the production of base chemicals, and energy conversion devices such as fuel cells. Membranes are also used in medical devices such as haemodialysis, blood oxygenators, and controlled drug delivery products. Membrane separation processes are being increasingly integrated with conventional technologies as hybrid membrane systems to make products energy efficiently and with minimum environment impact.

Four developments are widely considered to be responsible for transferring membrane science from the laboratory to commercial reality since 1960:

(a) development of high-efficiency cross-flow membrane elements (modules) with large surface areas;
(b) creation of advanced materials with controllable capabilities to separate molecularly similar components (e.g. gases, salts, colloids, proteins);
(c) tailoring membrane morphology for controlling microscopic transport phenomena; and
(d) manufacturing membrane elements economically and reliably.[1]

Continuing advances in development of new membranes with better thermal, chemical, and improved transport properties have led to many new possible applications. Development of newer membrane modules in recent years provided a key stimulus for the growth of the membrane industry.

The market for membrane applications is determined by product quality and by the availability of new and better membranes and membrane processes, e.g. heterogeneous membranes made from polymers and ceramics

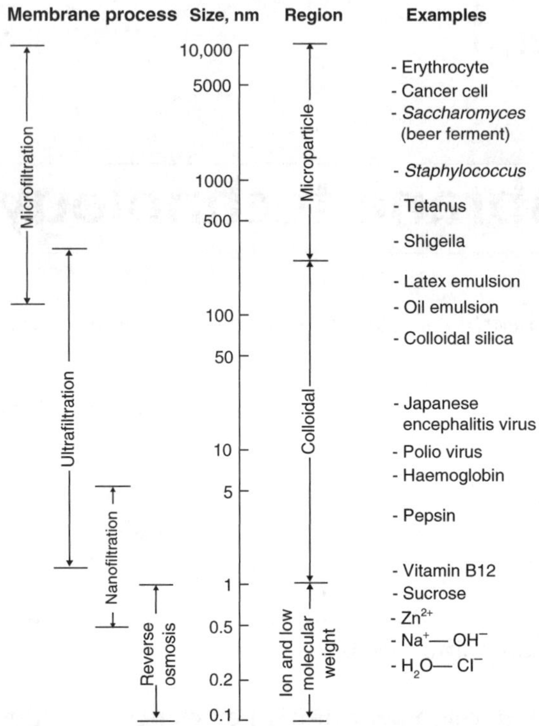

Figure 1.1 Membrane process designation by solute size

are being developed for handling non-aqueous feed streams.[1] Overall market for membranes and membrane systems is expected to approach $10,000 million by 2010. The largest market segment for membrane technology is in the biomedical sector followed by water purification (Table 1.1) with highest sales and moderate profits.[2] In the case of gas separation, pervaporation, membrane reactors, and bipolar membranes, the sales are smaller but the profits are higher.

1.2 Historical development

The first recorded study of membrane phenomena appears to have been by the Frenchman Abbe Nollet in 1748. He placed spirit of wine in a vessel, the mouth of which was closed with an animal bladder and immersed in water. Because it was more permeable

Table 1.1 Membrane equipment sales in 2000

Market segment	Sales (US $M)	Annual growth (%)
Haemodialysis/filtration	2,200	8
Blood oxygenator	350	2
Water desalination	350	10
Wastewater purification	400	10
Oxygen/nitrogen separation	100	8
Food processing	200	10
Biochemical industry	150	15
Electrochemical industry	150	8
Analytical/diagnostic	150	10
Miscellaneous	350	10
Total	4,400	>8

Source: Strathman

to water than to wine, the bladder swelled and sometimes even burst, demonstrating semi-permeability for the first time. Dutrechet introduced the term 'osmosis' in the 1820s to characterise the spontaneous flow of liquid across a permeable barrier. The first synthetic membrane was prepared by Fick in 1855, made from nitrocellulose. In 1861, Graham reported the first dialysis experiments with synthetic membranes. He also demonstrated that rubber films exhibited different permeabilities to different gases. During the next 30 years, Traube and Pfeffer prepared artificial membranes, and their work with osmotic phenomena led to the well-known van't Hoff osmotic pressure relationship.[3] Bechold coined the term 'ultrafiltration' in 1906, and Michaels pioneered modern ultrafiltration (UF) in the 1960s. Many of these membranes could be used as very fine particle or molecular filters. The microporous filters developed by Zigmondy and others around 1910 were asymmetric with a fine porous skin on the feed side and open structure on the permeate side. The tightest of the UF membranes were made from cellophane or regenerated cellulose. Membrane filters were first commercialised in 1927 by the

Table 1.2 Membrane milestones

Event	Scientist	Year
Osmosis	Abbe Nollet	1748
Laws of diffusion	Fick	1855
Dialysis, gas permeation	Graham	1861, 1866
Osmotic pressure	Traube, Pfeffer, van't Hoff	1860–1887
Microporous membranes	Zigmondy	1907–1918
Distribution law	Donnan	1911
Membrane potential	Teorell, Meyer, Sievers	1930s
Haemodialysis	Kolff	1944
Skinned membrane	Sourirajan and Loeb	1959
Membrane transport models	Kedem, Katachalsky, Lonsdale, Merten, Pusch, Sourirajan	1960–1970
Spiral-wound membrane element	Westmoreland, Bray	1965–1970
Hollow-fibre RO membranes	Mahon, Hoehn, and Milford	1965–1970
Thin-film composite membrane	Cadotte and Rozelle	1972

Source: Lonsdale[3]

Sartorius Company in Germany using the Zigmondy process. Reverse osmosis (RO) was first observed and studied in the 1920s. However, it remained unnoticed until rediscovered by Reid and his co-workers 30 years later. The first practical phenomenon of haemodialysis was demonstrated by Kolff in the 1940s. Membrane milestones are given in Table 1.2.

Until 1945, microporous membranes were used primarily for removing microorganisms and particulate matter from gaseous and liquid streams, for diffusion studies, and for estimating the size of macromolecules (or particle) and shape (spherical or linear). In the period after World War II, Juda and McRae pioneered electrodialysis (ED) using ion-exchange membranes. Soon after, ED systems were commercialised for brackish water desalination. In the early 1950s, RO (also called 'hyperfiltration') investigation for desalinating seawater was initiated by the Office of Saline Water to meet future water demands since microporous membranes were not suitable for desalination because of their large pore size. In the mid-1950s, Reid and Berton discovered polymeric membranes that exhibited high salt rejections, but they could not cast thin membranes without imperfections. The membranes were too thick, (about 6.0 μm), and exhibited water fluxes that were too low to be practical; the flux through a semi-permeable membrane is inversely proportional to the thickness of the membrane.

The first breakthrough came in 1959 when Sourirajan and Loeb discovered a method to make a very thin cellulose acetate (CA) membrane using the phase-inversion method.[4] This technique produces homogenous membranes with an asymmetric (or anisotropic) structure. The membranes were subsequently found to be 'skinned' when examined under an electron microscope by Riley in 1964.[3] The membranes consisted of a very thin, porous salt-rejecting barrier of CA, integrally supported by a fine CA porous substrate. The asymmetric membrane can be visualised as a pancake; when you are cooking observe that the top layer is skin-like after it is flipped up, and the now bottom side has large pores, as shown in Figure 1.2.

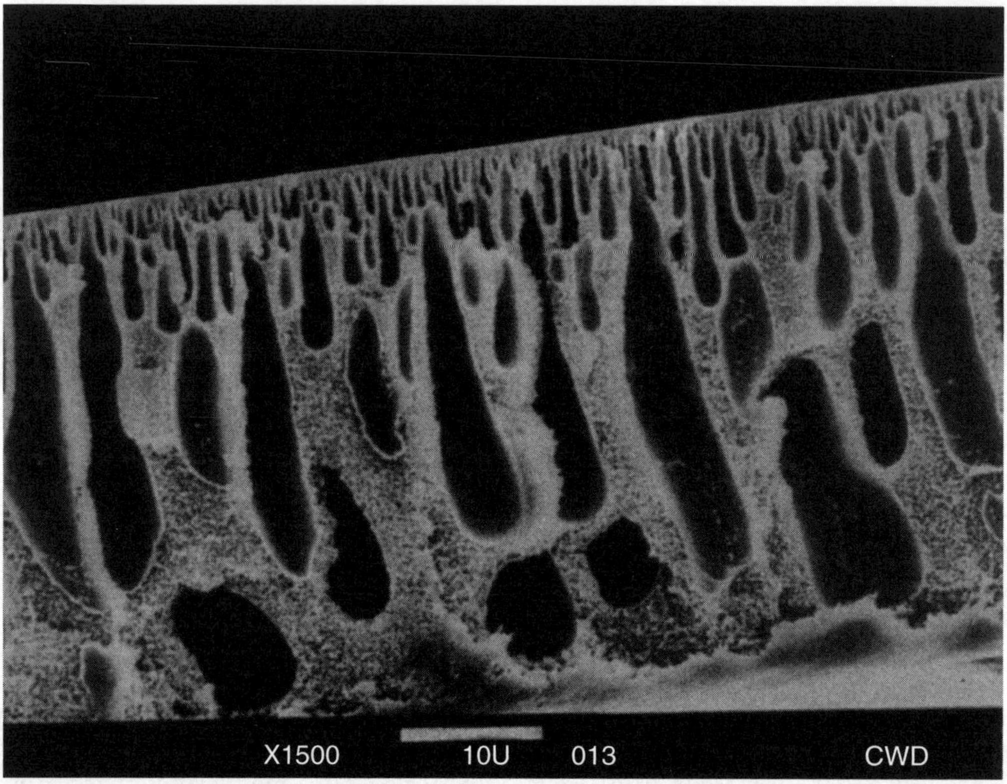

Figure 1.2 Electron micrograph of a UF asymmetric membrane showing the skin layer on the top (feed side) and macrovoids in the interior

These early Loeb–Sourirajan (L–S) membranes exhibited water fluxes that were more than 10 times those observed by Reid, and with comparable salt rejection.[5] The membrane was capable of producing a flux of 5–11 gallons per sq. foot per day (8–18 $l/m^3/h/m^2$) of 0.05% NaCl water from a brine containing 5.25% NaCl, under a pressure of 100–135 bar g. The high flux of these asymmetric membranes is due largely to their extreme thinness and porosity. The entire thickness of the membrane is about 0.1 mm, while the thickness of the active surface layer is only 30–100 nm.

Within a few years of the invention of the L–S membrane, development of RO membrane modules technique was realised. Haven and Guy developed tubular RO membranes in the mid-1960s. In the late 1960s, Westmoreland and later Bray invented the spiral-wound module, which was more efficient than the tube-in-shell module. The spiral-wound membrane can be viewed as a plate-and-frame arrangement that has been rolled up. The original module had a single leaf of membrane whereas modern spiral-wound modules contain multi-leaf membranes.

At the same time scientists and engineers at Dow and DuPont were examining new RO membrane materials and module designs. Mahon and others at Dow developed cellulose triacetate (CTA) hollow-fibre (HF) membranes. Although technically successful, this approach was not competitive commercially. The HF polyamide (PA) membranes developed by Henry Hoehn and George Milford at DuPont in the late 1960s turned out to be quite competitive with the L–S type CA spiral-wound membrane. The PA hollow fibres are solution spun into very fine hollow fibres. The inside diameter of the fibre was 42 μm and the outside diameter was 85 μm. The fibres had a skinned (asymmetric) structure on the shell side. The water flux was well below that exhibited by CA membranes; the aromatic PA membrane was less water permeable but more salt retentive than the CA membrane. However, this low flux was compensated by the several hundred thousand of these fibres that were incorporated into modules containing several thousand square feet of membrane area. The development of skinned hollow-fibre RO membranes led to the development of hollow-fibre UF and microfiltration (MF) membranes shown in the photograph in Figure 1.3.

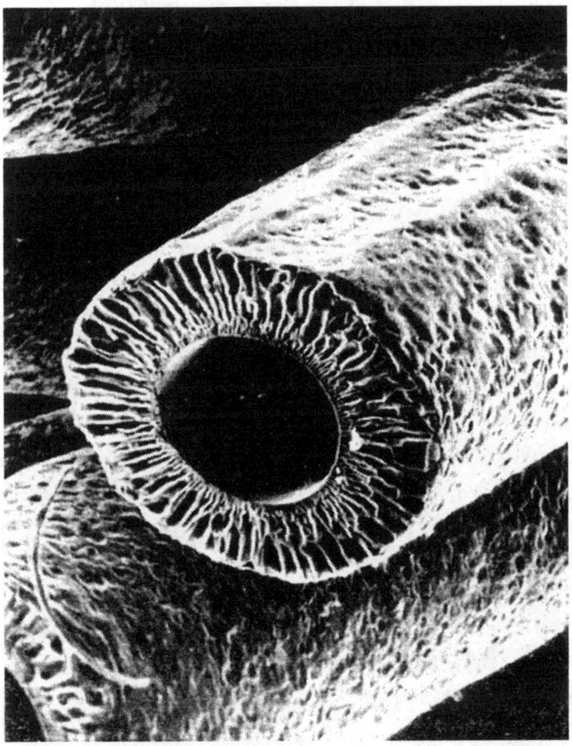

Figure 1.3 Cross section photograph of a UF hollow-fibre asymmetric membrane with the skin layer on the tube side. Source: Lonsdale[3]

Cadotte and Rozelle developed the so-called thin-film composite (TFC) RO membranes from a new class of membrane materials in the early 1970s.[6] It was made by depositing a very thin layer of a salt-rejecting PA polymer onto the surface of a suitable fine porous substrate such as polysulphone UF membrane by an interfacial polymerisation reaction. In 1976, Cadotte produced TCF membranes that showed high flux and high rejection of divalent ions but high permeability to aqueous chloride solutions.[7] These loose RO membranes were given the name 'nano-filtration' membranes by Petersen in 1984. Thus emerged another membrane process that is 'a most fascinating extension of reverse osmosis technology' in the words of Petersen.

The invention of the RO membrane stimulated both commercial and academic interest in other membrane processes and applications. For example, Henis and Tripodi, the co-inventors of HF gas separation membranes, made industrial gas separation economical by drawing on the experience of developments in RO membranes and modules.[8] Technical milestones for the commercial development of various membrane processes are given in Table 1.3.

Packaging of membranes in suitable element or modules that began in the 1960s resulted in several membrane modules by 1980: spiral-wound modules with CA and thin-film composite (TFC) polyamide membranes, PA hollow-fibre (HF) membranes, and UF polysulphone HF membranes. These membranes and modules found their own niches in the market place; broadly speaking, CA spiral-wound membranes for industrial and municipal water treatment, TFC spiral-wound RO membranes for water desalination and high purity water production, hollow-fibre RO membranes for seawater desalination, and hollow-fibre UF and MF membranes for industrial applications.

Table 1.3 Commercial development of membrane processes

Membrane process	Country	Year	Application
Microfiltration	Germany	1920	Laboratory use
Ultrafiltration	Germany	1930	Laboratory use
Haemodialysis	Holland	1950	Artificial kidney
Electrodialysis	USA	1955	Desalting
Reverse osmosis	USA	1965	Water desalination
Ultrafiltration	USA	1971	Concentration of macromolecules
Gas separation	USA	1979	Hydrogen recovery
Pervaporation	Germany, Holland	1982	Dehydration of organic solvents
X-flow microfiltration	Australia, USA	1980	Water treatment
Nanofiltration	USA	1986	Water softening
Electrodeionisation	USA	1987	Demineralisation

A historical review of membrane technology would be incomplete without a brief discussion of the subject of mechanism of water transport through RO membranes. The two leading transport mechanisms are the preferential sorption–capillary flow (PSCF) model and the solution–diffusion (SD) flow model. The PSCF model was developed in the 1960s by Sourirajan based on the assumption that the RO membrane surface has pores whereas the SD model assumes the membrane surface is pore free. Subsequently, it was shown that RO membranes have a two-size (bi-modal) distribution of pores in the RO membrane surface; predominantly small pores ($\leq$10 Å) thought to exist in perfect or ideal membranes; and occasional large ($\geq$100 Å) pores that are attributable to the inevitable existence of defects in the skin layer.[9]

The SD model was originally employed to explain the permeation of gases, vapours, and organic liquids through dense (non-porous, homogeneous) but defect-free membranes by Ferry.[10] The solution–diffusion mechanism involves molecular scale interactions of the permeating solutes (dissolved species) with the membrane. The model assumes that each component is sorbed by the membrane at one interface, transported by diffusion across the membrane through the voids between the polymeric chains (the so-called free volume), and desorbed at the other interface.

To summarise, the development and remarkable success of commercial membrane processes for liquid separations – reverse osmosis (RO), nanofiltration (NF), ultrafiltration (UF), and microfiltration (MF) – would not have been possible without the discovery and subsequent development of high-flux and extremely thin (skinned) CA membranes by Srinivas Sourirajan and Sidney Loeb, and culminating in the development of thin-film composite PA membranes by John Cadotte. Much of the newer and current membrane technology for NF, UF, pervaporation (PV), and gas separation (GS) systems was derived from the work on RO in the 1950s and 1960s.

1.3 Membrane separation characteristics

Membrane separation is characterised by simultaneous retention of species and product flow through the semi-permeable membrane.

Membrane performance is based on its high selectivity and flux; good mechanical, chemical, and thermal stability of membrane material, minimal fouling during operation, good compatibility with the operating environment, and defect-free production. Membranes responsible for such selective behaviour under mild operating conditions consist of porous or non-porous polymers, and ceramic or metal films with anisotropic or symmetric structures with homogenous, heterogenous, or composite materials, ionic membranes as well as liquid films with selective carrier components.[2]

1.3.1 Membrane separation basics

The membrane process

A membrane is defined as a permselective barrier between two homogeneous phases. Schematic diagrams of membrane separation processes are shown in Figure 1.4. The membrane processes are continuous steady-state operations consisting of three streams; feed,

PROCESS	SEPARATION CONCEPT	MATERIALS PASSED	DRIVING FORCE
Micro–filtration	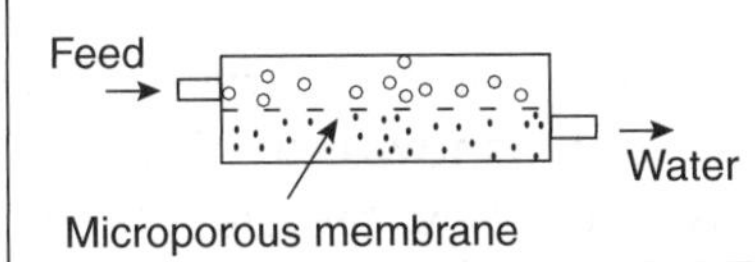 Feed; Water; Microporous membrane	Water and dissolved species	Pressure difference, Typically 10 psi
Dialysis	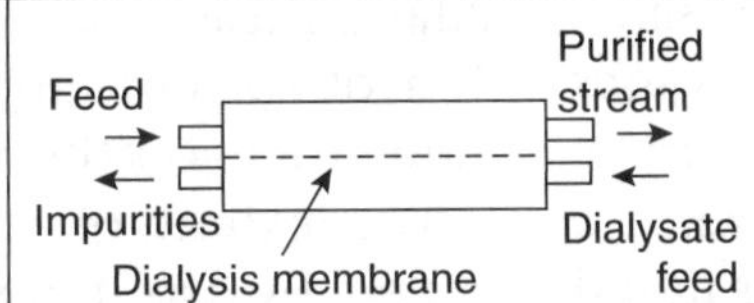 Feed; Impurities; Purified stream; Dialysate feed; Dialysis membrane	Ions and low–molecular–weight organics (Urea, etc,)	Concentration difference
Electro–dialysis	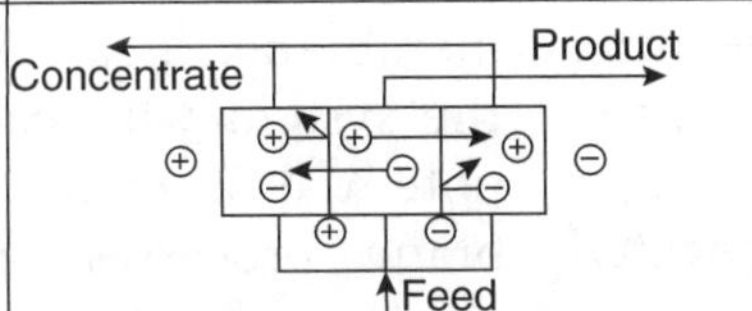 Concentrate; Product; Feed	Ions	Voltage, Typically 1–2V/Cell pair
Reverse osmosis	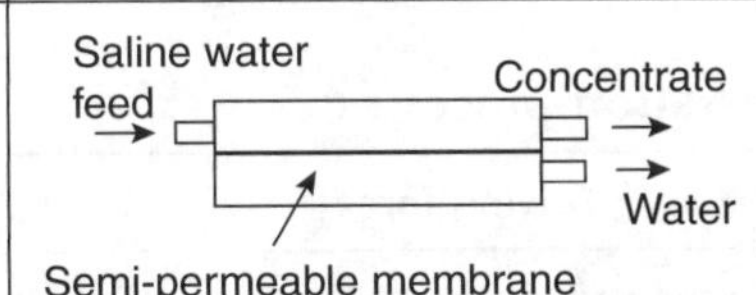 Saline water feed; Concentrate; Water; Semi-permeable membrane	Water	Pressure difference, Typically 100–800 psi
Ultra–filtration	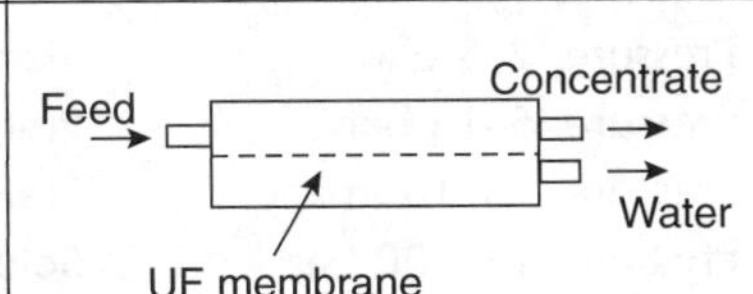 Feed; Concentrate; Water; UF membrane	Water and salts	Pressure difference, Typically 10–100 psi
Gas separation	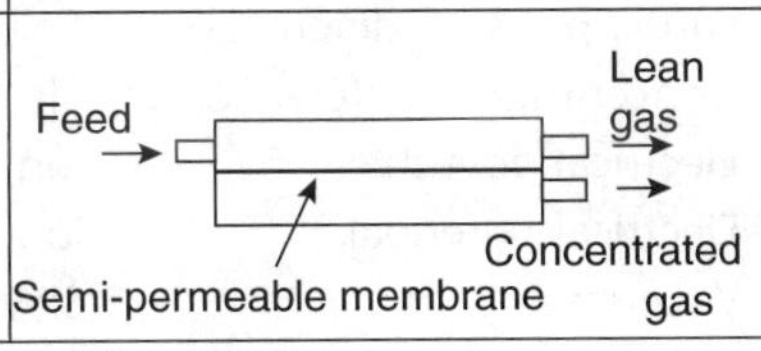 Feed; Lean gas; Concentrated gas; Semi-permeable membrane	Gases and vapours	Pressure difference, 1–100 atm

Figure 1.4 Membrane separation processes. Source: Lonsdale[3]

product (permeate), and reject (retentate). The dashed line in the process boundary is the semi-permeable barrier that selectively allows the passage of some components but not others, and it allows some components to pass through more rapidly than others. For example, in the case of RO, the membrane allows the passage of water but rejects almost all the ions and salts resulting in a concentrated retentate on the feed side of the membrane, and product on the other side of the membrane. Thus, the most permeable component gets enriched in the permeate stream while the least permeable component gets enriched in the retentate stream.

No one membrane process is suitable for every fluid stream owing to the nature of feed and product requirements as well as the nature of the membrane process and the membrane designed for that process (see Table 1.4 and Figure 1.4). Even for the same membrane process, the performance varies depending on the type of feed. For example, when a UF membrane is used to sterilise a water supply or to concentrate a viscous liquid such as a fruit puree under the same operating conditions of feed flow rate and pressure, the permeate flow is 70–80% lower in the case of the latter due to the concentration of rejected solutes on the membrane surface. Thus there is a clear distinction between the intrinsic membrane characteristics and its performance in a particular application. The main membrane processes are discussed in detail later in the chapter.

Membrane transport

The transport properties of solutes through a membrane is determined by the permeability of the membrane and by a driving force. The semi-permeable membranes allow preferential passage of certain substances under the influence of a driving force. In most cases, the driving force is either a pressure difference or a concentration difference between the feed side and the product side. Sources of various driving forces are shown in Figure 1.4 and Table 1.4. Other flow factors affecting membrane transport are axial diffusion of solute, longitudinal convection of solvent, and a flow-induced drag on the particle on the membrane surface.

The extent of the driving force is determined by the potential gradient across the membrane. If no external forces are applied, the system will reach equilibrium when the potential difference becomes zero. The membrane processes are steady-state processes

Table 1.4 Classification of membrane separation processes

Process	Pore size	Driving force	Transport mechanism
Microfiltration	0.05–10 μm	Pressure, 1–2 bar	Sieving
Ultrafiltration	0.001–0.05 μm	Pressure, 2–5 bar	Sieving
Nanofiltration	<2.0 nm	Pressure, 5–15 bar	Preferential sorption–capillary flow
Reverse osmosis	<1.0 nm	Pressure, 15–100 bar	Preferential sorption–capillary flow
Gas separation	Non-porous	Pressure, 15–100 bar	Solution–diffusion
Pervaporation	Non-porous	Partial pressure difference	Solution–diffusion
Dialysis	10–30 Å	Concentration difference	Sieving–diffusivity difference
Electrodialysis	MW < 200	Electrical potential	Ion migration
Electrodeionisation	MW < 200	Electrical potential	Ion migration
Membrane contactor	Porous, hydrophobic	Vacuum	Sieving

and not equilibrium processes such as traditional separation processes as distillation, evaporation, and crystallisation. When the driving force is constant, the flow through the membrane is constant once steady state has been established. The relationship between permeate flow or flux (J) and the driving force (X) is given by $J = k \times X$ where, k is the proportionality factor, and X the potential gradient per unit membrane thickness.

Mass transport through a membrane has been described by several semi-empirical mathematical models including Fick's, Hagen–Poisseuille's, and Ohm's laws. For pressure-driven membrane separation processes (RO, NF, UF, MF, GS), the transport relationship is given by:

$$J = K.\Delta P/t \qquad (1)$$

where, J is the transmembrane flux, K the membrane permeability constant, ΔP the pressure difference across the membrane, and t the membrane thickness. The importance of the thickness of the membrane is evident from equation (1) since flux is inversely proportional to thickness. The Loeb–Sourirajan RO membranes produced by the phase-inversion technique have an effective skin thickness of 0.1–0.2 μm that makes it possible to achieve acceptable fluxes (2–20 $l/m^2/h$) at reasonable feed pressures (30–60 bar g) for brackish water desalination. Ultrafiltration membranes with a similar but less tight skin structure (pore size 2–100 nm vs. <0.5 nm for RO membranes) have fluxes on the order of 15–150 $l/m^2/h$ at feed pressures of 1–5 bar g. The ability to minimise thickness without introducing defects relies upon controlling the membrane morphology during fabrication.

The flux also depends on the physical make up of the membrane and the pore size. Liquid flux for membrane processes is shown in Figure 1.5. The data show how the permeation rate varies with the size of the species and pore size (implicit in type of membrane). The ordinate represents the flux of water per unit pressure gradient. Since the pore radius of an RO membrane is less than 0.5 nm, water molecules whose radius is about 0.1 nm, can pass through the membrane freely while dissolved ions and organic solutes (e.g. sucrose) cannot. These solutes are either rejected at the membrane surface, or are more strongly attracted to the solvent water phase than to the membrane surface. The 'preferential sorption' of water molecules at the solvent–membrane interface, which is caused by the interaction force working between the membrane, solvent, and solute is responsible for the separation.[9]

As the pore size decreases and tends towards a non-porous skin structure, the transport mechanism changes from convective flow through pores to solution–diffusion in the membrane polymer. The latter is the transport mechanism in GS and PV. A membrane process property related to flux is product recovery defined below. In liquid separations, especially RO and NF, it is one of the key performance indicators.

$$\%\text{Recovery} = \frac{\text{Product flow}}{\text{Feed flow}} \times 100$$

Membrane selectivity

Membrane separation is governed both by the chemical and physical nature of the membrane material. Separation occurs because of differences in size and shape, chemical properties, or electrical charge of the substances to be separated. Microporous membranes control separation by size and discrimination. In the case of porous membranes, species small enough to enter the pores are transported by convective flow, and the membrane does not actively contribute to selectivity of the process while non-porous membranes rely on sorption and diffusion through the membrane. Furthermore, membranes do not have a sharp

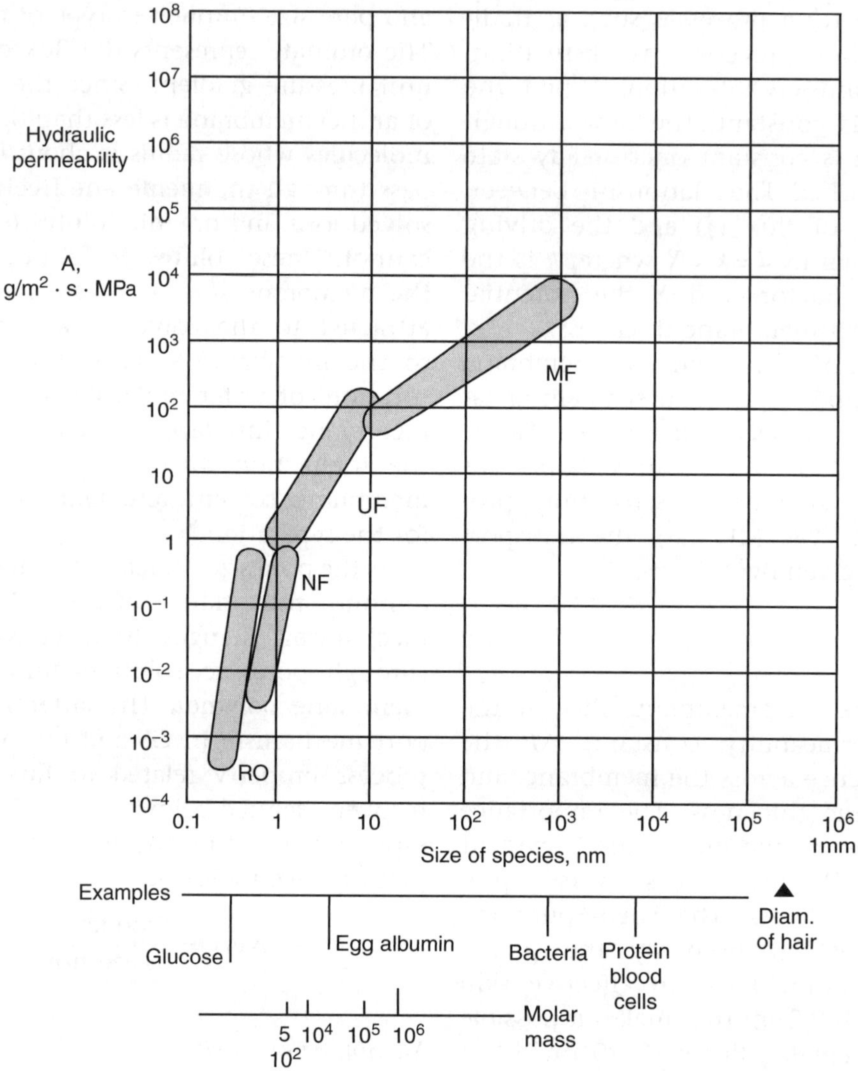

Figure 1.5 Solvent flux vs. solute size and solute type for various separation processes

separation between species. In the case of UF membranes, for example, a rule of thumb is that the nominal molecular cut-off rating of the membrane selected should be an order of magnitude lower than the average size of the solute to be rejected or retained. Synthesis of UF membranes with a sharp cut-off has been an elusive goal since the 1960s.[11] Membrane selectivity is defined by the rejection coefficient, which is a reliable indicator of the separating ability of a membrane process:

$$\%\text{Rejection} = \frac{\left\{\begin{array}{l}\text{Feed/Bulk solute concentration}\\ -\text{ Product solute concentration}\end{array}\right\}}{\text{Feed/Bulk solute concentration}} \times 100$$

For UF and MF processes, retention coefficient defined below is a better measure of selectivity because the build-up of solute particles on the membrane surface (concentration polarisation) affects solute passage through the membrane:

$$\%\text{Retention} = \frac{\left\{\begin{array}{l}\text{Membrane surface solute concentration} \\ -\text{Permeate solute concentration}\end{array}\right\}}{\text{Membrane surface solute concentration}} \times 100$$

Membrane performance is a trade-off between membrane selectivity and membrane productivity. The greater the selectivity, the smaller the permeate flux or productivity. This relationship for various RO membranes used with dilute NaCl solution is shown in Figure 1.6. The shaded regions refer to different feed concentrations and to different types of membranes. The data is fairly independent of the feed concentration but is a function of the physical and chemical properties of the membrane.

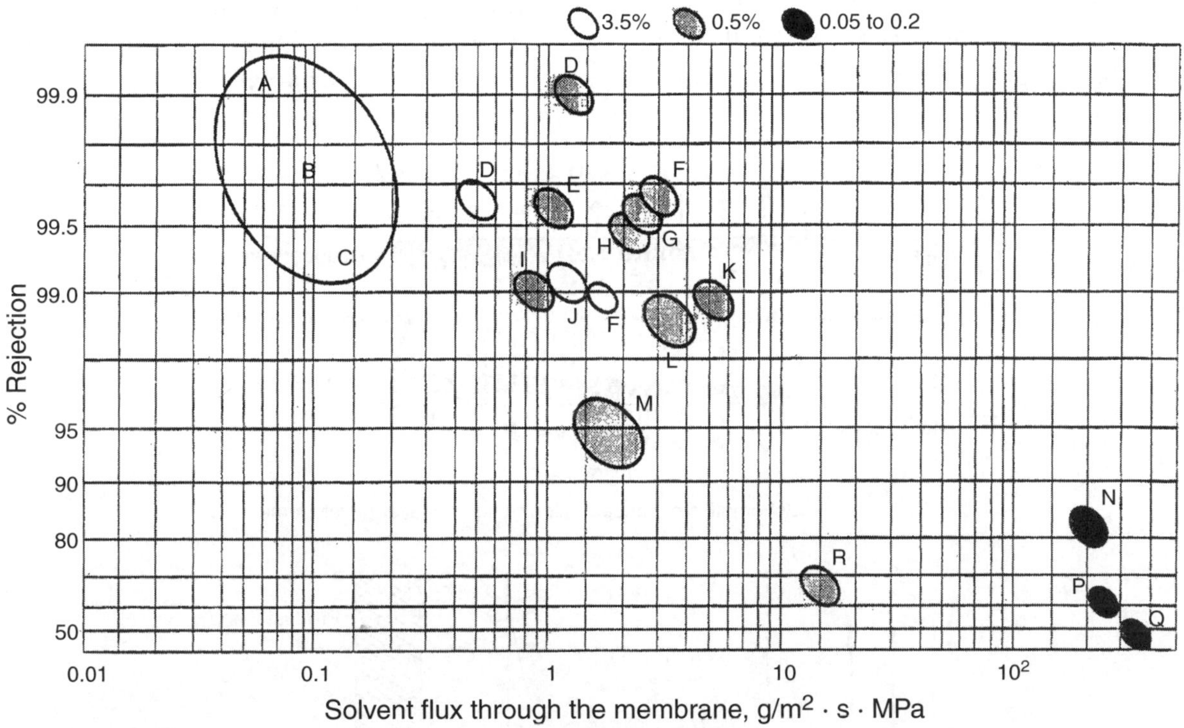

Membranes from various manufacturers: A, Hollosep–cellulose triacetate hollow fibre membrane (Toyobo); B, sulphonated polysulphone composite hollow fibre membrane (Albane International); C, B10–aromatic polyamide hollow fibre membrane (Du Pont); D, PEC-1000–composite flat-sheet membrane (Toray); E, NS-200–composite polyfurfuryl alcohol membrane; F, FT-30–composite polyamide flat-sheet membrane (Film Tec/Dow); G, NTR-7199–composite polyamide/polyurea flat-sheet membrane (Nitto Denko); H, TFC-803 (PA-300)–composite polyetheramide flat-sheet membrane (Fluid Systems/Signal; I, NS-100–composite polyurea flat-sheet membrane; J, TFC-801 (RC-100)–composite polyetherurea flat-sheet membrane. (fluid systems/Signal); K, NTR-7197–composite polyamide/polyurea flat-sheet membrane (Nitto Denko); L, B-15—asymmetric polyamide flat-sheet membrane (Du Pont); M, asymmetric cellulose acetate flat-sheet membrane (FilmTec/Dow); P, Romembra SU-composite polyamide flat-sheet membrane (Toray); Q, NF-50–composite flat-sheet membrane (FilmTec/Dow); and R, NTR-7250–composite polyvinyl alcohol flat-sheet membrane (Nitto Denko). Species: Sodium chloride Temp. 25°C

Figure 1.6 RO membrane performance data for various polymers. Source: E.K. Lee, Encyclopedia of Physical Science and Technology, Academic Press, 1987

Concentration polarisation

In pressure-driven liquid-phase processes such as RO, NF, UF, and MF, solutes and particles transfer to the membrane surface with the solvent by convection. The retention of rejected species at the membrane surface inevitably results in the formation of a secondary or a dynamic membrane due to concentration polarisation (CP). The solute mass balance based on the steady-state solute concentration profile is shown in Figure 1.7 is as follows:

Rate of convection towards the membrane
= Rate of diffusion back into solution +
Rate of permeation

i.e. $JC = D.\mathrm{d}c/\mathrm{d}x + J.C_\mathrm{p}$

The likely mechanisms for inducing back transport of solutes are:

(a) molecular (Brownian) diffusion,
(b) shear-induced diffusion, and
(c) interaction-induced migration (due to electro-kinetic and surface forces.[12] Diffusive back transport is the most common back-transport mechanism,[13] and is represented by the mass transfer coefficient, $k = D/\delta$ where D is solute diffusivity and δ is the boundary-layer thickness. The possible effect of slip flow velocity on diffusive back transport is discussed in Chapter 5.[14,15]

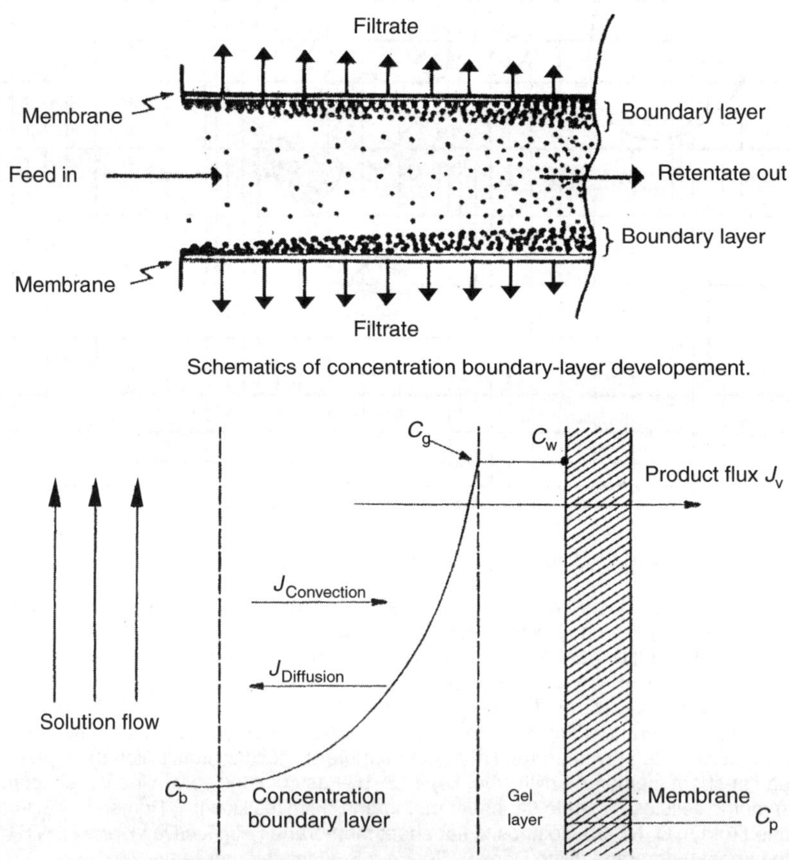

Figure 1.7 Concentration polarisation model schematic

According to the Film Theory Model:

$$JC - J.C_p = D.dc/dx \quad (2)$$

where, J is the solvent flux, C the feed solute concentration, C_p the permeate solute concentration, D the solute diffusivity, and x length co-ordinate. Integration over the boundary layer ($x = 0$ to δ, $C = C_w$ to C_b) gives for a 100% rejection ($C_p = 0$):

$$J = k \ln(C_w/C_b) \quad (3)$$

where, C_w and C_b are the concentrations of the solute at the membrane surface and in the bulk, and as shown in Figure 1.7. In the case of <100% rejection, equation (3) is modified to:

$$J = k \ln(C_w - C_p/C_b - C_p) \quad (4)$$

The polarisation modulus, C_M is defined as:

$$C_M = C_w/C_b = \exp(J/k) \quad (5)$$

The magnitude of C_M depends on the solute diffusivity. Both solute flow and concentration polarisation increase as C_M increases. Hence any technique that decreases C_M such as turbulence and surface shear would increase solvent flux through the membrane as also shown by the slip-flow model in Chapter 5.

According to equation (3), the flux is directly related to the boundary layer mass transfer coefficient, k. Both k and C_w can only be determined experimentally. In the case of k, boundary layer mass transfer is calculated using the well-known relationship between dimensionless numbers: Sherwood, Schmidt and Reynolds numbers.[16] A comprehensive recent review is provided by Fane.[12] Wall concentration, C_w is measured by extrapolating the semi-log plot of J vs. C_b in equation (3) to x-intercept (zero flux) shown in Figure 1.8 when the flux is independent of pressure so that C_w becomes constant (Region III in Figure 1.9).

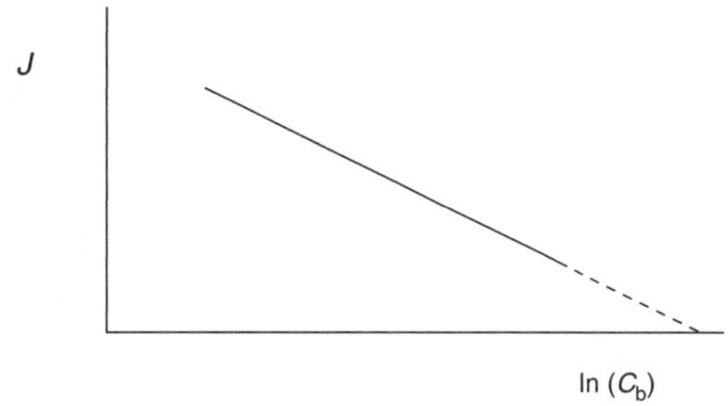

Figure 1.8 Flux (J) vs. feed/retentate concentration (ln C_b)

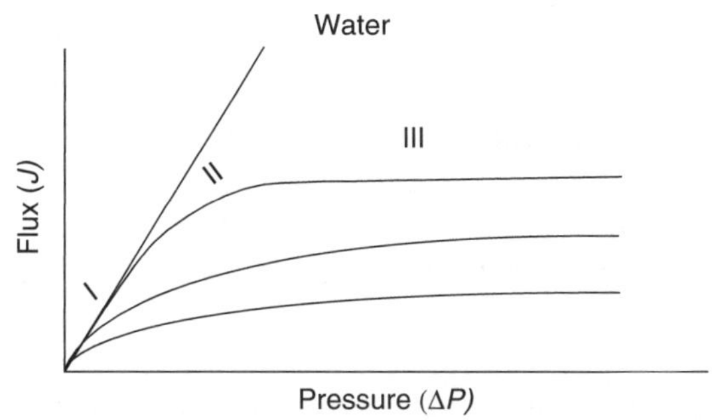

Figure 1.9 Flux (J) vs. pressure (ΔP)

Concentration polarisation results in substantial decline in flux, e.g. when the permeate line is closed the flux drops to zero, and C_w reverts to C_b.[12] The secondary membrane or gel layer formed during concentration polarisation does not adhere strongly to the membrane surface, and the effect can be minimised by turbulence in the membrane channel. However, concentration polarisation often leads to fouling, which is irreversible and can only be rectified by cleaning. In the case of biofouling, it is often difficult to restore flux in spite of repeated fouling. Fouling is discussed in detail later in this chapter and in Chapter 2.

1.3.2 Membrane materials and properties

Synthetic membranes can be made from a large number of materials. A list of common polymers used in membrane separations is given in Table 1.5. Membranes should combine high permeability and high selectivity with sufficient mechanical stability. Not all materials are suitable for all pressure-driven membrane processes with the exception of cellulose acetate and its derivatives. Preferably a membrane should have both hydrophilic and hydrophobic characteristics. In the case of porous UF and MF membranes, polymer (or other material) is selected based on the processing requirements, fouling tendency as well as thermal and chemical stability of the membrane. In the case of non-porous GS and PV membranes, the material choice is based on flux and selectivity. Several methods of manufacturing synthetic membranes are given in Table 1.6. Each method produces different membrane morphology; different porosity, pore size distribution, and ultrastructure. Membrane formulation techniques are beyond the scope of this book. These are discussed in detail in several texts.[17–19]

Polymeric membranes

Polymers provide a range of properties important for separations, and modifying them can improve membrane selectivity. Since it is desirable to use a polymer, which is chemically and thermally stable, materials with high glass transition temperatures, T_g, high melting points and high crystallinity are preferred. T_g, is a property that defines the state of a polymer; a glassy state below T_g, and a rubbery state above T_g. Glassy polymers have stiffer polymer backbones and, therefore, let smaller molecules, such as hydrogen and helium, pass

Table 1.5 Characteristics of common polymeric membrane materials

Polymer	Structure	T_g (°C)	Melting temp (°C)
Polythylene	$-(CH_2)_n-$	−60 to −90	137–143.5
Poly (vinylidine fluoride)	$-(CH_2CF_2)_n-$	−40	160–185
Polypropylene	$-(CH_2CH(CH_3))_n-$	−10	167–170
Polycarbonate	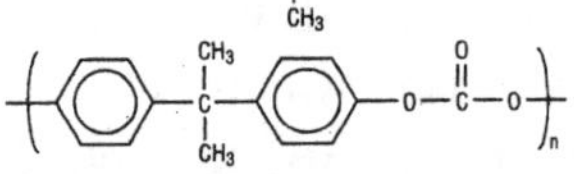	150–155	240
Teflon	$-(CF_2)_n-$	−113	327
Cellulose acetate	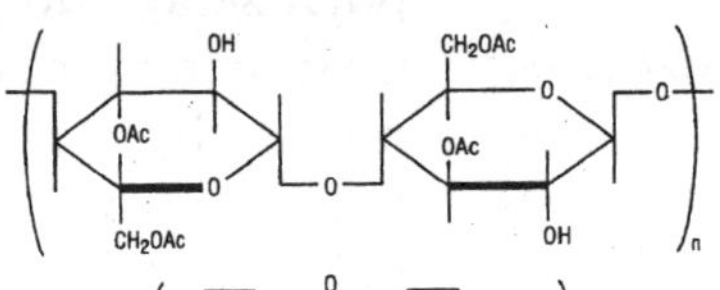	69	230
Poly (ether sulphone)	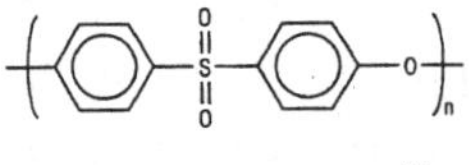	225	NA
Polysulphone	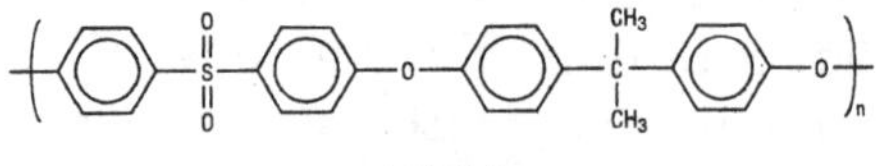	190	NA
Poly (vinyl alchol)	$-(CH_2CH(OH))_n-$	65–85	228–256
Polyacrylonitrile	$-(CH_2CH(CN))_n-$	80–104	319
Poly (phenylene sulphide)	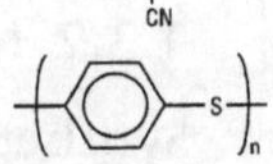	85	285

NA: Not available.
Source: Scott

Table 1.6 Synthetic membrane manufacturing methods

Process	Materials
Phase inversion:	
Solvent evaporation	CA, PA;
Temperature change	PP, PA;
Precipitant addition	PS, nitrocellulose
Stretching sheets of partially crystalline polymers	PTFE
Irradiation and etching	Polycarbonate, polyester
Moulding and sintering of fine grain powders	Ceramics, metal oxides, PTFE, PE

faster while large-sized hydrocarbons permeate the membrane more slowly. In the case of rubbery polymers, the permeabilities of hydrocarbons are higher than the much smaller gas molecules. A rubbery polymer is an amorphous polymeric material that is above its softening or glass transition temperature, T_g under the conditions of use. The degree of crystallinity has a large influence on the mechanical and transport properties of a polymer. Crystalline polymers show a high chemical and thermal stability and are usually used for MF membranes. However, as a polymer becomes more stable, it becomes more difficult to process into a membrane as it is no longer soluble in suitable solvents. The molecular structure and thermal characteristics of membrane materials are given in Table 1.5.[20] Amorphous glassy polymers such as PS are used for UF membranes as they are convenient for control of the small pore size required and display a high T_g. Polymers, such as polypropylene (PP) and polyamides (PA) are semi-crystalline, consisting of an amorphous and a crystalline fraction.

Aromatic polyamides are superior to aliphatic polyamides because of their superior mechanical, thermal, chemical, and hydrolytic stability, and superior selectivity properties. Aromatic PA is widely used for RO and GS membranes, whereas aliphatic PA such as nylon-6, nylon 6-6, and nylon 4-6 are used for MF membranes. Cross-linked aromatic PAs form the skin layer of the thin-film composite (TFC) membranes. The best known and used is interfacially polymerised PA formed from 1,3-diaminobenzene and 1,3,5-benzenetricarboxylic acid chloride. The residual negative charge resulting from incomplete reaction and subsequent hydrolysis and ionisation of the third acid-chloride group helps in reducing colloidal fouling.[3,9,18]

The most common cellulose esters in use is a blend of cellulose acetate (40.1% acetyl) and cellulose triacetate (43.2% acetyl) even though they are limited in terms of chemical, thermal, and mechanical stability. They undergo hydrolysis when operated beyond a pH range of 4–6, and are biodegradable. Cellulose acetate gets slowly hydrolysed in water, which reduces the acetyl content in the membrane, and affects its performance adversely.[9,18] They swell in polar organic solvents, plasticising them. Despite these limitations, CA is used in several membrane separations, including dialysis (regenerated CA); microfiltration (cellulose nitrate (CN) and cellulose triacetate (CTA)); ultrafiltration (CTA); reverse osmosis (CTA); and gas separation (CA). Although the cellulose polymer is very hydrophilic, it is not water soluble because the polymer is quite crystalline, and because of intermolecular hydrogen bonding between the hydroxyl groups.

The interfacial properties, such as interfacial tension and adsorption are considered to be the most important physical properties of the membrane material because the solid membrane phase is in contact with a liquid or a gas phase. Physical properties refer to the degree of change in the physical structure of the material in a given environment such as the degree of swelling of a polymeric membrane in an organic solvent, and the change in crystallinity at high or low temperatures. Interfacial properties in the form of adsorption between the solute and the membrane surface due to van der Waals interactions, hydrogen bonding, electrostatic effect, charge transfer effects, and dipole moments play a critical role in the performance of UF and MF membranes. For example, the precipitous drop in permeation flux with hydrophobic membranes is the result of irreversible adsorption of particles such as protein to the membrane surface (fouling) whereas cellulosic hydrophilic membranes, which swell in water, present a more 'water-like' surface to the proteins.

Despite the excellent chemical and thermal stability of hydrophobic polymers (e.g. polytetrafluoroethylene (PTFE) and polyvinylfluoridine (PVDF) membranes are used in MF of organic solutions), stable hydrophilic polymers are more effective as membrane materials because of their reduced adsorption tendencies. In general, hydrophilic membranes, such as CA, polyethylene (PE), and inorganic alumina are less susceptible to fouling than hydrophobic membranes made from PVDF, polysulphone (PS), polyether sulfone (PES), and polypropylene (PP).

In the case of RO membranes, the Hilderbrand's solubility parameter (SP) of a polymeric material can be used to select suitable polymeric materials. It has also been used for the study of PV membranes. The solubility parameter expresses the nature and magnitude of the interaction force working between molecules: between the permeant molecule and the molecules of the membrane material.[9] The SP of polymers, from which RO membranes with excellent performance have been prepared are between 24 and 35. In this range the membrane material hydrophobic/hydrophilic ratio is well balanced; a low SP reflects hydrophobicity of the polymer, whereas hydrophilicity is associated with a high SP value.

Inorganic membranes

There are four general categories of inorganic membranes: ceramics, sintered metals, glass, and zeolite.[18,20,21] These membranes have become viable alternatives to polymeric membranes for some UF and MF applications because of their outstanding thermal and chemical stability. Ceramic UF and MF membranes are microporous sieves in which separation takes place on the basis of the size and speed of a particle through a tortuous path of the pore. Capillary action, adsorption phenomena, and surface charge, all play roles in retention and separation. Although ceramic membranes can be manufactured with a sharp pore size distribution, they are limited in their size-exclusion capability.

Ceramic UF and MF membranes are made generally by sintering and sol-gel processes. Sintering is limited to pore sizes of about 0.1 μm. UF membranes are generally prepared by the sol-gel technique. γ-alumina (γ-Al_2O_3) and α-alumina are the most common ceramic membrane materials. Others include zirconia (ZrO_2), titania (TiO_2), silicon carbide (SiC), and glass (SiO_2). Alumina-based MF membranes and porous carbon substrates are made into UF membranes by depositing a layer of ZrO_2 on the surface. Characteristics of some of these membranes are given in Table 1.7.

Because of their greater rigidity, ceramic microfilters accommodate fluxes 5–10 times greater than those of asymmetric polymeric membranes, and can be backwashed frequently without damaging the membrane skin layer. Ceramic membranes offer applications in high temperature applications whereas polymeric membranes are restricted to temperatures less than 90°C. Ceramic membrane elements also have a greater resistance to cleaning chemicals, and can be steam sterilised and autoclaved repeatedly, which makes them well suited for biopharmaceutical processing. Another advantage of ceramic

Table 1.7 Characteristics of some ceramic membranes

Characteristic	Membrane property	
Substrate	α-Alumina	α-Alumina
Membrane	α-Alumina	α-Alumina
Pore sizes[a] (μm)	0.2, 0.45, 1.0	0.2, 0.45, 0.8, 1.2, 1.5, 3.5
Cross section	Circular	Hexagonal
Length (cm)	88	85
Diameter (cm)	1.6	2.8
Number of lumens	19	19, 37
Lumen diameter (mm)	2.0, 2.7, 4.0, 6.0	3.0, 4.0, 6.0
Available surface area (m^2)	0.14	0.20, 0.30
Pressure limit (bar)	21[b]	>100[c]
Water permeability[d] (m^3/h·m^2·bar)	4.0	3.0

[a] Smaller-pore ultrafiltration ceramic membranes are also manufactured.
[b] Operating pressure.
[c] Burst pressure.
[d] At 20°C for a 0.2 μm pore membrane.

membranes is a significantly longer life, up to 10 years. For most polymer membranes, typical life varies from one year for hydrophilic membranes, two to three years for hydrophobic membranes, and up to four years for fluoropolymers. Ceramic membranes are, however, much more expensive than polymeric membranes and are brittle.

Other inorganic membranes include:

(a) zeolite membranes, which have very narrow pore sizes, and are very good prospects for GS and PV membranes;
(b) sintered metal membranes; and
(c) organo-mineral membranes that are intermediate between polymeric and ceramic membranes.

Mineral membranes are obtained by sintering metal powders or stainless steel filaments with diameters between 1.5 and 80 μm.[18–20] They can be more expensive than ceramic membranes, and are limited in selection by their pore size range. Organo-mineral membranes are hybrid membranes, e.g. UF organo-mineral membranes are composed of a polymeric matrix (mostly PS) in which zirconia grains are dispersed finely as a filler material. The organo-mineral membranes have a considerably higher flux than traditional polymeric membranes.

Charged membranes

Charged membranes are highly swollen gels containing polymers with a fixed ionic charge. Ion exchange (IX) membranes are usually homogeneous (symmetric) produced by polymerisation of functional monomers and cross-linking agents, or made by chemical modification of polymers.[18,21] The membrane polymer is usually hydrophobic, for example polystyrene, and to a lesser degree, polyethylene, and polysulphone. In the case of polymer electrolyte membrane fuel cells, high-temperature fluoropolymer-charged membranes are used. The IX membranes must not only be highly permselective but must also have low electrical resistance.

IX membranes are either cation exchange membranes (CEM) or anion exchange membranes (AEM). CEM are usually made from polystyrene copolymerised with divinylbenezene and sulphonated to attach SO_3^{2-} ion exchange groups.[18,21] AEM are polystyrene–divinylbenzene polymer-based membranes with quarternary amines as the positive ion exchange carrier groups.

IX membranes are used in several membrane processes including electrodialysis (ED), electrodeionisation (EDI), fuel cells, electrodialytic water dissociation (water splitting), diffusion dialysis, Donnan dialysis, and electrolysis. Donnan exclusion is an important rejection mechanism for charged membranes, for example charged NF membranes.

Molecular weight cut-off (MWCO)

MWCO has emerged as one of the most useful tools for characterising UF membranes. The early UF membranes were used for purification of biological solutions for retaining macromolecules such as proteins. Since macromolecules are characterised by their molecular weights, the ability of UF membranes to retain these macromolecules is based on their molecular weight. Thus, the term MWCO came into being for characterising UF membranes. It is arbitrarily defined as the molecular weight at which 90% of the macromolecular solute is rejected by the membrane.

The main limitations for defining MWCO precisely are as follows:[11,18,19,21] First, UF membrane's pore size distribution is not 'sharp' as alluded to earlier in the chapter; Second, most UF membranes being hydrophobic adsorb proteins on the surface and inside the pores due to membrane–protein interaction resulting in higher retention (a) due to reduced pore size, and (b) due to the secondary film formed on the membrane surface. Adsorption is controlled by many factors such as concentration, pH, ionic strength, and temperature as discussed later in this chapter; Third, it is difficult to pick the 90% point on the S-shaped retention vs. molecular weight curve; and Fourth, it is difficult to select the marker protein since polymers of the same molar mass have different molecular size, e.g. linear polyacrylic acid passes through a membrane that retains a globular protein of the same molecular mass. In addition, the shape of the protein changes in the vicinity of the membrane surface. Hence, the feed concentration must be kept to minimum to prevent concentration polarisation.

The definition is not absolute because the macromolecule to be retained may or may not be spherical. However, it is a useful tool for pre-screening and selecting UF membranes for a particular application. The rules of thumb for good separation are (a) the molecular mass must differ by a factor to ten to separate two solutes of different molecular weights, and (b) the MWCO rating of the membrane must by at least one-half of the solute to be retained. Another factor that affects is separation of a macromolecule is its molecular shape. For example, high rejections are achieved for globular proteins as compared to flexible polymers with extended chains, such as polystyrene, polyethylene glycol and dextran. In the case of very dilute solutions of flexible polymers, however, the rejection decreases to zero with pressure due to shear-induced deformation.[18]

Bubble point

Bubble point is used to characterise the pore size of MF membranes.[18,21] It is also used to test the integrity of membrane assembly including membrane seals. According to the capillary theory, the smaller the diameter of the pore filled with water, the harder it is to blow water out of the pore. The bubble point pressure is a direct measurement of the largest pore in the membrane. Based on the surface tension of the fluid in the membrane pores and the wetting angle of the membrane polymer, the diameter of the pore can be calculated using the Young–Laplace equation, $\Delta P = (4\gamma \cos \theta)/d$ where ΔP is the transmembrane pressure (Pa), γ the surface tension (N/m), θ the liquid–solid contact angle, and d the pore diameter. For a fluid wetting the membrane completely, $\cos\theta$ is 1. The test method is also used to ensure the sterility of injectable drugs in pharmaceutical manufacturing.

1.4 Membrane processes

Membrane processes have been widely adopted over the last 30 years in spite of inherent limitations such as fouling, thermal and chemical resistance, and maximum achievable purity. One reason for their wide success is the emergence of hybrid membrane processes for several applications, some of which are discussed in Chapters 3 and 4. The principal characteristics of membrane processes are listed in Table 1.4. All processes except dialysis, ED and EDI are pressure-driven, and all except pervaporation and membrane distillation involve no phase change.

1.4.1 Reverse osmosis

Reverse osmosis is a process for removing solutes such as dissolved ions from a solution. When a semi-permeable membrane separates water from a concentrated solution of ions, the water moves through the membrane into the concentrated solution due to osmosis to even out the water concentration or equilibrate. Water continues to flow until sufficient osmotic pressure builds up on the concentrated solution side to prevent further flow of water up the concentration gradient. When pressure greater than the osmotic pressure is applied as illustrated in Figure 1.10, the flow is reversed. The process is called 'reverse osmosis' that results in a concentrated salt solution on the feed side of the membrane and a near pure solution on the other side of the membrane; in effect water gets squeezed out and flows out through the membrane under pressure. The effective water flow through the membrane at constant temperature is defined by the following equation:

$$J_w = A(\Delta P - \Delta\pi) \qquad (6)$$

where J_w is the membrane water flux, A the membrane permeability coefficient for water, ΔP the hydraulic pressure differential across the membrane, and $\Delta\pi$ the osmotic pressure differential across the membrane. In practice, the membrane is permeable to some low molecular weight solutes (rejection is always less than 100%). Hence, the real osmotic pressure is $\sigma\Delta\pi$ where σ is the reflection coefficient. The value of A is approximately in the range of $3(10^{-3})$–$6(10^{-5})$ $m^3/h{\cdot}m^2{\cdot}bar$ for RO membranes and $3(10^{-3})$–$2(10^{-2})$ $m^3/h{\cdot}m^2{\cdot}bar$ for NF membranes.[18] Osmotic pressure depends on solute concentration, solution temperature, and the type of ions present. For dilute solutions (applicable to membrane cases), the osmotic pressure, π is approximated using the van't Hoff relationship,

$$\pi = \nu_i c_i RT \qquad (7)$$

where π is the osmotic pressure, ν_i the number of ions formed when the solute dissociates (e.g. ν_i is 2 for NaCl, and ν_i is 3 for $BaCl_2$), c_i the molar concentration of the solute, R the gas constant, and T the absolute temperature. Thus, $\nu_i c_i$ is total concentration of ions in $kmol/m^3$.

The solute flux through the membrane is given by the following equation:

$$J_s = B(\Delta C_s) \qquad (8)$$

where J_s is the solute flux, B the solute permeability coefficient, and ΔC_s the solute concentration difference across the membrane. The value of B is in the range of $5(10^{-3})$–$1(10^{-4})$ m^3/m^2 h for RO membranes with NaCl as the solute. While the water flux increases linearly with pressure difference (equation (6)), the solute flux is independent of pressure and is a function of the concentration difference (equation (8)). Hence, when the membrane is operating near the osmotic pressure of the feed, the salt passage is not diluted by the product water. During normal operation, when the permeate water flux drops, it follows, the permeate salt concentration will increase.

The effect of operating variables on RO membrane performance is depicted in Figure 1.11. The qualitative graphs 'i' through

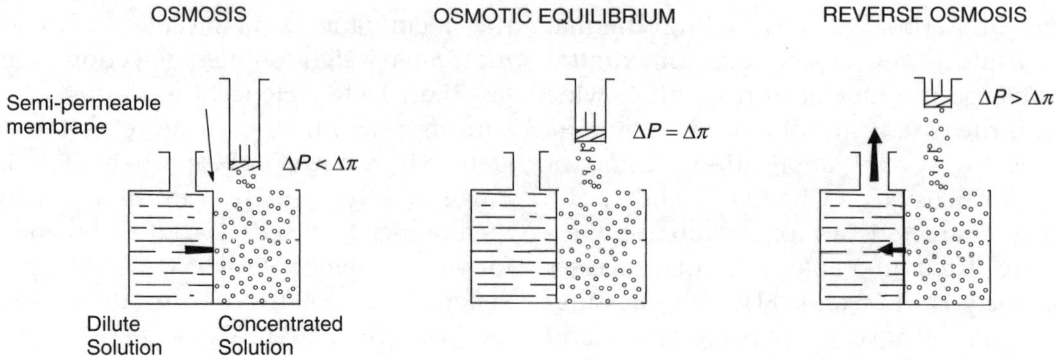

Figure 1.10 Osmotic phenomena. At osmotic equilibrium, the osmotic pressure ($\Delta\pi$) across the membrane is exactly counterbalanced by the hydrostatic pressure (ΔP) applied to the concentrated solution. Source: Lonsdale[3]

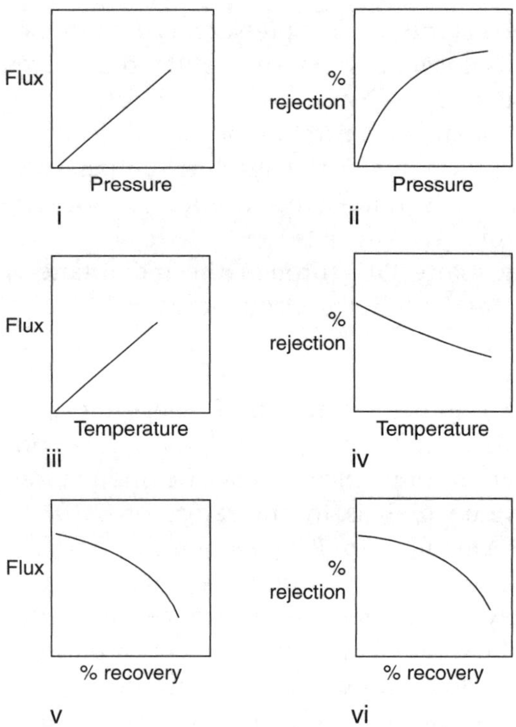

Figure 1.11 The effect of operating variables on RO membrane performance. Source: Eykamp; Cabibbo et al.

'vi' are summarised as follows:[21,22] in graph 'i' water flux increases with increase in feed pressure at constant π based on equation (6); in graph 'ii' product quality increases with feed pressure since water flux increases but solute flux stays constant based on equation (8); in graph 'iii' water flux increases with temperature as viscosity drops faster than the rise in π at 25°C (a rule of thumb is that flux increases 3% per °C rise in temperature); in graph 'iv' product water quality decreases (higher solute concentration in permeate or lower solute rejection) with rise in temperature due to higher π and because solute flow has a higher activation energy than water flow; in graph 'v' water flux declines as water recovery increases because increase in solute concentration in the feed/reject stream results in higher π, and also higher concentration polarisation; and in graph 'vi' product water quality decreases with an increase in water recovery due to higher concentration polarisation, and possibly precipitation of sparingly soluble salts when they exceed their solubility limits.

The most important RO membranes are homogeneous asymmetric cellulose acetates and polyamides (PA) made by the phase-inversion process, and cross-linked TFC polyamides. Both CA and PA membranes possess an economically viable combination of high rejection and water flux.[8] TFC membranes are more expensive than CA membranes but have a higher flux and, hence operate at a lower feed pressure than CA membranes.

The TFC membranes are not biodegradable, and have a higher rejection of silica, nitrates, and organic compounds than CA membranes. Because of the above performance attributes, the TFC membranes are now used exclusively in high-purity water plants.

The composite membranes consist of a dense ultrathin barrier layer, 0.2 μm thick, chemically bound on top of a polysulphone UF membrane support. A variety of organic polymers are used for the top skin layer. The data in Table 1.8 shows that these membranes are either positively charged or negatively charged.[23] The very high salt rejection of polyamide TFC membranes without sacrificing much higher flux performance is due to its ionic character, which results in co-ion repulsion. The negatively charged TFC membranes have an iso-electric point between 4 and 5. Below a pH of 4.0, the membranes have a net positive charge, and above 5.0, a net negative charge. Hence, during RO plant operation, the feed water pH is usually maintained above 7.0 and 8.5 to enhance rejection.

Table 1.8 Classification of composite RO and NF membranes

Class	Name	Manufacturer	Charge
Polyamides			
Heterocyclic/ aromatic	NTR-729HF	Nitto Denko	Negative
	NTR-739HF	Nitto Denko	NA
	UTC-60	Toray	NA
	UTC-20HF	Toray	NA
	NF-40	Dow/Film-Tec	NA
	NF-40HF	Dow/Film-Tec	NA
Aromatic	NTR-759HF	Nitto Denko	Negative
	CPA2	Hydranautics	NA
	UTC-70	Toray	Negative
	BW-30	Dow/Film-Tec	NA
	NF-50	Dow/Film-Tec	NA
	NF-70	Dow/Film-Tec	NA
Aromatic/ Alicyclic	A-15	DuPont	Negative
Polyureas			
Aliphatic/ aromatic	NTR-719HF	Nitto Denko	Positive
	IFCI	Hydranautics	NA
	TFC Series	UOP	NA
	PA 300	UOP	Positive
	UTC-40HR	Toray	NA
Polyether			
Sulphonated aromatic	NTR-7400	Nitto Denko	Negative

NA: Not available
Source: Matsuura

There are two main types of TFC membranes used, PA and polyether urea (PEU). Unlike PA membranes, PEU membranes are neutral to slightly positive. PEU (and CA) membranes, unlike aromatic PA membranes, have a smooth surface and are, therefore, less prone to fouling.[24] For RO feed water such as surface and municipal water, which are treated with cationic polymeric coagulants, PEU membranes and not negatively charged PA membranes are used to minimise fouling. Further, due to their superior pH stability (2–12 range), more rigorous chemical cleaning and chemical sterilising methods can be deployed. Hot water sanitisable TFC membranes are now routinely used in the pharmaceutical, food and beverage industries ensuring RO systems are sterilised up to 80°C instead of with chemicals. High flux, high rejection, and lower energy consumption composite membranes have been developed for use in applications such as high-purity water systems, and in large RO installations where fouling is not a likely problem. Similarly, low-pressure TFC membranes are being used in many brackish water applications. A typical spectrum of composite RO membranes for different applications is given in Table 1.9.

One of the few drawbacks of the PA membranes is that they are highly susceptible to degradation by chlorine with extended exposure (1,000 ppm-h) due to chain deformation and alteration of the average inter-chain

Table 1.9 Typical TFC membranes for various applications

Membrane type	Remarks
Seawater RO membrane[1]	Seawater conversion to industrial water and drinking water. Purify high TDS brackish water (>10,000 ppm).
Brackish water RO membrane[2]	Production of high-purity water, concentration of food and heavy metal recovery from well water, river water, lake water, or industrial wastewater.
Low-Pressure RO Membrane[3]	Lower pressure brackish water membrane. Ideal for small drinking water systems due to low energy consumption.
Fouling-resistant RO membrane[4]	Improved fouling-resistant characteristics possibly due to a hydrophilic character and a net negative charge. Applicable where fouling tendency is high.
Household RO membrane[5]	For removing solutes larger than 10^{-4} μm carcinogens such as THM, heavy metal ions, bacteria, and germs in household water.
NF membrane[6]	Water softening; >98% rejection of divalent ions. Removal or concentration of dyes, paints, and liquid foods.

Test conditions:
[1] 55 bar g; 32,000 ppm NaCl; 25°C; 6.5–7.0 pH; 10% recovery.
[2] 15.5 bar g; 2,000 ppm NaCl; 25°C; 6.5–7.0 pH; 15% recovery.
[3] 10 bar g; 2,000 ppm NaCl; 25°C; 6.5–7.0 pH; 15% recovery.
[4] 15.5 bar g; 2,000 ppm NaCl; 25°C; 6.5–7.0 pH; 15% recovery.
[5] 4 bar g; 250 ppm NaCl; 25°C; 6.5–7.0 pH; 15% recovery.
[6] 15.5 bar g; 2,000 ppm $MgSO_4$; 25°C; 6.5–7.0 pH; 15% recovery.
Source: CSM membrane catalogue

displacement within the PA membrane skin structure.[25] The net result is increased permeability and decreased selectivity, and eventual degradation. Intolerance to chlorine also means that PA membranes are unsuitable for treating wastewaters that contain oxidants.

Development of a chlorine-resistant PA membrane, therefore, has been identified as one of the critical development areas of membrane technology and is discussed in Chapter 5. RO feed water must be dechlorinated when using PA membranes. Unfortunately, absence of chlorine can result in biofouling that is often irreversible. Newer membrane module designs that eliminate the use of brine seals and membranes that are hot water sanitisable at 80°C are being increasingly used in bio-applications.

A list of RO applications is given in Table 1.10. RO plays a major role in providing potable water, defined as <1,000 ppm total dissolved solids (TDS) by the WHO or <500 ppm TDS based on the U.S. EPA criterion. RO has extensive and well-proven application in desalination, both seawater (50–100 bar g; 35,000 ppm TDS) and brackish water (15–25 bar g; 1,000–10,000 ppm TDS) and dairy processing. Brackish water desalination for drinking water is the largest application where it competes with electrodialysis (ED). The RO process is currently the optimum choice for desalination of low-salinity water (26,000–30,000 ppm TDS) independent of plant capacity.[21] The highest rejection RO membranes are those that can make potable water from seawater using multi-staged, single-pass RO units. The industry standard is 99.4% rejection of NaCl. The CTA hollow-fibre membranes are rated even higher. The membrane modules currently used for

Table 1.10 List of RO applications

Industry	Applications
Desalination	Potable water production
	Seawater
	Brackish water
Water reclamation	Municipal wastewater
Ultrapure water	Semiconductor industry
High-purity water	Pharmaceuticals
	Medical usage
	Beverages
Utilities; power generation	Boiler feed water
Chemical process industries	Process water reuse
	Effluent water reuse
	Water/organic separation
	Organic liquid mixtures separation
Metals and metal finishing	Mining effluent treatment
	Plating rinse water
	Reuse and recovery of metals
Food processing	Dairy processing
	Sweeteners concentration
	Juice and beverage processing
	Light beer and wine production
	Waste stream processing
Textiles	Dyeing and finishing chemicals recovery
	Water reuse
Pulp and paper	Effluent disposal and water reuse
Biotechnology/ medicine	Fermentation products recovery and purification
Hazardous waste water	Removal of environmental pollutants from water.

seawater desalination are CTA hollow-fibre and TFC spiral-wound.

Boiler feed water is another major application. RO membranes reduce the conductivity to 1–10 μS/cm depending on feed water TDS and RO system design that makes it economical to polish RO product water with ion exchange or EDI. Similarly, high-purity water for nuclear power plants is produced by RO membranes.[21,22] Increasingly RO systems are being used to reclaim wastewater at power plants and industrial sites for recycling, and for water pollution control by concentrating waste streams, as discussed in Chapter 3.

High-purity water production for pharmaceuticals, beverage, and microelectronics by RO membranes are discussed in Chapter 3. In the microelectronics plants, it takes about 4.5 m^3 of high-purity water to wash and rinse a 15-cm wafer during the fabrication of microchips. A large semiconductor plant can use 250–300 m^3/h of this ultrapure solvent, which is produced by a combination of several unit operations; MF, RO, EDI, ion exchange, ultra-violet sterilisation, membrane degasification, and UF. Similarly, RO is the key unit operation in the production of USP (United States Pharmocopea) grade water.

RO is used extensively in the dairy industry for pre-concentrating milk from 10 to 25 (wt.%) solids prior to evaporation thereby lowering the cost of evaporation, without denaturating the proteins and with minimal wastewater. Other dairy applications include concentration of cheese whey and milk, desalting of salt whey, and treatment of waste streams. Whey concentration by RO is limited to 20% total solids (concentration factor $= 4x$) by the solubility limits of lactose.[26] RO is used for concentrating sugar solutions, fruit and vegetable juices, and beverages for retaining salts and low-molecular weight flavour compounds. However, high osmotic pressure and flavour leakage through the membrane can be a problem.

1.4.2 Nanofiltration

Nanofiltration (NF) is closely related to RO, and is sometimes called 'loose RO'.[6] Like reverse osmosis, NF is a kinetic process and not an equilibrium process, and the basic principles of RO process discussed above are applicable to NF also. NF membranes have a surface charge due to ionisable groups, e.g. carboxylic or sulphonic acid groups. The equilibrium between the charged membrane and the bulk solution is characterised by an electric potential called the Donnan potential. Ions smaller than the pore size are rejected because of Donnan exclusion.[18]

Most of the NF membranes are PA thin-film composites as listed in Table 1.8. As in the case of RO membranes, the PA rejecting barrier is a fragile film (0.2 μm thick) on a polysulphone UF membrane. It is usually negatively charged so that anion repulsion is the primary determinant of solute rejection. The rejection of divalent ions is >98% as compared to <50% rejection of monovalent ions. The average pore size of NF membranes is 2 nm, which is between the size for RO and UF membranes. The size of the solute molecules that are separated from water, and the range of operating pressures are also between those for RO and UF. NF membranes are, therefore, ideally suited for rejecting organics such as lactose, sucrose, and glucose with molecular weights above 200 Da, and for rejecting multivalent ions. For a comprehensive review of NF and its applications, refer to the recent publication.[7]

Nanofiltration has several advantages over RO and UF processes for the following reason:[7,27]

- Its ability to selectively reject low molecular weight organic compounds and divalent compounds, it is a better alternative to RO and UF for treating wastewater streams, e.g. desalting cheese whey, removing heavy metals, and separating dyes and colour compounds in the textile industry.
- In the treatment of bleaching effluents from pulp and paper plants, it is cheaper than RO, which has lower permeate flux and higher energy cost. It is more efficient than UF, which is not effective in rejecting low molecular weight toxic chlorinated compounds.
- Separation of salts with monovalent ions from organics in the MW range of 200–1,000 Da are ED and ion exchange. NF is superior because it can simultaneously concentrate the high MW compounds and remove the monovalent salts.

The largest application of NF is in water softening. Besides removing hardness, NF is also used for removing organic colour compounds and trihalomethane (THM) precursors such as humic acids. In Florida, where shallow aquifers have a colour intensity similar to surface waters, and also have a high THM formation potential, NF is the preferred technology. Not surprisingly, the EPA has recommended NF for its Disinfectant and Disinfection By-Products Rule (the so-called DBRs) requirements. NF membrane elements are designed for removing TOC and THM precursors with a medium to high salt passage for surface and groundwater applications where good organic removal is required with partial softening in order to maintain a minimum level of hardness for organoleptic properties and preservation of water distribution lines.

Since NF membranes reject polyvalent anions to a much greater degree than monovalent ions, e.g. low rejections of monovalent salts (e.g. NaCl) and high rejection of divalent salts (e.g $BaSO_4$) and organics allow low operating pressures to be used for selective removal of multivalent ions from brine solutions such as seawater, organics from offshore oil platforms, or removing oil from oily wastewater.[28] NF membranes also have higher solvent flux, and operate at much lower pressures than RO: 40–45 $l/m^2/h$ flux at 5 bar g for NF vs. 20–25 $l/m^2/h$ at 15 bar g for RO.

The major source of drinking water in most countries is groundwater, which is increasingly contaminated with chemicals such as nitrates, and organic molecules such as halogenated hydrocarbons and aromatic compounds. Thin-film composite aromatic PA and PEA membranes are the most effective in removing trace amounts of these small organic and carcinogenic molecules from sources of drinking water supply.[29]

Other likely applications of NF include:[6,7,18]

- Cleaning up of contaminated groundwater
- Treatment of effluents containing heavy metals and oils
- Selective salt rejection at offshore oil platforms
- Pulp and paper wastewater treatment to remove colour
- Electroless copper plating
- Cheese whey production
- Yeast production
- Pharmaceuticals
- Food processing.

1.4.3 Ultrafiltration

Ultrafiltration (UF) membrane process can be defined as between NF and MF (Table 1.4) with pore sizes ranging from 1 to 100 nm (0.001 to 0.1 μm). Both UF and MF membranes are porous in nature, but UF membranes always have an asymmetric structure, as shown in Figures 1.2 and 1.3 with a much denser skin layer, i.e. smaller pore size and lower surface porosity that produces higher hydrodynamic resistance. The thickness of the top layer is generally less than 1.0 μm. Flux through a UF membrane is defined by Darcy's law of flow through porous materials as:

$$J = A(\Delta P) \tag{9}$$

where A is the membrane permeability constant and ΔP the transmembrane pressure. The value of A is a function of membrane porosity, pore size, and membrane thickness, and varies from 0.5 m^3/m^2.day.bar for dense membranes to 5 m^3/m^2.day.bar for more open membranes.[18] New UF flat membranes have a water flux greater than 1 mm/s. However, the flux drops to 100 μm/s when the membrane is in a device. In actual process applications, the flux is much lower. Some typical UF membrane design fluxes are given below:[21]

- Potable water, 125 l/s
- Water, 108 l/s
- Whey, 17 l/s
- Polyvinyl alcohol, 6 l/s.

The model most frequently used for describing flow through pores is based on the well-known Hagen–Poiseuille equation for laminar convective flow through straight capillaries with $A \approx \varepsilon r^2$:

$$J = \frac{\varepsilon r^2.\Delta P}{8\mu\tau.\Delta x} \tag{10}$$

where, J is the flux in cm/s, ε the porosity, r the pore radius, ΔP the applied pressure, μ the dynamic viscosity, τ the tortuosity factor (=1 for cylindrical pores), and Δx the membrane thickness. The unit of J is cm/s. There is a fairly good correlation between pore size and water flux based on the reported literature.[26] The direct proportionality between flux and the applied pressure is only true for water. For typical aqueous solutions used in UF and MF separations, flux drops off drastically due to concentration polarisation and fouling as discussed earlier.

The solute flux, J_s through a pore is equal to the sum of convective and diffusive affects:

$$J_s = K_c V C_s - K_d D dC_s/dZ \tag{11}$$

where K_c is the hindrance factor for convective transport, V the filtration velocity, C_s the average solute concentration, K_d the pore hindrance factor for diffusive transport, D the solute diffusivity, and Z the pore axial dimension. The rate of solute transport and rejection are adversely affected by non-hydrodynamic interactions between the solute and the pore wall, viz. electrical interactions and solute–solute interactions.[17]

Typical performance curves for UF are shown in Figures 1.8 and 1.9. Figure 1.8 shows a plot of flux as a function of log of feed/retentate concentration and is discussed further in Chapter 2. Figure 1.9 shows flux as a function of pressure; at low transmembrane pressure (I), flux is directly proportional to pressure (equation (9)). The magnitude of flux is same as water when the viscosity is the same. As the pressure is increased beyond a certain threshold value, the flux becomes constant and independent of pressure (III). Region II curve reflects the start of concentration polarisation, and is the ideal operating range. However, it is very difficult to design feasible equipment to operate in this region.[21]

Owing to the low or negligible osmotic pressure of macrosolutes, UF operates at very low pressures, typically between 2 and 5 bar g. Further, due to the larger pores in the membrane skin, the UF membranes have an order of magnitude higher flux than RO membranes. UF performance parameters are calculated using the equations given in Section 1.3.

UF membrane separation depends upon membrane pore size, solute–membrane interactions, shape and size of the macromolecule, and concentration polarisation. For maximum separation efficiency, there should be a 10-fold difference in the sizes of the species to be separated. In addition, since all liquid separation membranes have a certain pore size distribution[23] (e.g. there is a bi-modal distribution in the case of RO membranes predominantly small pores, ≤10Å, and occasional large pores, ≥100Å that are attributable to the inevitable existence of defects in the skin layer) molecular weight cut-off (MWCO) of UF membranes should be at least one-half that of the smallest solute to be removed.

The UF membranes are manufactured by the phase-inversion process. The most widely used polymer is polysulpone (PS), but increasingly other polymers are being used; CA, PES, polyacrylonitrile (PAN), and PVDF polymers (Table 1.5). Typically, CA-based membranes have a higher flux at equivalent rejections than PS membranes. Although, CA membranes are less prone to fouling, PS membranes are necessary for many applications because of their higher stability. In the food industry, where steam sterilisation is required, PES is used instead of PS. Among the newer membranes in use are polyimide (PI) polymeric membranes. Polyimide UF membranes are promising because of their resistant to many organic solvents[30] such as hexane, benzene, methanol, acetic acid, acetone, ethyl ether, ethoxy ethanol, and chlorinated hydrocarbons.

The share of the world's membrane and systems market of UF is almost one-half that of RO. A list of UF applications is given in Table 1.11. Industrial UF was developed primarily for the treatment of wastewater to remove particulate and macromolecular matter. Its applications now include water treatment, food processing, biotechnology, and chemicals processing. UF is used for removing macromolecules, colloids, colloidal silica, emulsified oil, endotoxins, pyrogens, viruses, and bacteria.

Major UF food applications are in the dairy industry.[19,21,26] Dairy applications have the largest share of membrane capacity in the world with applications in whey processing – treatment of cheese whey for recovering milk proteins – and cheese-making. Membrane systems are also used for the production of whole and skim milk concentrates, and

Table 1.11 UF and MF process applications

Concentration
- Proteins – enzymes, milk proteins, egg white.
- Polymer latices – styrene butadiene, polyvinyl acetate, polyvinyl chloride.
- Starch and pectin.
- Yeast production, mammalian cell harvesting.

Recovery
- Electrodeposition paint from wash and rinse baths.
- Textile sizing agents – polyvinyl alcohol, carboxymethyl cellulose.
- Lignin from Kraft black liquor in pulp and paper processes.
- Indigo dyes, printing inks.

Clarification/stabilisation
- Fruit juices and wine – removal of haze components.
- Beer – removal of cellular debris and bacteria.
- Sugar refining – removal of polysaccharides, proteins, colloidal impurities.
- Sterile filtration of biologicals – removal of bacteria and viruses.

Effluent treatment
- Oily wastewater – from washing operations.
- Bleach effluents – pulp and paper industry.
- Leather and tanning effluents.

Purification
- Blood fractionation – plasmapheresis, haemofiltration.
- Separation – antibiotics or vaccines from fermentation broths.
- Buffer exchange – purification of pharmaceuticals by solvent diafiltration.

Water treatment
- High-purity water – final filtration in semiconductor, pharmaceutical, and beverage applications.
- Potable water – removal of bacteria and viruses.
- RO pre-treatment

Source: Zeman and Zydney, Copyright® 1996 from Microfiltration and Ultrafiltration by L.J. Zeman and A.L. Zydney. Reproduced by permission of Routledge/Taylor & Francis Group, LLC

in the manufacture of lactose-reduced milk products.

In the beverage industry it is used to improve product yield and quality. UF systems remove proteins, suspended colloids, polyphenolic compounds, starch, pectin, microorganisms, from natural juice providing a brilliantly clear juice (permeate) that is stable even after extended storage. UF is commonly used for product concentration (i.e. solvent removal) in bioprocessing; concentration of cell-free fermentation broths containing complex biological compounds such as monoclonal antibodies as well as biomolecule/product recovery from very dilute solutions. UF systems are also used for virus removal in the production of therapeutic proteins and vaccines, and for antibiotics recovery.

Other major industrial applications include electroplating rinse water processing for paint recovery, treatment of oil/water emulsions, treatment of industrial wastewater containing heavy metals, oil and grease, as well as textile and pulp and paper industries.[19,21,27] Electrodeposition is one of the most successful large-scale commercial applications of UF, which is an ideal process for recovering the electropaint and eliminating significant loss of paint with the aid of hollow-fibre membranes.

UF is used to treat wastewater to meet the municipal effluent discharge standards and to recover valuable component for recycle as in the case of oily wastewater. The sources of oily wastewater include metal works, textile industries, automotive, aeronautical and animal/vegetable fats industries, and industrial laundromats. Hydrophilic UF membranes (e.g. polyacrylonitrile and PES) are used for recovering oil from oily wastewaters instead of hydrophobic membranes (e.g. PVDF) that would get fouled with oil resulting in loss of flux. The oil droplets are completely retained by the UF membrane although the membrane is permeable to free (soluble) oil. UF permeate of highly stable emulsified feed stream can be discharged to a municipal wastewater system. The concentrated recovered oil is less than 5% of the original waste volume. Similarly, tubular UF membranes are used for treating industrial laundry wastewater to recover heavy metals such as lead and chromium to remove oil prior to effluent discharge.[19,21,26]

Newer applications include potable (drinking) water production for removing colloids, microorganisms, and particulates along with organic impurities such as insoluble fats, oils and grease, and high-purity water production for microelectronics and pharmaceutical plants. Typically UF systems are used in high-purity water plants as the final polishing step. Double-skinned PES hollow-fibre membranes are being increasingly used in semiconductor plants providing high particle retention. These membranes have a skin or barrier layer both on the inside and the outside of the fibre. This morphology provides greater structural support reducing possible particle shedding. Further, the two skin layers also ensure sterile conditions by preventing migration of any bacteria (colonised in the porous support layer) downstream. Most UF systems for water applications operate in Region I (Figure 1.9) discussed earlier because there is virtually no concentration polarisation.

1.4.4 Microfiltration

Microfiltration (MF) is the oldest membrane process (Table 1.2) using dead-end mode. In this mode, unlike cross-flow MF, the entire feed is passed through the membrane, and there is no reject recycle stream. The basic flux relations and separation principles for UF process discussed above are applicable to cross-flow MF. The UF polarisation model discussed in Section 1.3 is also applicable to MF except that the diffusivities of larger particle and colloids are an order of magnitude smaller than those for macromolecules.

The pore size of MF typically varies from 0.1 to 10 μm (Figure 1.1). However, the separation mechanism is not a simple sieve mechanism where, the particles, whose sizes are smaller than the pore size flow freely through the pore while the larger particles are rejected. In many cases the particles to be separated are adsorbed onto the surface of the pore, resulting in a significant reduction in the size of the pore. Particles also deposit on top of the membrane, forming a cake-like secondary filter layer. Hence, the particles that can be separated by MF membranes are often much smaller than the pore size.

Several commercial microporous polymeric membranes are listed in Table 1.5. The choice of the membrane depends on several factors: chemical and thermal resistance to the process conditions, sharp separation, wettabilty of the membrane, tendency to adsorb hydrophobic materials, and resistance to cleaning. The most common polymeric materials are PTFE, PVDF, PP, PS, CA/CN, CTA, PE, polycarbonate, polyester, polyether imide, and nylon 6. Of these only PTFE, PVDF, and PP have excellent to good chemical stability. Even though hydrophilic CA/CN and CTA membranes have limited chemical stability, they are best suited for treating high fouling feeds using tubular membranes.

MF membranes are prepared by sintering, track-etching, stretching, or by phase inversion (Table 1.6). They are usually symmetric unlike UF membranes. MF membranes with well-defined pores are now manufactured in inorganic materials such as alumina and zirconia. α-alumina-based ceramic membranes are available in multi-tube sheet and monolith honeycomb modules.[19,21,26] Ceramic MF membrane modules cost 3–4 times the polymer membrane modules. Some of this high initial cost is offset by the lower operating costs because of much longer life of ceramic membranes. Characteristics of some commercial ceramic membranes are given in Table 1.7.

A list of MF applications is given in Table 1.11. MF is primarily used for clarifying liquids while UF is used for fractionating solutions. In bioengineering, MF is used for removing microorganisms from fermentation products (e.g. antibiotics) during downstream processing, and for recovering penicillin.[26,31] Examples of downstream processing include clarification of fermentation broths, sterile filtration, cell recycle in continuous fermentation, harvesting mammalian cells, cell washing, mycelia and lysate recovery, enzyme purification, and vaccines. In the beverage industry, MF is used to remove yeast from alcoholic beverages and to remove turbidity from fermentation products. In the food industry, it is used to recover bacteria for disinfecting milk, and for separating fat from milk.[26] Back washable microfiltration is preferable to conventional filtration to remove suspended solids, as discussed in Chapter 2. Cross-flow MF can also be used to remove long-chain traces of saturated fat from food oils such as sunflower oil.[27]

Microfiltration is being used in municipalities for tertiary wastewater treatment and for producing potable water. Cross-flow MF can reduce algal loading from 400,000 algae cells/ml to less than 50 algae cells/ml. The PP hollow-fibre membrane units use gas pulse backwashing every 15 min to maintain the flux thereby reducing the frequency of cleaning the membranes to 4–6 weeks. Backwashing enables the MF system to run on feed water with turbidity up to 500 NTU. Backwashing is possible because the membranes are symmetric with a homogenous pore structure, i.e. there is no thin skin (1–4 μm) that can be damaged by high backpressure.

Submerged cross-flow MF (SCMF) membrane systems have been introduced by Kubota, USFilter and Zenon in the last decade for water and wastewater treatment (Table 1.12). SCMF is a low cost (no membrane pressure vessel), low maintenance process well suited to low pressure UF and MF, and membrane bioreactors (MBR).[32,33] MBR technology is discussed later in this chapter.

In the SCMF process, HF or flat sheet membranes made from PP or PVDF polymer are immersed vertically in a water tank open to atmosphere pressure. Water enters each membrane cell and is drawn through the outside of the membrane to the inside either through gravity, by a filtrate pump, or by suction producing filtered permeate.[27,32] Concentration polarisation is controlled by bubbling air from the bottom of the tank. The membranes are backwashed intermittently using filtrate and air. Periodic chemical cleaning is performed when the maximum transmembrane pressure is reached. Unlike standard membrane systems, the submerged membrane operates under a slight vacuum.

Table 1.12 Properties of various membrane modules

Characteristics	Plate and Frame	Spiral-wound	Tubular	Hollow-fibre	Submerged	High shear
Packing density (m^2/m^3)	Moderate (200–500)	High (500–1000)	Low–moderate (70–400)	High (500–5,000)	Moderate	Low–moderate
Energy usage	Low-moderate (laminar)	Moderate (Spacer losses)	High (turbulent)	Low (laminar)	Low	Moderate
Fluid management and fouling control	Moderate	Good (no solids) Poor (solids)	Good	Moderate–good	Moderate–poor	Very good
Standardisation	No	Yes	No	No	No	No
Replacement	Sheet (or cartridge)	Element	Tubes (or element)	Element	Element (or bundle)	Sheet (or cartridge)
Cleaning	Moderate	Can be difficult (solids)	Good–Physical cleaning possible	Backflush possible	Backflush possible	Good
Ease of manufacture	Simple	Complex (automated)	Simple	Moderate	Moderate	Complex

Source: Fane

Low operating pressures provide increased membrane life, reduced replacement costs and energy requirements and fouling (particles are not forced into the membrane pores under high pressure).

Several leading commercial MF applications for water treatment include:

- Potable water. Large-scale applications for producing potable water by removing *Cryptosporidium* and *Giardia* cysts with the aid of back washable 0.2 μm cross-flow MF (CMF) and submerged cross-flow MF (SCMF) technology.
- RO pretreatment: CMF provides excellent pre-treatment for RO systems eliminating conventional filtration and clarifiers.
- Tertiary water treatment: CMF treatment of secondary treated wastewater followed by RO and ultraviolet radiation or ozonation to produce drinking water.
- Water reclaim. Large-scale applications with the aid of back-washable 0.2 μm CMF and SCMF processes.
- Oil-water treatment: MF and UF reduce the oil content of water from 10–30 (wt.%) to less than 5 ppm.

Dead-end MF

Dead-end filtration has been used most commonly for laboratory and medical filtration. Unlike established cross-flow membrane processes – RO, NF, UF, MF – the fluid to be filtered is fed perpendicularly to the filter element and all the fluid passes through the filter element with retained solids forming a cake layer on the element surface. In the case of charged dead-end microfilter cartridges, the separation is based on two mechanisms: pore blockage/surface retention due to size, and surface/pore adsorption due to electrostatic interaction. In the absence of a charge, the pore size rating is nominal (80–90% retention) whereas in the presence of a charge, the pore size rating is absolute. The flux in dead-end mode is given by:

$$J = \Delta P/(R_m + R_c) \qquad (12)$$

where, J is the flux, ΔP the pressure drop across the membrane and cake, R_m the membrane resistance, and R_c the cake resistance. The filter element must be replaced when the limiting pressure drop is reached. The cake build-up can be reduced by

(a) periodically backwashing the filter;
(b) installing a media filter or a depth filter upstream of the MF; or
(c) using cross-flow operation.

The advantages of dead-end MF include a very low capital cost and the recovery of a highly concentrated feed whereas cross-flow MF has a high capital cost and is limited to solids content <10%. Given the pore size range of dead-end MF membranes (0.1–10 μm), low cost and simple operation, it is not uncommon to see microfilter cartridges cover the entire pore size range in a single high-purity water plant. Newer pleated cartridge filter designs with 100% higher surface area shown in Figure 1.12 have been introduced. The new cartridges have flow rates exceeding 20 l/min per 25 cm length which is more than double that of traditional pleated cartridges.

1.4.5 Dialysis and electrodialysis

Dialysis

Dialysis pertains to the transport of a solute across a membrane by diffusion resulting from a concentration difference. Since concentration difference, ΔC is the sole driving force, the difference should be large and the membrane should be thin to reduce the diffusion path.[3,18] The process is slow compared with pressure-driven membrane processes. Unlike UF or RO where solvent passes

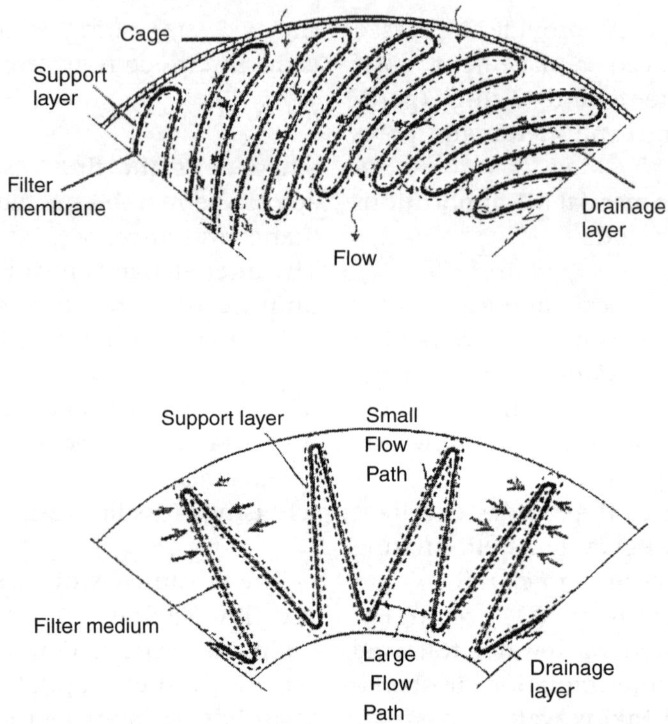

Figure 1.12 Schematic of dead-end cartridge membrane filters: Top – deep-pleated filter with increased surface area (Pall Corp.); Bottom – traditional pleated filter with less surface area and non-uniform flow

through the membrane, it is the solute that passes through the membrane. The dialysis membranes (regenerated CA or PS) can be considered as swollen non-porous films. The membrane pores must be very small to prevent convective transport of the solution resulting from a small pressure difference across the membrane. Separation in dialysis is governed by the small pores and diffusion, and therefore small molecules diffuse faster than large ones. The solute flux across the membrane is given by the relation, $J = U\Delta C$ where U is the sum of resistances: membrane and the two film resistances. In continuous flow systems, ΔC may be taken as the log mean concentration difference across the membrane.

Dialysis has the largest market segment of all the membrane processes (Table 1.1).[2] The biggest market application of dialysis is in artificial kidneys for haemodialysis to treat patients suffering renal failure. Since the process is very gentle, there is less damage to blood cells than there would be in a pressure-driven process.[3] Haemodialysis units are usually hollow-fibre devices with a membrane area of 0.5–1.5 m^2. The haemodialysis process is described as follows: Blood is drawn from the patient and passed through the lumen of the hollow fibres, while water is passed through the shell side of the unit. Since water is loaded with salts, it has the same osmotic pressure as blood cells than there would be in a pressure-driven process. Urea diffuses from the blood into the water, and is removed quite selectively but is mixed again with salts. The urea-loaded salt solution is disposed of as waste.[3,18]

In certain pharmaceutical and animal cell culture applications that require addition and removal of small molecular weight products simultaneously while retaining high molecular weight solutes, sterile dialysis membrane modules have possible applications. Dialysis is used to remove electrolytes from colloidal suspensions to render the latter more stable, e.g. it is used to recover sodium hydroxide from certain industrial wastes that are contaminated with organic substances. The sodium and hydroxide ions pass through the permeable membrane cell walls into the surrounding water. The water is evaporated to recover the sodium hydroxide while the organic waste in the cells is disposed off.

Electrodialysis

The dialysis principle is used for desalting liquids using electrodialysis (ED). Essentially ED employs charged membranes and uses electrical energy to flow the ions against concentration gradient causing separation and purification. For example, put brackish water both inside and outside an ED cell, and place electrodes in the water outside the cell. When a current is applied, the ions within the cell are caused to flow through the semi-permeable membrane by ionic conduction, and concentrate outside. The cations flow toward the cathode and the anions flow toward the anode on the basis of electrochemistry – electrochemistry deals with the relationships between electrical and chemical phenomena. The water within the cell is demineralised (see Figure 1.13). ED modules consist of a large number of thin, continuous flow cells to make the process efficient for large-scale usage; the membrane stack is composed of hundreds of parallel cells where the ions are depleted and concentrated in alternate cells, as shown in Figure 1.13. Besides desalting of brackish water (ED is not suitable for the desalination of seawater), ED is used extensively for deashing whey, and the desalted product is a useful baby food

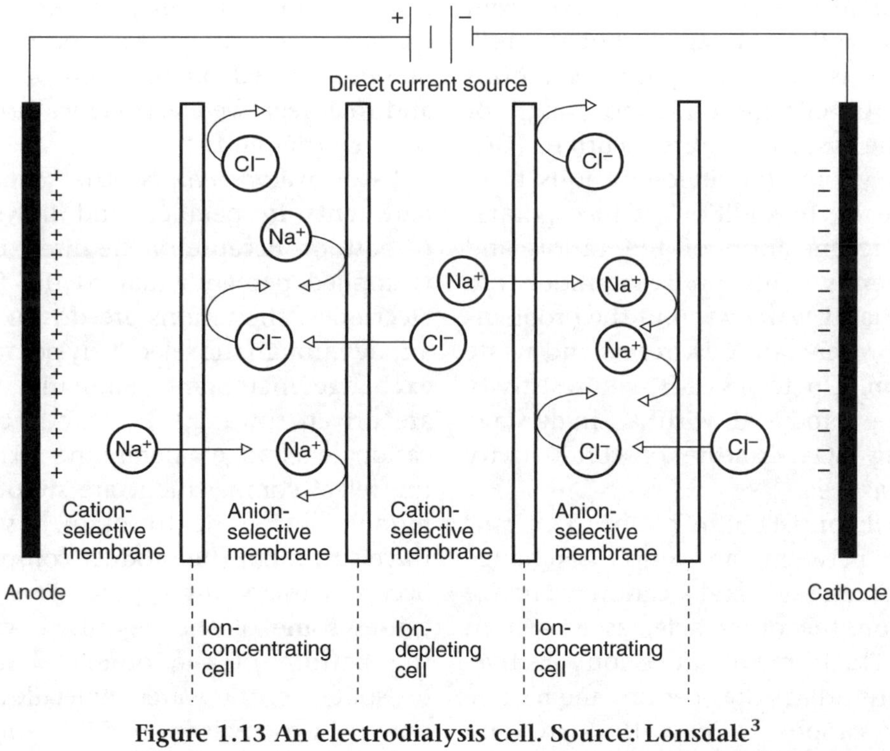

Figure 1.13 An electrodialysis cell. Source: Lonsdale[3]

additive. It is useful in separating electrolytes from non-electrolytes. ED is also used to remove specific pollutants such as heavy metals, nitrates, and radium from contaminated waters.

ED reversal (EDR) units have been deployed to control fouling and reduce cleaning cycles. EDR is the process of reversing the polarity of the ED stack intermittently from several minutes to several hours.[21] When the polarity is reversed the diluate compartments become concentrate compartments and vice versa. Although an expensive process, it reduces concentration polarisation, scaling and fouling, and addition of feed-water additives such as antiscalants. ED energy requirements and costs are illustrated in Chapter 5.

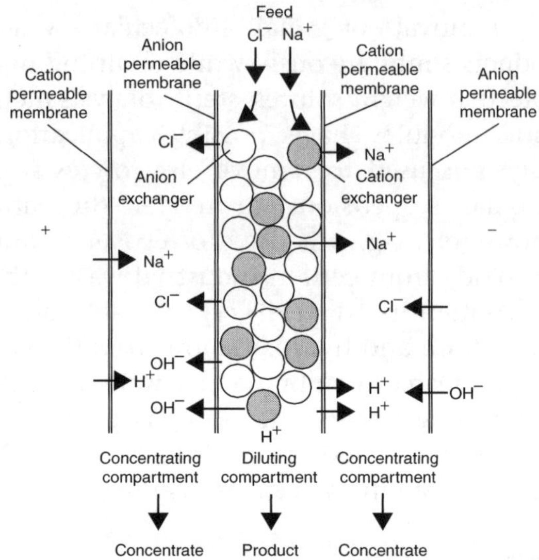

Figure 1.14 Schematic diagram of an EDI cell. Source: USFilter/IonPure

1.4.6 Electrodeionisation

Electrodeionisation (EDI) combines electrodialysis (ED) with ion exchange (IX), i.e. 'electrodialysis-moderated ion exchange'. ED is generally used to reduce the salt level from 1,000–10,000 mg/l to a few hundred mg/l, whereas EDI is used to purify solutions containing 10–100 mg/l down to 1 mg/l or less. As in the case of ED, direct current (DC) is the driving force for removing ions from the feed stream. In addition, it incorporates continuous regeneration of the cation and anion IX resins. This process produces a consistent quality water without the problems and costs of regenerating IX resins and waste neutralisation. Up to 18.0 MΩ-cm resistivity water can be produced with a single-stage system using RO permeate of conductivity $<1.0\,\mu S/cm$ as feed.[34]

An EDI cell consists of mixed-bed IX resins sandwiched between an anion exchange membrane on one side and a cation exchange membrane on the other side, as shown in Figure 1.14. The IX membrane is ion selective, which means that the cation membrane allows only cations to pass through, and anion membrane allows only anions to pass through. Since the membrane polymer is hydrophobic, it is not permeable to water. An EDI module consists of multiple cell pairs (up to 240 cell pairs) stacked end-to-end. DC power is applied to a positive electrode (anode) located on one end of the module, and to a negative electrode (cathode), located on the other end.[34]

Feed water enters the product compartments in parallel, and flows from top to bottom. Because of the direct current that is applied perpendicular to the flow by the electrodes, the anions are driven toward the anode through the selectively permeable anion exchange membrane. Similarly, the cations are driven toward the cathode through the cation exchange membrane. The ions in the reject compartment are swept out of the module and sent to drain. As the water travels through the product compartments, it becomes increasingly pure. The DC voltage causes some of the deionised water toward the bottom of the product compartments to 'split,' i.e. the water molecules are broken down into hydrogen (H^+) and hydroxyl

(OH^-) ions. The H^+ and OH^- ions react to regenerate the mixed-bed resins at the bottom of the compartments continuously, thus, producing highly pure water. The polished water exits the EDI module as product. The electrochemical regeneration of the resin produces high-quality water consistently, without the use of chemicals for regenerating the resins.

EDI is used as a hybrid membrane process in the semiconductor, pharmaceutical, and beverage industries for producing high-purity water in tandem with RO as well as providing high-purity water for power plants replacing mixed-bed ion exchange polishers downstream of the RO unit.

1.4.7 Gas separation

Membrane gas separation (GS) has evolved rapidly in the last 25 years. Gas diffusion through non-porous membranes occurs due to a concentration gradient or for gases it is directly proportional to pressure. Hence, membrane GS is often described as a pressure-driven diffusion process. Separation is achieved because of differences in the relative transport rates of the feed components. A membrane will separate gases only if one or more components pass through the membrane more rapidly than others (Figure 1.15). Gas separation membranes are unique in that they are able to separate two or more gases (fluids) unlike liquid separation membranes which separate dissolved and suspended solids from the fluid.[35,36]

The membranes used in most commercial applications are non-porous in structure where separation is based on the solution–diffusion. This mechanism involves molecular scale interactions of the permeating gas with the membrane polymer. The model assumes that each component is sorbed by the membrane at one interface, transported by diffusion across the membrane through the

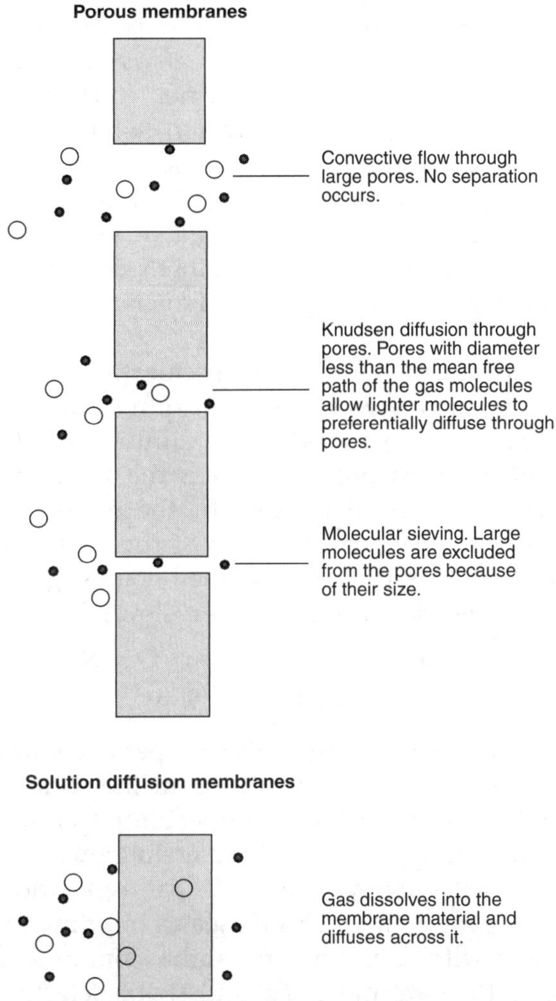

Figure 1.15 Schematic of membrane gas separation mechanisms. Source: Membrane Handbook, Ho and Sirkar, 1992

voids between the polymeric chains (the so-called 'free volume'), and desorbed at the other interface. According to the solution–diffusion model, the flux of gas through a membrane is given by,

$$J = P(p_f - p_p)/t \tag{13}$$

where P is gas permeability, p_f and p_p are the partial pressures of the gas on the feed side and the permeate side, respectively, and t is the thickness of the membrane. For commercial membranes, t is $\leq 0.2\,\mu m$. P is a property of the membrane material and is independent of membrane thickness. It is equal to gas solubility (S) in the polymer × diffusivity of the gas (D) in the polymer. Solubility is thermodynamic in nature and is affected by polymer–penetrant interactions as well as excess inter-chain gaps in glassy polymers. The diffusion coefficient is kinetic in nature, and is determined by the polymer–penetrant dynamics.

Gases can have high permeation rates due to high solubilities, high diffusivities or both. Generally speaking, the diffusion coefficient becomes smaller whereas the solubility increases with an increase in the molecular size of the gas. On the basis of their relative permeabilites through the membranes, gases have been classified as 'fast' or 'slow':[35]

H_2O, H_2, He, H_2S, CO_2 O_2 Ar, CO, N_2, CH_4
'Fast' 'Slow'

Thus, gases with higher permeabilites enrich on the permeate side of the membrane, while gases with lower permeabilities enrich on the feed/reject side of the membrane.

Virtually every polymeric gas separation membrane is either entirely glassy or primarily glassy with a minor crystalline component since they are more size and shape selective than rubbery polymers.[35] A rubbery polymer is an amorphous polymeric material that is above its softening or glass transition temperature, T_g under the conditions of use. Gas molecules permeate through a rubbery polymer quickly because the binding force between molecular segments of the polymer is not strong, and segments can move relatively easily to open a channel through which even large molecules can pass. Further, solubility often dominates diffusion characteristics for transport in rubbery polymers. Thus, volatile organic compounds (VOCs) which have a strong affinity (high solubility) for polymeric materials, permeate through a rubbery polymeric membrane much faster than the much smaller oxygen and nitrogen. Membranes prepared from rubbery polymers such as polydimethylsiloxane (PDMS) and composite PEI/silicone rubber membranes are, therefore, effective in removing VOCs from air. The permeability of rubbery membranes is 1000 times higher than the permeability of glassy membranes.[35,36]

A glassy polymer is an amorphous polymeric material that is below its softening or glass transition temperature, T_g, under the conditions of use. Because the chain segmental motions in glassy polymers are restricted as compared to rubbery polymers, glassy polymers are able to discriminate between extremely small differences in molecular dimensions of common gases (0.2–0.5 Å). Unlike rubbery polymers, transport in glassy polymers is not governed by solubility but by the size of the molecule and diffusion coefficient. Thus, since hydrogen molecules have the smallest size (high diffusion coefficient), membranes prepared from glassy polymers are used effectively for hydrogen separation. Similarly, in CO_2/CH_4 separation, carbon dioxide permeates through glassy membranes much faster than methane, partly because a CO_2 molecule is slightly smaller than a CH_4 molecule, and partly due to the stronger affinity of the polymeric materials to CO_2.

Typical examples of glassy polymeric membranes in commercial use are CA, PS, and PES. Aromatic PS is among the most successful GS membrane polymers. However, rigid polymers, such as PEI, and polyimide (PI), with higher T_g and greater resistance to organic solvents are required. PI membrane is highly chemically durable and is also thermal resistant.[30] Therefore, it is possible to eliminate special pre-treatment steps such as scrubbing, adsorption, absorption, and cooling. Polyimides can withstand temperatures up to 300°C. Due to the limitations of materials used in modules, however, 150°C is the upper limit of operating temperature. Higher temperature translates into fewer modules since the permeation rate of gases increases with the operating temperature. PI membranes are also highly resistant to widely used solvents such as methanol, ether, benzene, toluene, xylene, and gasoline. Perhaps the most important property is that PI membranes have a higher selectivity at a comparable permeability than other membranes for gas separations involving He/CH_4, CO_2/CH_4, H_2/CH_4, and O_2/N_2.[30,35] This is a major improvement given that, in general, membranes with higher permeability coefficients possess lower selectivity.

PI membranes are currently in use to separate hydrogen, helium, and carbon dioxide from various gases. Of these, H_2 separation/recovery is the largest market for PI membranes. One important application is in the separation of CH_4 from N_2. In most polymers used for gas separation membranes, the permeability of CH_4 is greater than that of N_2. However, in the case of aromatic PI, the reverse is true. Although the solubility of CH_4 in PI is higher than in other polymers, N_2 has a substantially higher diffusivity in PI than in other polymers. Advanced organic and inorganic membranes and materials being investigated include polymers of intrinsic microporosity (PIMs), microporous PVDF, perovskite and palladium alloy membranes.[37] PIM membranes have displayed both high permeability with high selectivity for various gas mixtures. The major commercial and promising applications of membrane GS are delineated below:[35–37]

- **Nitrogen production (90–99.5% purity).** This is likely to be one of the largest applications of GS membranes. The main applications include nitrogen blanketing of tanks and vessels, purging of process vessels and pipelines before shut downs for safety uses in the chemical process industry, and controlled atmosphere storage of fresh produce. The latter requires dry nitrogen for the preservation of perishable food. Membranes preferentially permeate oxygen, carbon dioxide, and water leaving a non-permeate (retentate) stream enriched in dry nitrogen, which is used to maintain an inert atmosphere during storage and transportation of fresh produce.
- **Hydrogen recovery.** Hydrogen is an expensive chemical feedstock. Hence, recovery of hydrogen from mixtures of hydrocarbon gases increases process efficiency and reduces operating costs. Hydrogen is recovered from naphtha during the synthesis of substitute natural gas (SNG), and from the purge gas from an ammonia synthesis reactor. The purge gas in ammonia synthesis contains nitrogen, argon, and helium. Unless recovered, it is purged with inert gases or used only for its fuel value. Installation of GS polysulphone hollow-fibre membranes in a high-pressure purge gas loop in ammonia plants to recover H_2 was the first major commercial breakthrough for gas membrane technology. Other commercial applications include recovery from tail gases from catalytic reforming, hydrocracking, and hydrodesulphurisation. Membrane GS has an edge over other technologies used in hydrogen recovery; cryogenic distillation is energy and cost intensive, and pressure swing adsorption is complex.
- **Synthesis gas ratio adjustment.** Synthesis gas is a mixture of hydrogen and carbon monoxide. It is generally produced from natural gas by steam reforming or by partial oxidation/steam reforming. It is the feedstock for manufacturing various chemicals including acetic acid, methanol, ethylene glycol, acetic anhydride, acetaldehyde, ethanol, and oxoalcohols. As produced, the H_2/CO ratio is >2.0 whereas

a ratio of between 1 and 2 is required for the production of the above chemicals. Membranes such as PI and PS are used to adjust the H_2/CO ratio by virtue of the higher diffusion rate of hydrogen.

- **Carbon dioxide removal.** Separation of carbon dioxide from nitrogen and low-molecular weight hydrocarbons is a multi-staged process because of the low selectivities. Several applications have achieved commercial success including the removal of CO_2 from natural gas and from biogas, and in enhanced oil recovery. In the treatment of natural gas, membranes are used to remove impurities such as CO_2, H_2S, and moisture to produce pipeline-grade gas. Similarly, in the treatment of biogas (produced from anaerobic digestion of sewage, municipal waste, agricultural waste, and landfills), these 'fast' gases permeate rapidly across the membrane yielding methane-rich retentate.
- **Dehydration.** The applications include the production of dry nitrogen and moisture-free natural gas, and methane from biogas.
- **Helium recovery.** Helium is abundant in natural gas resources. Because of its low boiling point and inert character it is used in many manufacturing facilities. Spiral-wound and hollow-fibre membranes have been commercialised for helium recovery from natural gas and for purification of crude helium contaminated with air.
- **Removal of VOCs.** Removal of various organic and other relatively high boiling species from vent gas streams is one of the largest growth areas of GS membranes. Membrane GS is able to remove much larger organic molecules compared to gases such as nitrogen, which usually make the bulk of vent streams. Even though the diffusion coefficient of the organic molecules are smaller than the permanent gases, they have a high permeability rate due to their much higher solubilities in siloxane-based rubbery polymers such as PDMS and composite PEI/silicone rubber membranes.

1.4.8 Pervaporation

Pervaporation (PV) combines the evaporation of volatile components with their permeation through a membrane. It involves the use of a liquid feed to produce a vapour permeate and a liquid reject. The feed is usually supplied at above 100°C and at a pressure slightly above 1 bar. The permeate pressure is maintained at vacuum well below atmospheric pressure. Vapourisation occurs as the permeating species pass through the membranes. Mostly asymmetric composite hydrophilic membranes, such as composite poly vinyl alcohol (PVA)/PS (or a polyacrylonitrile) are used.[35,36] Membranes made from a thin layer of PDMS cast on PAN have been found to be useful in separating polar compounds from non-polar ones in dealcoholisation of liqors. Flat-sheet and spiral-wound modules are commonly used.

The major applications of pervaporation (PV) include the removal of small amounts of water from organic solutions (e.g. drying of isopropanol/water, butanol/water, and ethanol/water azeotropes to produce a relatively pure organic chemical), drying of organic liquids by removing low concentrations of water (a few %) from nearly pure organics, and the removal of small amounts of organics from water (e.g. in wastewater cleanup). Ethanol purity levels of 99.85% are attained in separating ethanol/water mixtures using a distillation/PV hybrid process. It has been used to remove alcohols from fermentation broths where continuous removal of ethanol is required so as to not inhibit the production activity of micro-organisms. Similarly, PV units are used for recovering isopropanol from water with a purity of 99% alcohol starting with a feed containing 94% alcohol and 6% water. Water formed during the esterification reaction must be removed for reusing the alcohol.

Currently, the commercial sales of PV systems are very small because of unfavourable economics resulting from: (a) reheating of permeate between stages is required to sustain a meaningful flux, and (b) large membrane surface area is required due to a low driving force with pressure drops are limited to 1 bar.

1.4.9 Membrane contactors

Membrane contactors (MC) is a phase contacting device for use in gas absorption and stripping (degassing) processes as well as in biomedical gas transfer processes.[18,27,36] The function of the membrane is to facilitate diffusive mass transfer between contacting phases such as liquid–liquid, gas–liquid, and gas–gas. The membrane phase contactor uses polyolefins, e.g. PP microporous HF membranes, which are packed densely in a high surface area module.

Since membranes are hydrophobic and have small pores (0.05–0.1 μm), water does not pass through the membrane pores easily. The pressure required to force water to enter the pore is called the breakthrough pressure. The breakthrough pressure for a polypropylene membrane with a pore size of 0.05 μm is greater than 10 bar g.

A MC module contains thousands of microporous hollow fibres, which are knitted into a fabric that is wound around a distribution tube with a central baffle, as shown in Figure 1.16. The baffle ensures the water is distributed across the fibres, and also results in reduced pressure drop across the contactor. The HF are packed densely in a membrane module with a surface area of up to 4,000 m^2/m^3. The liquid flows outside (shell side) the hollow-fibre membrane, while vacuum is applied on the inside of the fibre (tube side), and forms a film across the pores of the membrane. Mass transfer takes place through this film and the pores due to the difference in the gas partial pressure between the shell side and tube side. Since the membranes are hydrophobic, they are not wetted by water, thereby, effectively blocking the flow of water through the membrane pores. The membrane provides no selectivity. Rather its purpose is to keep the gas phase and the liquid phase separated. In effect, the membrane acts as an inert support which allows intimate contact between gas and liquid phases without dispersion. Vacuum on the tube side of the membrane increases the mass transfer rate as

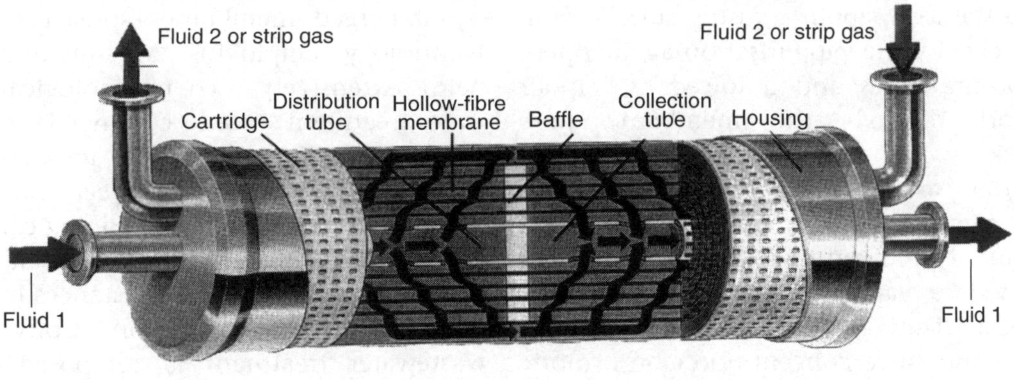

Figure 1.16 A Liqui-cell membrane contactor module

in a vacuum tower. The efficiency of the process is enhanced with the aid of nitrogen sweep gas flowing on the permeate side of the membrane.

Membrane contactors are used commercially for stripping dissolved oxygen from ultrahigh-purity water for semiconductors manufacture. Membrane contactors have also been commercialised in the beverage industry for carbonation (adding CO_2). Membrane contactors are in operation worldwide with most being used for gas stripping. Blood oxygenation is the earliest application of MC. A large number of microporous hollow-fibre oxygenators are in use in hospitals. One interesting potential application is in aerobic fermentation processes, which require O_2 for the cells in the fermentation broth, and simultaneously require the removal of CO_2 formed during fermentation.[38]

Membrane distillation (MD) and osmotic distillation (OD) are extensions of MC for applications when concentration of aqueous solutions with non-volatile solutes such as salts and sugars, e.g. in water desalination.[18,27] In the case of MD, two liquids at different temperatures are separated by a hydrophobic (non-wetting) microporous membrane. The temperature gradient generates a vapour pressure difference that results in vapour molecules permeating through the microporous membrane from the high vapour pressure side to the low vapour pressure side. Separation occurs by the vapourisation at the openings of the pores and followed by vapour transport through the membrane pore network.

In the case of OD, the permeate side of the membrane has a low vapour pressure solution, e.g. a concentrated salt solution, to provide a vapour pressure driving force. A hybrid UF–RO–OD process for concentration of fruit juice concentration by osmotic distillation has been realised.[27,39] Osmotic distillation being an athermal process allows for juice concentration without loss of flavour or product deterioration.

1.4.10 Other membrane technologies

Membrane reactors

Molecular separation along with simultaneous chemical transformation has been made possible with membrane reactors. The selective removal of reaction products increases conversion of product-inhibited or thermodyanamically unfavourable reactions, e.g. in the production of ethanol from corn.[26] Enzyme-based membrane reactors were conceived by UF pioneer Alan Michaels.[40] Membrane biocatalytic reactors are used for hydrolytic conversion of natural polymeric materials such as starch, cellulose, proteins, and for the resolution of optically active components in the pharmaceutical, agrochemical, food, and chemical industries.

Catalytic membrane reactors have been realised with the development of high-temperature operation membranes. Membrane reactors can produce hydrogen for fuel cell power plants from methanol or natural gas by steam reforming more efficiently, and with a higher quality as compared to conventional technologies;[27] the membrane reactor produces hydrogen while selectively removing hydrogen during conversion using solid oxide membranes.

Submerged membrane bioreactor (SMBR) technology employing MF membranes is being extensively used for biological wastewater treatment since its commercialisation in the 1990s by Kubota, USFilter and Zenon.[32,33] Besides hollow fibres, flat sheet and tubular membranes are also used. Biological wastewater treatment is used to remove all organic and nitrogen containing substances by anaerobic or aerobic conversion. Conventional wastewater treatment is composed of two reactors; an aerated and a non-aerated reactor.

In the SMBR process, the biological reactor is integrated with the immersed membrane unit, which is placed in the nitrification area. The air sparged along the membrane surface creates turbulence at the membrane surface, cleaning, or scrubbing it. Membrane bioreactors are able to remove more than 95% of the chemical oxygen demand (COD) while the reduction of biological oxygen demand (BOD) was even higher. The SMBR process is further discussed in Chapter 2.

Dynamically formed membranes (DFM)
DFM were developed at Oak Ridge National Laboratory in the 1960s for water desalination, but have been more effectively used to remove dyes from textile manufacturing process effluent. These membranes are formed by flowing materials such as zirconia colloid and zirconia polyacrylic complex in low concentrations (1,000 mg/l) over porous carbon or sintered stainless steel tubes. The performance of the membrane depends mainly on the characteristics of the membrane forming material; permeate flux and membrane retention are strongly dependent on the particle shape and size distribution. Membrane supports with pore sizes varying from 1 to 10 µm are typically used. If the membrane becomes damaged or clogged, it is removed or regenerated; hence, the term 'dynamic membrane'.[8,41] Other applications of DFM include lime softening treatment, toxic heavy metals removal, clarification of turbid waters, and for treating rinse water from semiconductor manufacturing plants.

A variation of the DFM is the supported liquid membrane used in rejuvenating or regenerating RO membranes.[42] Because the ultra-thin skin layer of RO membranes is fragile, it degrades with time by the use of strong chemical cleaning agents, abrasion, enzyme activity, aging, or manufacturing defects. This degrades membrane performance with a noticeable loss of salt rejection. The performance is recovered by adding polymers and/or surfactants in very low concentrations (25–100 mg/l) to the feed and recirculating the mixture until the rejection is restored. Regeneration requires that a relatively stable liquid membrane be created, i.e. show near-term permanency of more than 200 h. This is possible if the additives have the following properties:

- are good dispersants on low-energy surfaces such as synthetic polymers; tannic acid and lecithin (a lipid) are such materials;
- have polar and non-polar groups well spaced or, preferably, on opposite sides of the backbone; then hydrophobic groups can strongly adsorb to the membrane without being masked by the hydrophilic groups. Water dispersible polyesters were found to be very stable;
- have the capability to strongly adsorb either through physical adsorption or through electrostatic interaction. For example, positively charged quaternary ammonium chloride is highly effective in regenerating cellulose acetate membranes.
- can hydrogen bond inter-molecularly to the membrane material. Strong solute–membrane interactions are exhibited in systems with distinct hydrogen bonding.
- capabilities of the additive with functional sites of membrane matrix.

The most common regenerating chemicals used are polyvinyl acetate copolymer, polyvinyl methyl ether, tannic acid, and polyvinyl acetates.

Electrofiltration
Electrochemical effects can play a major role in increasing the performance of conventional MF and UF processes. According to the conventional Film Theory Model, the polarised layer is described as arising due

to the convective transport of particles to the membrane surface and their diffusive back-transport into the bulk solution, as shown in Figure 1.7. Charged particles experience an extra force, electrostatic repulsion, as they are brought closer together in the polarised layer. In the case of charged particles, therefore, flux augmentation due to electrostatic effects can be very significant for particles <100 nm in size. The main types of electrofiltration processes are summarised in Table 1.13.[43] Results of tests conducted with several MF membranes and colloidal solutions are summarised below:[44]

- Flux levels are an order of magnitude higher than in the absence of imposed force fields.
- Both electric and ultrasonic fields reduce membrane fouling caused by the deposition of colloidal material over a range of operating conditions.
- The rate of fouling is affected by electric field strength and frequency, concentration of colloids, particle size, particle shape, and the surface properties of the dispersed phase.
- Electric fields improve filtration by electrophoresis and other secondary electrokinetic effects like electroosmosis. Flux is enhanced with ultrasonic fields largely as a result a of cavitation.
- Synergistic effects were observed when electric and acoustic fields were applied simultaneously. The coupling mechanism is believed to be due to a combination of effects induced by cavitation and electrokinetic phenomena.
- The cross-flow velocity is an order of magnitude lower than in conventional cross-flow filtration.
- Assisted filtration in conjunction with low cross-flow velocities results in reduced power consumption and lower pumping costs, less degradation of shear sensitive streams, and reduced heat transfer requirements in recirculation systems for biological applications.

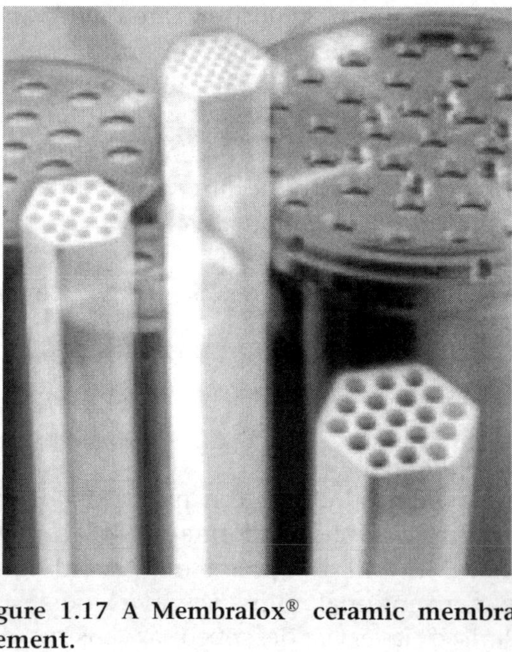

Figure 1.17 A Membralox® ceramic membrane element.

Polyelectrolyte-enhanced ultrafiltration (PEUF)
Large MW polyelectrolytes such as sodium polystyrene sulphonate (PSS) have also been used for removing multivalent cations. For example, PSS (MW = 85,000 Da) was used for softening water with 10,000 MWCO membranes.[45] The negatively charged colloidal PSS binds calcium and magnesium; rejections of up to 99.7% of hardness ions were obtained. The permeate flux was 70 $l/m^2/h$ at 4 bar g transmembrane pressure and 20°C, substantially higher than what is possible with RO and NF membranes. Similarly, cationic polyelectrolytes have been used to remove anions such as chromate.[46]

Activated carbon-enhanced filtration (ACEUF)
Ultrafiltration has been used to remove organic matter and THM precursors from surface water with the aid of powdered activated

Table 1.13 Electrically enhanced membrane filtration

Process	Electric field	Membrane	Remarks
Electrofiltration	Continuous	Polymeric, inorganic	Reduced concentration polarisation and fouling, improved selectivity*
Pulsed electrophoretic cleaning	Pulsed	Polymeric, inorganic	Reduced fouling, low power consumption, minimal changes in feed attributable to electrolysis
Electrolytic membrane cleaning	Pulsed	Electrically conducting	Reduced fouling, low power consumption; works in all conductivity ranges
Electrolytic membrane restoration	Pulsed	Polymeric, inorganic, or electrically conducting	Cleaning membranes under mild chemical conditions
Electroosmotic backwashing	Pulsed or continuous	Polymeric, inorganic	Reduce fouling, low power consumption, minimal changes in feed attributable to electrolysis

* Synergistic effects are observed when electric and acoustic fields are applied simultaneously.
Source: Bowen; Tarleton and Wakeman

carbon (PAC). Similarly, tubular ceramic MF membranes with a pore size of 0.2 μm were used to treat landfill leachates.[47] Leachate is a difficult water to treat because it contains high quantities of organic matter, colour, heavy metals, and suspended solids. The permeate flux was 180 $l/m^2/h$ at 1 bar g feed pressure, 70% higher than with no PAC treatment. The flux increased with increasing dosages of PAC, even though the suspended solids concentration also increased. There were two reasons for enhanced performance: (a) more colloids are adsorbed at higher PAC loadings, and (b) PAC acts as an abrasive and helps to remove the gel layer deposited on the membrane.

Micellar-enhanced ultrafiltration (MEUF)

Because of their unique aggregation and solubilisation properties, surfactant micelles have been used for removing low molecular weight organic compounds by UF.[45] MEUF combines the high selectivity of RO with the higher flux of UF. The basis of MEUF is that at concentrations above the critical micellar concentrations (CMC), surfactant molecules attach to each other forming organised aggregates or micelles. These large aggregates are then easily rejected by UF membranes. The micelles form in such a way that the hydrophobic portions of the molecules align towards the centre away from the surrounding hydrophilic aqueous solution. These aggregates solubilise organics such as volatile organic compounds and hydrocarbons.

The MEUF process is based on the formation of micelles and on the increased solubility, in a micellar solution, of an organic, which is sparingly soluble in water. The overall result is a two-phase system, (a) the dispersed or discontinuous phase consisting of micelles, and (b), the continuous or aqueous phase consisting of surfactant monomers. Both these phases exist in a dynamic equilibrium. This concept of phase separation suggests that the permeate concentration should be constant. Experimental data show that this is indeed the case.[48]

MEUF is also used to remove multivalent heavy metal ions with ionic surfactants. The ionic micellar surface has a high charge density and a high absolute electrical potential. Therefore, the heavy metal cations electrostatically adsorb onto or near the micellar surface formed by anionic surfactants such as SDS and sodium alkylbenzene sulphonate. Similarly, cationic surfactants, e.g. cetylpyridinium chloride has been shown to be effective in removing multivalent hazardous anions. Non-ionic surfactant micelles are larger, hence more effective, as discussed in Chapter 5.

Fractionation of macromolecules

Fractionation of proteins by UF has been an important area of membrane development. Because of factors such as

(a) the deformation of polymer chains induced by shear forces,
(b) concentration polarisation,
(c) molecular weight distribution, and
(d) membrane pore-size distribution, fractionation was only possible when the MW of two proteins to be separated differed by an order of magnitude.

With the use of charge-modified membranes, however, it has become possible to fractionate mixtures of proteins with MW close to each other,[18,43] e.g. myoglobin (MW = 17,500) and cytochrome C (MW = 12,400). Proteins have a net zero charge at their isoelectric points (pI). At a pH value higher than pI, proteins are negatively charged, and when the pH < pI, protein are positively charged. To illustrate the point, during the separation of myoglobin from cytochrome C (CyC), negatively charged UF membranes were used at a pH = 9.2 which is also the isoelectric point of CyC (pI = 9.2). Since myoglobin (pI = 7.09) has a net negative charge at this pH, it was rejected by the negatively charged membrane whilst CyC permeated completely through the membrane. Conversely, at a pH of 5.5, CyC was rejected by a positively charged membrane and myoglobin passed through the pores.

Similarly, amino acids with widely different pI values can be separated by varying the pH of the solution. It follows that below the pI of the protein, when the protein is positively charged, a positively charged membrane is less prone to fouling, resulting in a higher flux. Note, also that when the pH < pI, proteins foul hydrophilic membranes which are mostly negatively charged.

1.5 Membrane modules

1.5.1 Conventional modules

Characteristics of membrane modules such as ceramic filters shown in Figure 1.17, are summarised in Tables 1.12 and 1.14. The spiral-wound (SW) and hollow-fibre (HF) modules shown in Figures 1.18 and 1.19 are the most commonly used. The SW module is the most popular for large-scale RO, NF, and UF modules, whereas the HF modules are used in UF and MF applications. One of the earliest UF modules developed and used extensively in the dairy industry, it has largely been replaced by SW modules.

The tubular membrane (TM) modules shown in Figure 1.20 are 1.5–3 cm in diameter and have a very low surface area. Tubular membranes are usually preferred to HF and SW membranes for industrial wastewater applications; TM modules are less prone to severe polarisation and fouling since they are operated under turbulent flow conditions (Re > 3,000). The TM module is best suited for feeds with high suspended solids since it can handle particulates and can be physically cleaned with foam balls. Although, TM plants consume 50% more power than HF or SW plants, TM are relatively inexpensive ($200 compared to $1500 for a HF), and have a longer life (5 years compared to 3 years for HF).

Shear sensitive biological materials require gentle operating conditions; hence, turbulent flow conditions are not desirable. In contrast to TM modules, SW modules offer gentler flow fields above the membrane surface. Plastic mesh spacers are used in SW modules to promote localised turbulence above the surface of the membrane. The SW modules come in standard sizes of 2.5, 4, and 8 in. diameter × 40 in. long. The elements fit in a pressure vessel connected in series by O-rings up to six elements per vessel.

1.5.2 Recent developments

High-performance SW modules

In the production of water for pharmaceutical and beverage applications, full-fit modules are now used. Full fit means there is no brine seal that provided dead spaces and potential areas of bacterial growth (see Figure 2.16). Because of this design, the chemicals can be rinsed out faster. In addition, hot water sanitisable (up to 80°C) RO membrane elements are now available. Other new spiral-wound designs include energy-saving polyamide membrane (ESPA®) module commercialised by Hydranautics, ultra-low pressure RO membrane (TFC-ULP®) introduced by Dow/Film-Tec and high surface area Magnum® modules manufactured by Koch.

Ceramic modules

Ceramic membrane modules are tubular in design, as shown in Figure 1.17. They are the most expensive due to the low surface areas of the module (Table 1.14). Ceramic membranes, however, are good candidates for immobilised whole cell reactors, cell harvesting, and protein purification because of their excellent chemical, thermal, and mechanical properties, as discussed earlier. A novel approach to increase the surface area and, thereby, reduce the capital cost is to use honeycomb monolith ceramic filters containing up to 10 channels/cm^2 have been modified to produce UF and MF modules.[21,26] The filters are made of alumina with ZrO_2 layers with pore sizes between 40 and 500 Å.

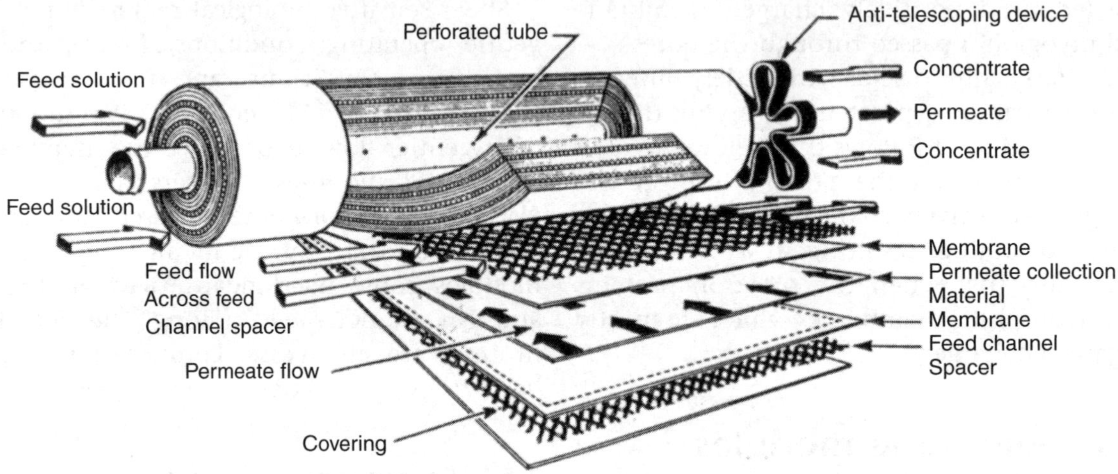

Figure 1.18 A spiral wound membrane module

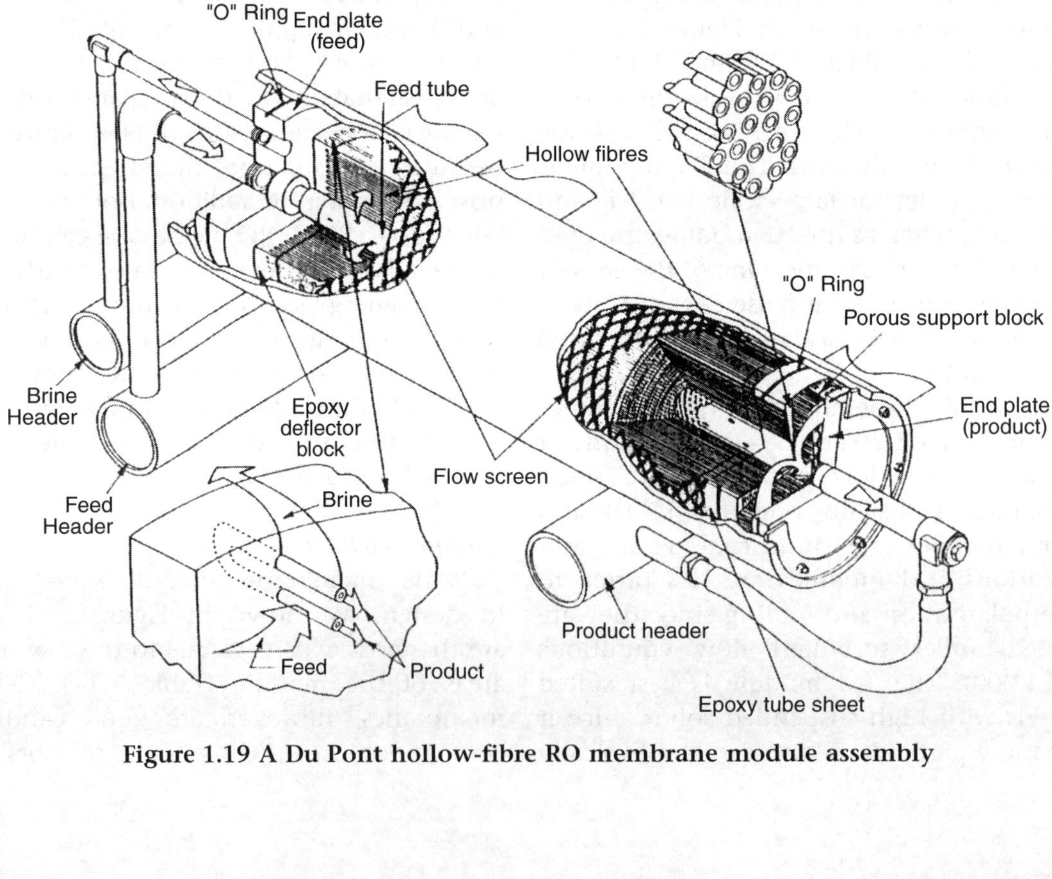

Figure 1.19 A Du Pont hollow-fibre RO membrane module assembly

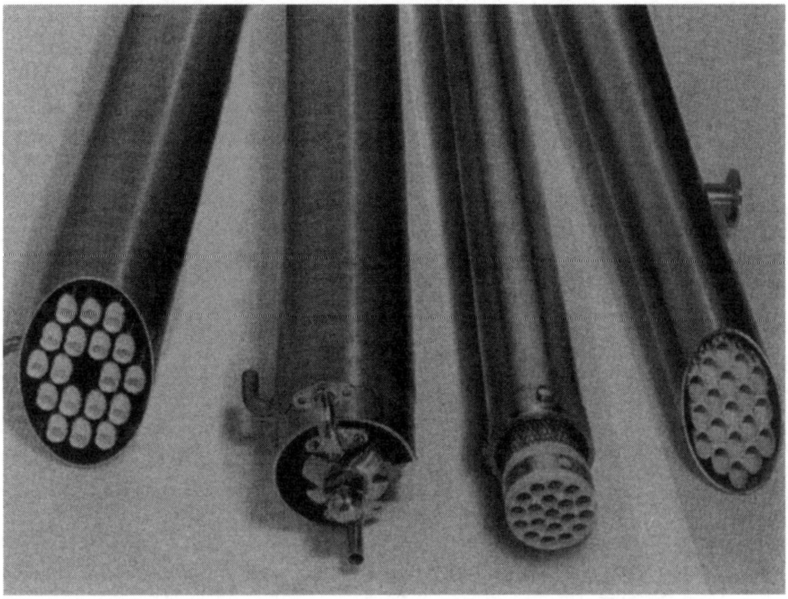

Figure 1.20 Typical tubular membrane modules

Typically, a 100 cm long × 15 cm diameter module has a membrane surface area of 10 m^2.

Short flow-path modules

A few years ago, during a visit to an industrial laundromat that was using tubular PVDF UF membranes to remove emulsified oil, grease and prioity pollutants, we observed that there was no permeate flow in the end section of membrane modules, the so-called 'far downstream' section. This came as no surprise to us; the solute concentration profile continues to grow with the length of the membrane, as shown in Figures 1.7 and 1.21. The severity of concentration polarisation with membrane length in UF modules was proven with the help of mathematical models in the 1970s.[14,15] Short-length UF membrane modules have, therefore, been recommended[16,42] in order to reduce concentration polarisation and fouling.

Short flow-path disc-tube (DT) RO membrane modules developed by ROCHEM for treating highly turbid waters and fouling feeds are now used extensively to minimise concentration polarisation, scaling, and fouling in certain applications. Typically, for RO systems the Silt Density Index (SDI) of feed water must be less than 3 to prevent excessive fouling. An empirical measurement, SDI is the method of choice for measuring the colloidal fouling potential of RO feed water (ASTM standard D-4189–82, 1987). Because of their unique design, DT modules can tolerate feed water with an SDI value as high as 16.[49] The salient features of the DT module shown in Figure 1.22 are:

- The extremely short flow path (7.5 vs. 100 cm for the shortest TM, SW, or HF module) limits concentration polarisation.
- High permeate flux; 70 l/m^2/h at 70 bar g.
- The delayed formation of the gel layer allows longer cleaning intervals and, thereby, extends membrane life. The flow is turbulent.
- The open channels allow efficient cleaning of the membranes.
- The packing density is high; 150 m^2 of membrane area/m^3 of module.

Table 1.14 Characteristics of various modules

Characteristics	Module type				
	Hollow fibre	Spiral wound	Tubular	Plate-and-frame	Ceramic
Relative expense	Low	Low	High	High	Very high
Packing density	~1800–3600	~300	~20	~150	~180
Parasitic pressure drops	High	Moderate	Low	Low	Low
Fouling propensity	High	High	Low	Moderate	Low
Ease of cleaning	Fair	Poor	Excellent	Good	Excellent
Typical feed stream filtration requirement[b]	$<5\,\mu m$	$<10\,\mu m$	Not required	$<10\,\mu m$	N/A

N/A: Not applicable.
[a] Includes the cost of membrane and modules.
[b] Applicable to RO and NF systems.

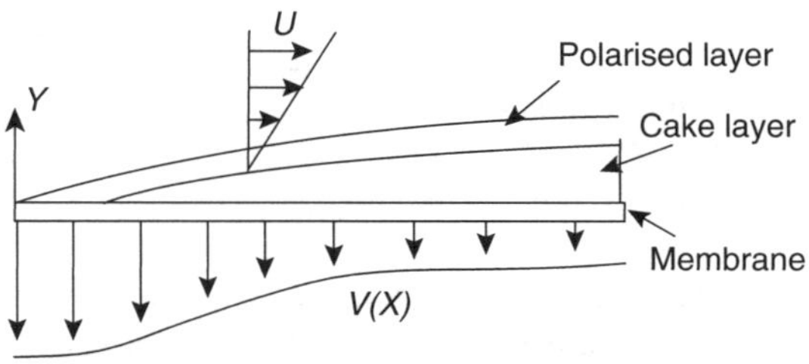

Figure 1.21 A schematic of concentration polarisation and permeate flux, *V*(*x*) profiles. *U* is axial velocity. Source: Schafer et al.

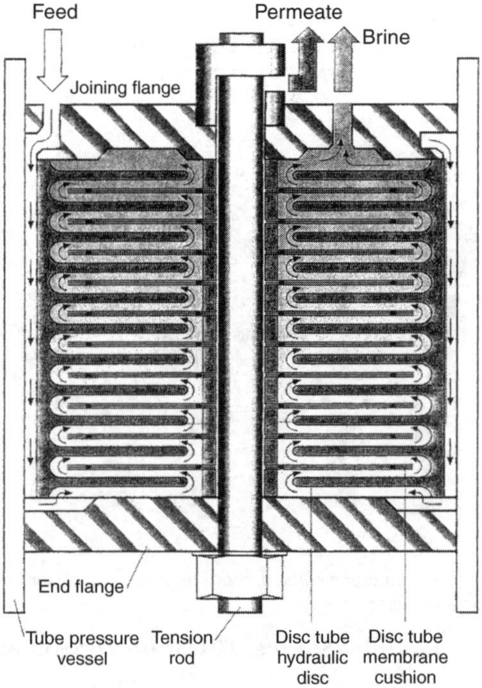

Figure 1.22 ROCHEM short flow-path membrane module. Each disc is ~8 cm long

Advanced PAF membrane module

A novel plate-and-frame (PAF) membrane and module design consists of a composite membrane, wherein a proprietary polymer is coated on a porous grooved support, as shown in Figure 1.23. The uniqueness of the design lies in the enhanced surface area achieved by forming a membrane on the channel walls, thereby, increasing the surface area 2–3 times, which is not the case with the existing PAF modules commonly used in pervaporation applications. By providing high surface area modules, the relative cost of PV could be reduced. Further, to increase surface shear and control fouling, short, narrow, non-uniform, multi-path channels are formed. These turbulence-promoting pathways, without any fouling prone constrictions such as used in SW modules, could go a long way in mitigating fouling in UF and MF membrane systems.

Flux-enhancing techniques

Several flux-enhancing techniques are summarised in Table 1.15.[12] Among the commercialised modules are high shear modules such as the vibratory VSEP module shown in Figure 1.24, the discover rotating disc module and the vortex flow filtration rotating cylinder module. These modules are designed to

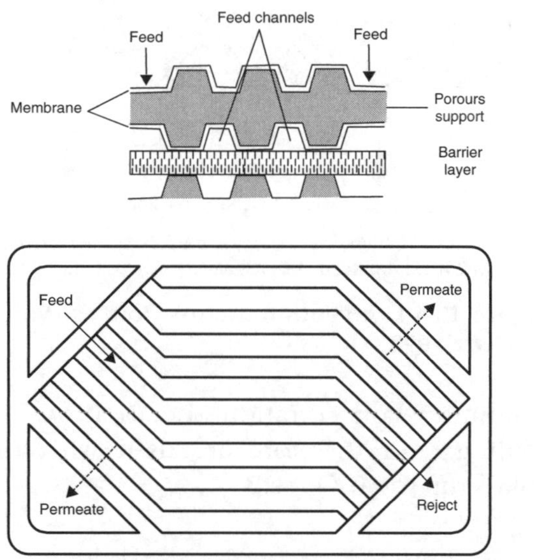

A novel membrane module design. Top: Cross-sectional view of a membrane-coated channel. Bottom: Channel flow pattern. High surface-area modules could reduce the relative cost of pervaporation. The short, narrow, nonuniform, multipath channels that create higher shear are formed. The turbulence promoting pathways could go a long way to mitigate fouling in ultrafiltration and microfiltration membrane systems.

Figure 1.23 Advanced module design concept schematic

Table 1.15 Flux-enhancing strategies

Process/technique	Effect/comments
Feed-channel spacers – spiral-wound modules and some flat-sheet modules	Eddies and increased shear
High shear – vibrating membrane	Generates high shear at membrane surface
High shear – rotating the membrane	Generates high shear and Taylor vortices
High shear – rotor above the membrane	Generates high shear at membrane surface
Dean vortices	Induced flow in curved channels
Pulsing the feed	Unsteady-state flow generates eddies
Baffles	Eddies. Enhances pulsations
Air bubbling	Bubbles generate eddies and shear transients
Back-pulsing	Permeate reversal removes deposits. Best for hollow-fibre MF and UF membranes

Source: Fane

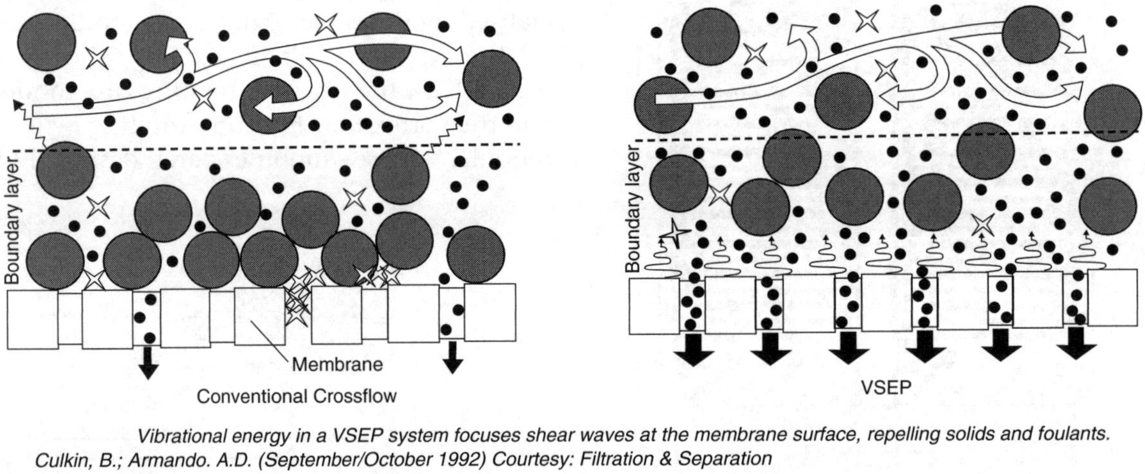

Figure 1.24 Conventional cross-flow vs. VSEP membrane module. Source: Filtration & Separation, Sep/Oct 1992

minimise concentration polarisation and fouling, and therefore are used with high solids (up to 50%) feeds.

1.6 Membrane fouling

1.6.1 The fouling phenomena

The performance of membrane processes is severely affected by fouling. Fouling causes a loss in water flux and quality, reduced operating efficiency, lost service time, premature membrane replacement (membrane costs vary from 20–30% of installed costs for RO plants to 50–75% for UF/MF plants) and higher operating costs. Typically, fouling factors for RO, UF, and MF are 30, 80, and 90%, respectively. Membrane fouling is frequently confused with concentration polarisation. The causes of fouling vary depending on the nature of the solute and solute–membrane interactions.

Fouling is often the result of a strong interaction between the membrane and the components in the feed stream, for example fouling by colloids, iron, and biomaterials can be especially severe. As a rule, a flux reduction that is reversible is due to concentration polarisation, whereas an irreversible flux reduction is due to fouling. Flux drop is usually recoverable following cleaning. Similarly, the precipitation of sparingly soluble salts leads to scaling that can be prevented by a well-designed pretreatment system, and minimised by defining operating conditions. Not surprisingly prevention and reduction in the rate of fouling and concentration polarisation are important factors in the design and operation of membrane systems; fouling results in higher operating costs due to frequent shutdowns for cleaning that in turn affects membrane life adversely. In addition, anticipated lower throughput warrants more membrane surface area resulting in higher capital and operating costs.

The flux of a solvent through a membrane is roughly proportional to the applied pressure as discussed earlier. For liquids other than pure water, however, this proportionality does not exist and the permeate flow may be as low as 20% of that of pure water. The permeate flux stabilises after an initial rapid decline. The steady-state flux value is a function of the nature of the feed (viscosity, solute concentration and diffusivity, pH, ionic strength) and operating conditions such as fluid shear rate at the membrane surface and transmembrane pressure drop.

Most fouling substances are hydrophobic and carry a surface charge.[18,19,26] Proteins such as bovine serum albumin (BSA) are readily adsorbed on the surface of hydrophobic membranes due to hydrophobic interactions, than hydrophilic solutes, e.g. dextran. Adsorption from only 10 mg/l BSA solutions can result in a 25% loss of flux with an accumulation of less than 10 mg/m^2 of protein (corresponding to a full monolayer of protein on the surfaces). The adsorbed layer is also more difficult to remove from a hydrophobic membrane than from a hydrophilic membrane. Similarly, during the removal of natural organic matter using NF membranes, the flux decline for the hydrophobic organic matter fraction was 60% compared to 20% for the hydrophilic fraction. Typical substances that foul membranes are given in Table 1.16.

Water purification RO and NF plants, therefore, rely on extensive feed water pretreatment to control fouling, as discussed in Chapter 2. Thus, water should not have any colloidal silica or iron, nor should the sparingly soluble salts in the reject stream above the membrane be allowed to exceed their solubility limits by limiting product water recovery. In addition, systems are designed to prevent biofouling; the slightly negative charge of bacteria and cell hydrophobicity result in the formation of a biofilm gel layer.[51] Usually, a rapid flux decline in RO systems indicates the presence of foulants rather than the mineral scale.

Table 1.16 Sources of membrane fouling

Substance	Extent and/or mechanism
Fe, Mn, Al hydroxides	Severe fouling, rapid kinetics
Mineral salts[1]	Form mineral scales when their solubilitiy is exceeded
Colloids	Electrically charged; SDI[2] and Zeta potential determine fouling
Microbiological	Forms a biofilm gel layer
Proteins	Fouling by hydrophobic and charge interactions
Polyelectrolytes	Fouling by charge interaction
Organic acids	Humic and fulvic acids cause severe fouling
Oil and grease	Hydrophobic membrane fouling
Suspended solids[3]	Cannot exceed 0.5 ppm

[1] $CaCO_3$, $CaSO_4$, $BaSO_4$, $SrSO_4$
[2] Silt Density Index
[3] Applicable to RO/NF

1.6.2 Fouling control techniques

Hydrophilic vs. hydrophobic membranes

Many of the membrane polymers today do not contain the desired properties; they are susceptible to fouling or are not durable. One way to change the membrane property is to modify the membrane before filtration either permanently (polymer blending, surface modification) or reversibly (dynamic membranes). Hydrophilic membranes, e.g. regenerated CA are less susceptible to irreversible fouling than hydrophobic membranes such as PVDF, PS, PES, and PP. The influence of the membrane material on the fouling of membranes shows that the flux of hydrophilic membranes is only marginally reduced, whereas the flux reduction of hydrophobic membranes is significant.[21,26]

In general, hydrophilic polymers such as cellulosics and aliphatic polyamides (nylon-4,6, nylon-6, nylon-6,6, nylon-11, nylon-12) have restricted chemical and thermal stability. Blending of two polymers, one to give the membrane sufficient chemical and thermal stability, the other to render the membrane more hydrophilic, therefore, has been used to prepare stable UF hydrophilic membranes,[18] e.g. a blend of hydrophilic polymer, polyvinylpyrrolidone (PVP) and hydrophobic PS and PES polymers.

Surface modification

Among the surface modification techniques recently introduced is surface fluorination. Fluorination of PS membranes nearly doubled the flux when treating potato starch waste (0.6% solids, mainly starch and proteins) with a 100,000 Da MWCO membrane.[38] Fluorination makes the membrane surface hydrophilic (the contact angle reduced from 90 to 56°) by adding fluorine and oxygen to the membrane polymer.

Polyamide TFC composite RO membranes are being increasingly used for water demineralisation. However, their application in treating wastewater effluents has been less successful because of severe fouling. When these membranes were fluorinated with hydrofluoric and fluorosilicic acids, the surface-modified membrane showed an order of magnitude increase in flux without any reduction in rejection.[52]

Like PS membranes, PVDF membranes have high chemical and thermal stability but are sensitive to fouling. Asymmetric PVDF membranes were coated with a hydrophilic layer (less than 1.0 μm) of polyether/polyamide block co-polymer to produce composite hydrophilic composite UF membranes. The performance was comparable to cellulosic membranes; lower susceptibility to fouling and reduced flux decline during ultrafiltration of wastewater oil emulsions containing 6 (wt.%) oil. Similarly, PVDF MF membranes, surface treated by grafting 2-methacryloyloxyethyl phosphorycholine (MPC), showed a flux decline of only 10% as compared to a decline of 86% for the uncoated membrane during the filtration of BSA solutions, as shown in Figure 1.25.[53]

Polymer modification

The development of modified charged UF membranes proved to be a major success for electrocoat paint recovery, one of the leading UF applications.[21,26] There are two types of electrocoating processes: anodic deposition where the item to be painted is positively charged and the paint is negatively charged, and cathodic deposition where the item to be painted is negatively charged and the paint is positively charged. In order to avoid fouling by charge interactions, positively charged membranes are used for cathodic paint recovery, and negatively charged membranes are used for streams for recovering anodic paints (see Figure 2.15).

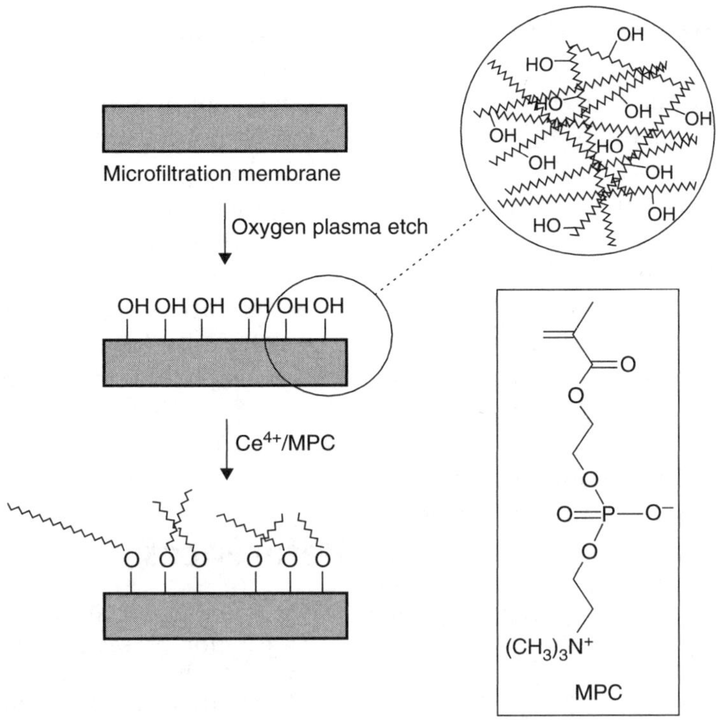

Figure 1.25 An example of surface treatment of a PVDF microfiltration membrane to minimise fouling. Source: Akhtar et al.

Ceramic/polymeric hybrid membranes

The advantages of the ceramic UF and MF membranes discussed earlier in this chapter are often offset by their low selectivity, which makes their use economically unfeasible for many applications. These membranes separate solutes from solvents primarily by site exclusion, and to a lesser extent, by interactions with the membrane surface. The selectivity of ceramic membranes can be improved by modification of the membrane surface either by polymer grafting or graft polymerisation. The latter has the advantage of providing a covalently bonded brush layer of high surface coverage with minimal diffusional limitations and steric hindrance effects.

A composite ceramic membrane was formed[54] by the graft polymerisation of a hydrophilic polymer, PVP onto the surface of silica membranes (pore size = 3.0 μm). The flux of an unmodified membrane during UF of an oil/water emulsion (4.7%), decreased with time as compared to the flux of the composite membrane. The decline in flux was caused by fouling and/or the immediate formation of an oil gel layer on the surface of the unmodified membrane. The modified membrane, in contrast, was not only more resistant to adsorption of oil, but also had a higher oil rejection. The performance of the modified ceramic membrane depends upon the configuration of the grafted chains in response to solvent–polymer interactions. Thus, the hydrophilic PVP polymer chains tend to expand or extend away from the surface in aqueous solutions, preventing oil adsorption on the membrane surface. Simultaneously, the hydrophilic polymer allows the passage of water molecules preferentially over oil.

1.7 Concluding remarks

Membrane technology has progressed remarkably in the last 25 years from an understandably slow start in the 1950s. Membrane separation science and the latest developments are discussed in this chapter. Research is continuing for addressing fouling problems by designing new modules and membranes to increase membrane flux and performance especially as possibilities of new applications are realised, and to tackle existing process problems.

Recently several membrane technology areas that require special attention for improving membrane performance and reducing operating costs of water treatment plants were identified by the U.S. National Research Council.[55] These high priority research topics included:

- Improving RO membrane permeability while maintaining existing salt rejections in order to operate at a lower feed pressure for a given module cost.
- Developing membranes with fouling-resistant surfaces, e.g. the development of polymers that resist biofouling – the spontaneous and unwanted adsorption of proteins, cells, and bacteria on surfaces – has been reported.[56] Coatings made from such polymers could help prevent fouling.
- Developing membrane modules and configurations that minimise concentration polarisation and upstream pretreatment for removing particulates, e.g. increasing cross-flow velocity, introducing Dean vortices and/or using backwashable UF or MF systems.
- Improving membrane selectivity or developing membranes with selective rejection capabilities.
- Improving membrane integrity by developing defect-free membranes as well as developing RO/NF membranes resistant to chlorine.
- Developing online sensors for determining the integrity of the membranes.
- Exploring membrane process design configurations and materials to reduce costs, e.g. dual membrane and hybrid membrane systems with process optimisation to enhance higher recovery and product purity at lower costs. Several hybrid systems are discussed in Chapters 3 and 4.

References

1 W.J. Koros, Evolving beyond the thermal age of separation processes: membranes can lead the way, AIChE J. 50 (10) (2004), 2326–2334.
2 H. Strathmann, Membrane separation processes: current relevance and future opportunities, AIChE J. 47 (5) (2001), 1077–1087.
3 H.K. Lonsdale, The growth of membrane technology, J. Memb. Sci. 10 (1982), 81–181.
4 S. Loeb, S. Sourirajan, Sea water demineralization by means of an osmotic membrane, Advances in Chemistry Series #38 (1963), 117–132.
5 S. Loeb, The Loeb-Sourirajan membrane: how it came about, ACS Symposium Series #153 (1981), 1–9.
6 R.J. Petersen, Composite reverse osmosis and nanofiltration membranes, J. Memb. Sci. 83 (1993), 81–150.
7 A.I. Schafer, A.G. Fane, T.D. Waite, eds., Nanofiltration, Elsevier: Oxford, UK (2004).
8 H.K. Lonsdale, The evolution of ultrathin synthetic membranes, J. Memb. Sci. 33 (1987), 121–136.
9 T. Matsuura, Synthetic Membranes and Membrane Separation Processes, CRC Press: Boca Raton, FL, USA (1994).
10 J.D. Ferry, Ultrafilter membranes and ultrafiltration, Chem. Rev. 18 (1936), 373–455.
11 A.S. Michaels, Ultrafiltration, in Progress in Separation and Purification, Vol. I. (ed. E.S. Perry), Interscience: New York, NY (1968), 297.

12 G. Anthony Fane, Module design and operation, in Nanofiltration (eds. A.I. Schafer, A.G. Fane, T.D. Waite), Elsevier: Oxford, UK (2004), 67–88.
13 W.F. Blatt, A. Dravid, A.S. Michaels, L. Nelsen, Solute polarization and cake formation in membrane ultrafiltration: causes, consequences, and control techniques, in Membrane Science and Technology (ed. J.E. Flinn), Plenum Press: New York, NY (1970), 47.
14 R. Singh, R.L. Laurence, Influence of slip velocity at a membrane surface on ultrafiltration performance – I. Channel flow system, Int. J. Heat and Mass Transfer 22 (1979), 721–729.
15 R. Singh, R.L. Laurence, Influence of slip velocity at a membrane surface on ultrafiltration performance – II. Tube flow system, Int. J. Heat and Mass Transfer 22 (1979), 731–737.
16 M.C. Porter, Concentration polarization with membrane ultrafiltration, Ind. Engn. Chem. Product Res. Dev. 11 (1972), 234–248.
17 R.E. Kesting, Synthetic Polymeric Membranes, John Wiley & Sons: New York, NY (1985).
18 M. Mulder, Basic Principles of Membrane Technology, Kluwer Academic Publishers: Holland (1997).
19 L.J. Zeman, A.L. Zydney, Microfiltration and Ultrafiltration, Marcel-Dekker: New York, NY (1996).
20 K. Scott, Section 2, in Handbook of Industrial Membranes, Elsevier: Oxford, UK (1995).
21 W. Eykamp, Section 22, Alternative separation processes, in Perry's Chemical Engineers' Handbook (eds. D.H. Perry, D. Green) 7th Edition, McGraw-Hill: New York, NY (1999).
22 S.V. Cabibbo, D.B. Guy, A.C. Ammerlaan, A. Ko, R. Singh, Reverse Osmosis Technical Manual, NTIS Publication: Springfield, Virginia, USA (1979).
23 T. Matsuura, Future trends in reverse osmosis membrane research and technology, in Reverse Osmosis (ed. Z. Amjad), Van Nostrand Reinhold: New York, NY (1993), 37–75.
24 M. Elimelech, X. Zhu, A.E. Childress, S.J. Hong, J. Memb. Sci. 127 (1997), 101.
25 J. Glater, M.R. Zachariah, ACS Symposium Series, 281 (1985), 345.
26 M. Cheryan, Ultrafiltration and Microfiltration Handbook, CRC Press: Boca Raton, FL (1998).
27 E. Drioli, M. Romano, Progress and new perspectives on integrated membrane operations for sustainable industrial growth, Ind. Eng. Chem. Res. 40 (2001), 1277–1300.
28 E. Park, S.M. Barnett, Oil/Water separation using nanofiltration membrane technology, Sep. Sci. & Tech. 36 (7) (2001), 1527–1542.
29 L.D. Nghiem, A.I. Schafer, Trace contaminant removal with nanofiltration, in Nanofiltration (eds. A.I. Schafer, A.G. Fane, T.D. Waite), Elsevier: Oxford, UK (2004), 480–520.
30 H. Ohya, V.V. Kudryavtsev, S.I. Semenova, Chapter 6, in Polyimide Membranes, Gordon and Breach: Amsterdam, The Netherlands (1996).
31 C.A. Heath, G. Belfort, in Advances in Biochemical/Biotechnology Vol. 47 (ed. A. Fiechter), Springer-Verlag: New York (1993).
32 J.M.K. Timmer, J.T.F. Keurentjes, Nanofiltration bioreactors, in Nanofiltration (eds. A.I. Schafer, A.G. Fane, T.D. Waite), Elsevier: Oxford, UK (2004), 416–433.
33 G.T. Daigger, B.E. Rittman, S. Adham, G. Andreottiola, Are membrane reactors ready for widespread application? Env. Sci. & Tech. 39(19) (2005), 59–68.
34 G.C. Ganzi, Electrodeionization for high purity water production, AIChE Symposium Series No. 261, Vol. 84 (1988), 73.
35 R.E. Kesting, A.K. Fritzche, Chapter 7, in Polymeric Gas Separation Membranes, John Wiley & Sons: New York (1993).
36 J.L. Humphrey, G.E. Keller II, Chapter 5, in Separation Process Technology, McGraw-Hill: New York, NY (1997).
37 Membranes for Gas Separations, Chem. & Eng. News, 83 (40) (2005), 49–57.
38 K.K. Sirkar, Membrane separation technologies: current developments, Chem. Eng. Comm. 157 (1997), 145–184.
39 P. Hogan, R.P. Canning, P.A. Peterson, R.A. Johnson, A.S. Michaels, Osmotic distillation: a new option, Chem. Eng. Progress, 94 (7) (1998), 49–61.
40 A.S. Michaels, Membrane technology and biotechnology, Desalination, 35 (1980), 329–351.
41 D.G. Thomas, Chapter 14, Dynamic membranes – their technological and engineering aspects, in Reverse Osmosis and Synthetic Membranes (ed. S. Sourirajan), National Research Council Canada: Ottawa, Canada (1977).

42 R. Singh, Surface properties in membrane filtration, Chem. Eng. Prog. 85 (6) (1989), 59–64.
43 W.R. Bowen, Chapter 8, in Membranes in Bioprocessing (eds. J.A. Howell, V. Sanchez, R.W. Field), Chapman & Hall: London, UK (1993).
44 E.S. Tarleton, R.J. Wakeman, Electro-acoustic crossflow microfiltration. Filtr. & Sep. May (1994), 274.
45 J.F. Scamehorn, S.D. Christian, R.T. Ellington, Use of micellar-enhanced ultrafiltration to remove multivalent metal ions from aqueous streams, in Surfactant-Based Separation Processes (eds. J.F. Scamehorn, J.H. Harwell), Marcel-Dekker: New York, NY (1989).
46 A. Tabatabai, J.F. Scamehorn, S.D. Christian, Water softening using polyelectrolytes-enchanced ultrafiltration, Sep. Sci. & Tech. 30 (2) (1995), 211.
47 C. Vishvanathan, S. Muttamara, S. Babel, R. Ben Aim, Treatment of landfill leachate by crossflow microfiltration and ozonation, Sep. Sci. & Tech. 29 (3) (1994), 315.
48 R. Singh, Investigation of ultrafiltration rejection of surfactant micelles by dynamic light scattering, Sep. Sci. and Tech. 31 (9) (1996), 1351–1356.
49 Th.A. Peters, Membrane Technology: Applications to Industrial Wastewater Treatment (ed. A. Caetano), Kluwer Academic Publishers: Boston, MA (1995), 175.
50 R. Bradley, Design considerations for reverse osmosis systems, in Reverse Osmosis (ed. Z. Amjad), Van Nostrand Reinhold: New York, NY (1993), 104–138.
51 H.-C. Flemming, Mechanistic aspects of reverse osmosis membrane biofouling and prevention, in Reverse Osmosis (ed. Z. Amjad), Van Nostrand Reinhold: New York, NY (1993), 163–209.
52 D. Mukerjee, A. Kulkarni, W.N. Gill, Flux enhancement of reverse osmosis membranes by chemical surface modification, J. Memb. Sci. 97 (1994), 231.
53 S.P. Nunes, L.S. Mauricio, K.-V.J. Peinemann, J. Memb. Sci. 106 (1995), 49.
54 S. Akhtar, C. Hawes, L. Dudley, I. Reed, P.J. Stratford, J. Memb. Sci., 107 (1995), 209.
55 National Research Council, Review of the Desalination and Water Purification Technology Roadmap National Academic Press: Washington, D.C., USA (2004).
56 Polymers Resist Biofouling, Chem. & Eng. News, 83 (21) (2005), 8.

Chapter 2

Water and membrane treatment

'Whisky is for drinking, water is for fighting'.

Mark Twain

2.1 Priceless water

Current shortages of potable water around the world and the looming water scarcity especially in the developing countries has spurred a host of news items in the media[1,2] and priority themes of international conferences[3,4] in the last few years. The United Nations declared the year 2003 as the 'International Year of Fresh Water'. The reasons for water scarcity are simple; the global water consumption rose six times between 1900 and 1995, which was more than double the rate of population growth. As the world's population is expected to rise to nine billion from the current six billion in the next 50 years, the demand for water will get only worse especially when one-third of the world's population that currently lives in water-stressed countries is expected to rise to two-thirds by 2025. Inadequate supply of good-quality water coupled with higher water demand due to rapid population growth and industrialisation in developing countries are among the major reasons for the worsening water situation.

Nearly 60% of illness around the world is due to contaminated water and lack of sewer treatment. According to the World Health Organisation (WHO), about 2.4 billion people do not have access to basic sanitation facilities, and more than one billion people do not have access to safe drinking water. Unclean water causes diarrhea, cholera, dysentery, guinea worm infection, typhoid, intestinal worm infection, and trachoma.[4] According to the WHO, four billion people get diarrhea every year that kills nearly 1.8 million people of which 90% are children under the age of five.

Although water is the most common substance in the world, only 3% is fresh water (97% is seawater), and only 1% is available for human consumption. According to many experts, even 1% is adequate since fresh water supply is infinite because of the natural water cycle.[1,4] However, the availability and application of water is uneven around the world, and the quality of available or useable water is decreasing due to increasing environmental pollution. Water pollution from industrial, agricultural, and other human activity is becoming a major environmental problem today. For example, due to excessive use of fertilisers in India's bread-basket state of Punjab during the last 40 years, the nitrate level in groundwater has exceeded the carcinogenic level of 50 parts per million (ppm) in many areas. Nitrates are a health concern because they are converted to nitrites which interfere with haemoglobins to exchange oxygen in blood. This can cause serious problems especially for fetus and children. Nitrate residues also accelerate eutrophication – enrichment of ponds and lakes by nutrients that results in oxygen depletion – of water bodies. Another major source of water pollution is chlorine, which is used in pulp and paper bleaching, metal processing, pharmaceutical manufacturing, textile dyeing and cleaning, corrosion control, photography, and water treatment.[4] Arsenic, a carcinogen at concentrations greater than 10 parts per billion (ppb), occurs naturally in groundwater in several countries. The problem is especially acute in Bangladesh where 60 million people are seriously affected by arsenic in natural drinking water.

Water demands and its quality depend on its usage, and varies from region to region. Water shortages have been attributed to varied reasons such as nearly 50% loss due to leaks in piped water systems, and inefficient usage of water (70–90% of available water) for agricultural irrigation. For example, intensive and indiscriminate pumping of groundwater is depleting the huge Ogallala aquifer in middle America eight times faster than it is being replenished. Similarly, excessive use of water required for high-yielding grain crops is depleting groundwater in the Punjab region of India. Water scarcity problems, however, can be solved by a combination of water conservation, reclamation, recycling, and desalination, as discussed in Chapter 3.

Industrial water management relies heavily on separation science, e.g. coagulation and flocculation are used to separate or remove suspended solids from water for clarification by accelerating their settlement rates during conventional treatment. Membrane filtration processes – UF and MF – are being used for treating surface water in lieu of coagulants such as aluminium sulphate (alum) and polymers thereby eliminating the sludge and its disposal. Separation techniques such as adsorption, ion exchange, membrane separation, and chemical coagulation and precipitation are being investigated for removing arsenic from groundwater.

RO membrane separation has been traditionally used for seawater and brackish water desalination, and production of high-purity water for food, pharmaprocessing, and industrial waste treatment, as discussed in Chapter 1. The development of NF membranes has opened up many areas of application including water softening, removal of disinfection by-product precursors (trihalomethanes), removal of total organic carbon (TOC), food processing, and industrial water treatment.[5] Low-pressure cross-flow MF and UF systems are becoming the separation processes of choice for municipalities in the US and Europe for removing turbidity and pathogens.[6–8] Both these membrane filtration processes are very effective and reliable in removing microbiolgical parasites, *Giardia* and *Cryptosporidium*. Increasingly hybrid membrane systems are being used for the reclamation of industrial and municipal water where UF or MF replaces conventional pretreatment for treating RO feed water, e.g. treating secondary treatment effluent with membrane filtration followed by RO membrane treatment for producing water for industrial use and aquifer recharge. Several case studies of hybrid membrane processes used in water treatment are discussed in Chapter 3.

There are many reasons why membrane filtration is becoming the preferred technology for water treatment complementing RO or NF membrane systems in a truly hybrid membrane plant:

- Increasingly cost competitive; the cost of RO membranes dropped over 90%

between 1980 and 2003 while the permeate production increased threefolds.[9]

- Proven integrity of membranes albeit with a limited life span of 3–5 years for polymeric membranes. Ceramic membranes last longer but are expensive.
- Development of high-frequency back-pulsed UF and MF processes for combating membrane fouling using high-integrity hollow fibre membranes.
- Development of submerged cross-flow MF systems, and membrane bioreactors for biological treatment.
- Reliable product water quality including removal of parasites and pathogens.
- Minimal use of chemicals thereby meeting stringent water pollution standards with ease.
- Reliable and well-established membrane plant operation.
- Automated control of systems resulting in reduced operating labour costs.
- Small footprint and modular design allowing for easy future expansion.

2.2 Water treatment

The types of impurities depend on the water source. The feed water could be groundwater, well water, surface water, seawater, brackish water, or municipal wastewater. Water composition of some typical waters is given in Table 2.1. Surface water usually contains suspended solids, colloids, and organics including high bacteria count, and has significant variable quality during the year. Groundwater and well water usually have low amounts of suspended matter but may contain iron, manganese, and hardness ions of calcium and magnesium.[10] Industrial wastewaters are regulated for suspended solids, free oil and grease, pH and biochemical oxygen demand, and heavy metals, and often contain mineral acidity, detergent compounds and trace amount of chemicals and heavy metals due to industrial pollutants. Membrane separation processes are being increasingly used to treat these waters for environmental pollution control.[11]

Once the feed water source has been determined, analysis of the feed water composition is necessary before a treatment system can be designed. Feed water constituents that must be analysed prior to designing a RO/NF membrane system as per ASTM Designation D4195–88 'Standard Guide for Water Analysis for Reverse-Osmosis Applications' are discussed in Chapter 5. Water treatment methods are summarised in Table 2.2.

In the case of municipal wastewater for reclamation, secondary treatment effluent is often the feed water for membrane plants. Preliminary and primary treatment remove heavy solids, fine suspensions, grease, and fats. Sometimes raw water is chlorinated for odour control. In the case of raw municipal wastewater, primary treatment removes 30–50% of the suspended solids. The remaining organic matter is removed in biological secondary treatment.[12,13] Typically RO/NF feed water is primary treatment effluent except when the feed water source is brackish water or seawater. RO/NF pretreatment is used to remove (a) certain dissolved solids and minerals, (b) organics, (c) virus and pathogens, (d) trace metals, (e) nutrients, and (f) removing remaining suspended solids for in order to minimise the potential of fouling and scaling. The water treatment unit operations (UOPs) discussed include:[10,12,14]

- Coagulation
- Softening
- Granular media filtration
- Activated carbon filtration
- Membrane filtration
- Deaeration–Decarbonation
- Chemical oxidation

Table 2.1 Typical feed waters composition

Item/Ion	River[1]	Lake[2]	Well[3]	Brackish[4]	Seawater
Cations					
Calcium	50*(125**)	38	114	129	400
Magnesium	14 (57.4)	8	19	89	1,252
Sodium	35 (76.3)	11	-	521	10,561
Anions					
Bicarbonate	158 (129.5)	117	170	161	140
Carbonate	1.2 (2)	0	0	-	0
Hydroxyl	-	-	-	-	-
Sulphate	97 (100.9)	26	29	1,260	2,650
Chloride	16 (22.6)	18	76	262	18,980
Nitrate	4.6 (3.7)	1.8	-	-	1.5
Fluoride	-	-	0.1	-	1.4
Boron	-	-	-	2.8	4
Other					
Silica	13 (2.1)	12	10	49	-
Iron	0.1 (0.01)	3.5	Trace	1.2	-
Manganese	-	0	0.01	-	-
pH	7.9	6.8	7.5–8.4	7.7	-
Turbidity	58	0–20	0	-	-
TDS	-	-	-	2,478	33,990
Alkalinity**	-	-	-	133	
Hardness	-	-	-	692	-

* Ion as mg/l; **Ion as $CaCO_3$
[1] Mississippi
[2] Lake Erie
[3] West Virginia
[4] Coalinga, California
Source: Cabibbo et al.

- Disinfection
- Electrodialysis
- Ion exchange (IX)

When RO product water needs to be polished to produce higher grade water, several or all of UOPs listed below are also required. These UOPs remove trace amounts of contaminants, such as ions, gases, particles, microorganisms, and organic compounds:

- Mixed-bed IX
- Electrodeionisation (EDI)
- Membrane degasification
- TOC ultraviolet radiation

Besides feed water characteristics and product water quality, capital and operating budgets, manpower, cost of utilities, space availability, and future expansion requirements, and the efficacy of the specific techniques are considered in developing a water treatment system for a given application.

2.2.1 Coagulation

Coagulation is a chemical process for removing particles in suspended or colloidal form. These particles do not settle out on standing and cannot be removed by conventional physical treatment processes. The settling

Table 2.2 Water constituents and treatment methods

Constituent	Chemical formula	Effect/problem	Treatment
Turbidity	None (expressed as Jackson Turbidity Units (JTU))	Imparts unsightly appearance to water. Deposits in water lines and process equipment. Interferes with most process uses	Coagulation, settling, and filtration
Colour	N/A	May cause foaming in boilers	Coagulation, and filtration, chlorination, activated carbon, anion exchange, NF
Hardness	Ca^{2+} and Mg^{2+} salts expressed as $CaCO_3$	Chief source of scale in heat-exchange equipment, boilers, and pipelines. Forms curds with soap, interferes with dyeing	Softening, demineralisation, surface-active agents, NF, RO, ED
Alkalinity	HCO_3^-, CO_3^{2-} and OH^- expressed as $CaCO_3$	Foaming and carryover of solids with steam. Embrittlement of boiler steel. Bicarbonate and carbonate produce CO_2 in steam, a source of corrosion in condensate lines	Lime and lime-soda softening, acid treatment, hydrogen zeolite softening, demineralisation, dealkalisation by anion exchange, degasification
Free mineral acids	H_2SO_4, HCl, etc. expressed as $CaCO_3$	Causes corrosion	Neutralisation with alkalis
Carbon dioxide	CO_2	Causes corrosion in water lines, particularly steam and condensate lines	Aeration, deaeration, membrane contactors, neutralisation with alkalis
pH	Hydrogen ion concentration defined as: $pH = \log(1/H^+)$	pH varies according to acidic or alkaline solids in water. Most natural waters have a pH 6.0–8.0	pH can be increased by alkalis and decreased by acids
Conductivity	None (expressed as microSiemens/cm)	High conductivity can increase the corrosive characteristics of water	Demineralisation, lime softening, RO, ED, NF
Nitrate	NO_3^-	Adds to solids content. Useful for control of boiler-metal embrittlement	Demineralisation, distillation, RO, ED
Silica	SiO_2	Causes scale in boilers and cooling water systems, which forms insoluble turbine blade deposits	Hot-process removal with magnesium salts, adsorption by highly basic anion-exchange resins, in conjunction with demineralisation
Sodium	Na^+	Adds to solids content of water. When combined with OH^-, causes corrosion under certain conditions	Demineralisation, RO, ED, distillation
Aluminium	Al^{3+}	Usually present as a result of floc carryover from clarifier; can cause operation deposits and contribute to scaling	Improved clarification and filtration
Iron	Fe^{2+} (ferrous) Fe^{3+} (ferric)	Discolours water on precipitation. Source of deposits in water lines, and boilers. Interferes with dyeing, tanning and papermaking	Aeration; coagulation and filtration, lime softening; cation exchange; contact filtration, surface-active agents for iron retention, greensand filtration
Manganese	Mn^{2+}	Same as iron	Same as iron
Oxygen	O_2	Causes corrosion of water lines, heat-exchange equipment, boilers and return lines	Deaeration, membrane contactors, sodium sulphite, corrosion inhibitors
Hydrogen sulphide	H_2S	Causes of 'rotten egg' odour and corrosion	Aeration, chlorination, strong basic anion exchange
Ammonia	NH_3	Corrosion of copper and zinc alloys by formation of complex soluble ions	Cation exchange with hydrogen zeolites, chlorination, deaeration

Source: Branan, Gulf Publishing Co., 1998

time for silt, for example is about three hours as compared to three years for colloids. Suspended solids and colloids resist agglomeration because of the similar electrical charge (usually negative charge) on their surfaces that creates a mutually repellant force. The charge repellant behaviour is depicted in Figure 2.1 and is explained by the concept of 'zeta potential' in Chapter 5.

Colloids are particles in the size range of 0.1 (10^{-8} cm) to 1 nm (10^{-7} cm). They are either hydrophobic (e.g. clays) or hydrophilic (e.g. poorly ionised organic acids). High-valence cation (Mg^{2+}, Al^{3+}) coagulants

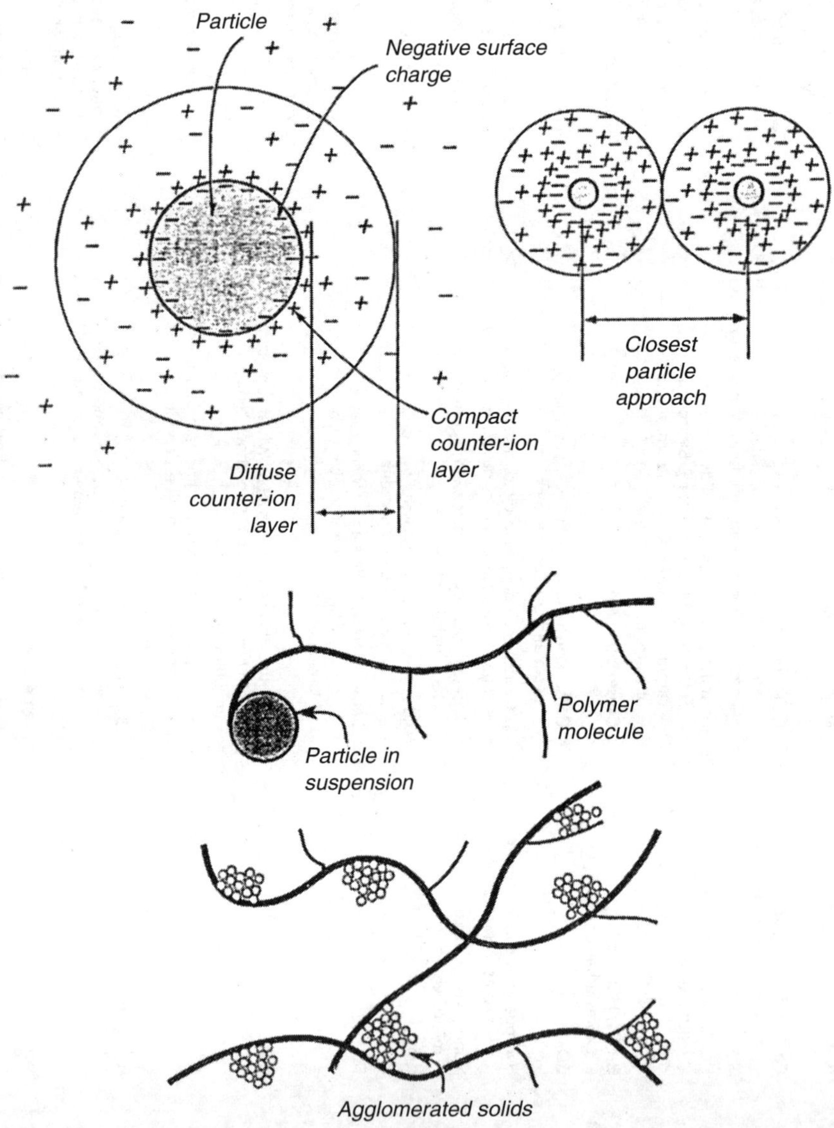

Figure 2.1 Repulsive force between suspended particles limiting or preventing agglomeration due to natural electric surface charge (top). Polymers attach to floc sites and enhance growth by chemical bridging (bottom). Chemical coagulants neutralise the charge, reducing the potential barrier and facilitating coalescence into floc

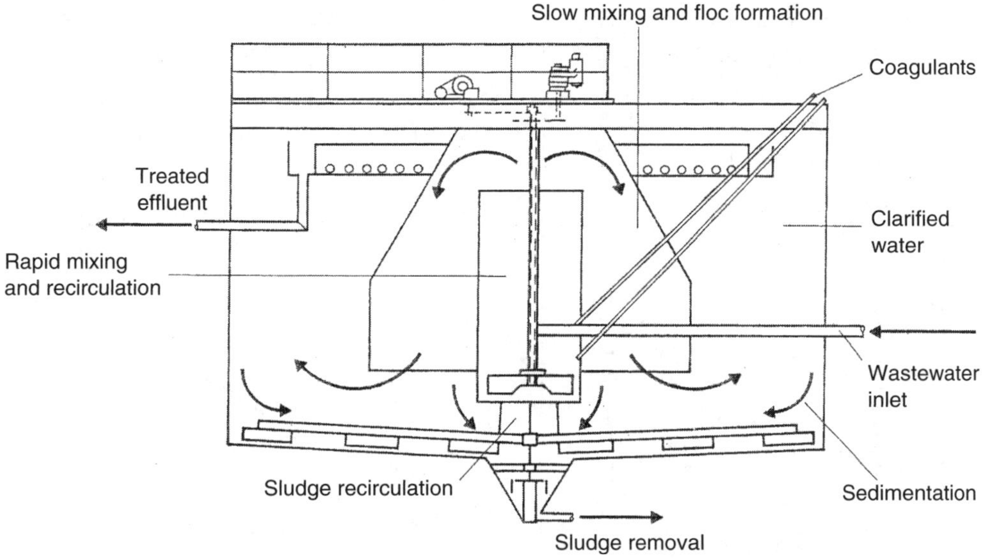

Figure 2.2 A reactor/clarifier designed for both coagulation and settling. Source: Eckenfelder, Reproduced with permission from McGraw-Hill

neutralise the negative charges thereby allowing the particles to come together or coalesce; the larger particles then precipitate and removed via sedimentation or filtration. A clarifier used for both coagulation and settling is shown in Figure 2.2. Chemical coagulants include aluminium sulphate (alum), ferric chloride, lime, and polymers. Low molecular weight cationic polymers and inorganic aluminium and iron salt (e.g. alum and ferric sulphate) are the most commonly used positively charged coagulants.[10,12,14]

Coagulation is a two-step process: first, the zeta potential is reduced to a level below the van der Waals attractive forces, and second, the micelles aggregate to form clumps which agglomerate the colloidal particles.[10,13] The mechanism of the coagulation process is shown in Figure 2.3. For an effective coagulation, alkalinity should be added first (bicarbonate provides alkalinity without raising the pH), alum or ferric salts are added next, and coagulant aids such as activated silica and/or polyelectrolyte for floc build-up and zeta potential control are added last. Activated silica is a short-chain polymer that binds together particles of microfine aluminium hydrate. Polyelectrolytes are high molecular weight polymers capable of forming bridges between particles or charge flocs. Large flocs (0.3–1 mm) are created when small dosages are added with alum or ferric chloride.

The reaction of alum $[Al_2(SO_4)_3{\cdot}18H_2O]$ when added to water in the presence of alkalinity is:

$$Al_2(SO_4)_3 \cdot 18H_2O + 3Ca(OH)_2 \rightarrow 3CaSO_4 + 2Al(OH)_3 + 18H_2O \quad (1)$$

The alum floc is least soluble at a pH of about 7.0; the floc charge is positive below pH 7.6 and negative above pH 8.2. Aluminium hydroxide is amphoteric and acts as either an acid or a base.

There are three types of polyelectrolytes: cationic, anionic, and non-ionic. Sometimes cationic polymers are added as a coagulant. The polymers bring the system to the

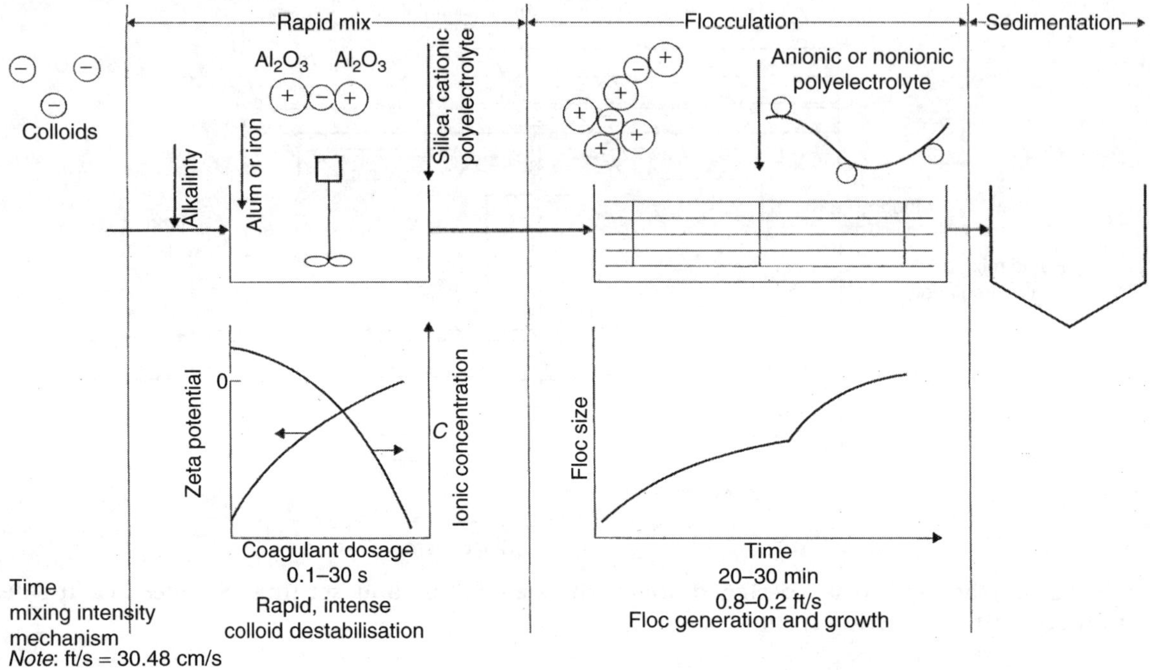

Figure 2.3 Mechanism of coagulation. Source: Eckenfelder, Reproduced with permission from McGraw-Hill

isoelectric point without a change in pH. Although polymers are many times more effective than alum, they are quite expensive. The dosage range for a cationic polymer is 2–5 mg/l vs. 75–250 mg/l for alum. The dosage range for anionic and non-ionic polymers is 0.25–1.0 mg/l.

Some clarification systems concentrate on producing only a fine or pin floc, which is then removed by in-line coagulation (coagulants are typically fed to the influent side of a media filter and allowed to mix in-line with the aid of an in-line mixer). In-line coagulation increases particle size and, therefore, filterability of solids that have carried over. Typical coagulant feed rates range from 0.5 to 20 ppm, with dosage for inorganic coagulants generally higher than polymeric coagulants. This process eliminates the large settling clarifier, and produces high-clarity water more quickly in smaller equipment.

2.2.2 Softening

Softening refers to removing calcium and magnesium hardness by chemicals. However, silica, alkalinity, and other constituents are also removed during lime softening. Other water softening methods include ion exchange and NF membrane separation.

Lime softening

Lime, sodium carbonate (soda ash), and/or sodium hydroxide (caustic soda) are added to water to convert soluble calcium and magnesium hardness to insoluble calcium carbonate and magnesium hydroxide in a contact vessel for 60–90 min.[10,12,14] Lime is not a true coagulant but reacts with bicarbonate alkalinity to precipitate calcium carbonate. Magnesium hydroxide precipitates at high pH levels. The solids are collected as sludge from the bottom of the settling basin. The sludge-free water is clarified by filtration to remove any turbidity and remaining solids thereby ensuring there is no carryover.

Calcium and magnesium occur as bicarbonates primarily. Lime and caustic soda breakdown the bicarbonate ions $(HCO_3)^-$ into water molecules and insoluble carbonate ions $(CO_3)^{2-}$ as follows:

$$Ca(OH)_2 + Ca(HCO_3)_2 \rightarrow 2CaCO_3\downarrow + 2H_2O \quad (2)$$

$$2NaOH + Ca(HCO_3)_2 + Ca^{2+} \rightarrow 2CaCO_3\downarrow + 2H_2O + 2Na^+ \quad (3)$$

Similarly, lime or caustic soda are used to provide hydroxide for precipitation of magnesium hardness as $Mg(OH)_2$. Hardness may also be present in non-carbonate form as sulphate, chloride, or nitrate. Calcium sulphate has reverse solubility and does not precipitate with heating. In the case of non-carbonate calcium compounds, carbonate for precipitation is provided by adding sodium carbonate (soda ash):

$$CaCl_2 + Na_2CO_3 \rightarrow 2NaCl + CaCO_3\downarrow \quad (4)$$

Cold lime softening is never complete because calcium carbonate and magnesium hydroxide are slightly soluble in water. Hot lime softening (HLS) with water heated to 100°C, on the other hand, significantly reduces silica in addition to hardness and alkalinity. Heating aids in the completion of the softening reaction, which in turn increases the efficacy of silica removal by providing more solids, particularly magnesium hydroxide that absorb silica. HLS is not, however, used for potable water applications. Lime softening is followed by filtration to remove any turbidity and solids. Inorganic coagulants are typically used to minimise carryover in cold process softening whereas anionic polymers are used with hot process softening. Effluent hardness from a cold process softener ranges from 35 to 80 mg/l; effluent hardness from a hot process softener is as low as 10–40 mg/l. In order to reduce the hardness further (1 mg/l), ion exchange softening is required.

Ion exchange softening

Water softening by ion exchange (IX) uses strong acid cation resins in the sodium form $(\text{—}SO_3Na)$ to remove scale-forming cations from solution.[10,12,14,15] IX softening involves the exchange of hardness ions such as calcium, magnesium, strontium, and barium for sodium ions to yield low hardness, or 'soft' water (see equation (19)). Hence, the softened water has a higher TDS. IX softening is typically used when the hardness is in the range of 50–500 mg/l[16], and for feed water flow rates of up to 60 m^3/h.

The exhausted resin is regenerated with a dilute NaCl (brine) solution. This removes calcium and magnesium in the form of their soluble chlorides and at the same time restores the resin to its original sodium form. The bed is rinsed free of undesirable salts and returned to service. The regeneration reaction may be written as:

$$Ca/Mg\text{—}R + 2NaCl \rightarrow Na_2R + CaCl_2/MgCl_2 \quad (5)$$

Weak acid cation (WAC) exchange softening involves the exchange of hardness ions such as calcium, magnesium, strontium, and barium for hydrogen ions. WAC resins remove all the alkalinity (hydroxyls, carbonates, and bicarbonates) as well as the hardness ions that are combined with it (calcium, magnesium, barium, and strontium). The WAC resins come in the hydrogen form $(\text{—}SO_3H)$. The exhausted resin is regenerated with dilute sulphuric acid. This removes the calcium and magnesium in the form of their soluble chlorides and at the same time restores the resin to its original hydrogen form. The bed is rinsed free of undesirable salts and returned to service. The regeneration reaction is:

$$Ca/Mg—R + H_2SO_4 \rightarrow H_2R + CaSO_4/MgSO_4 \quad (6)$$

The IX softeners normally operate at linear velocities of 14–20 m/h. About 8.5 lb (3.8 kg) of salt (NaCl) is required to regenerate 1 ft^3 (0.3 m^3) of resin, and remove approximately 4 lb (1.8 kg) of hardness. Hardness is given in grains/gallon or in ppm as $CaCO_3$. The reduction is directly related to the amount of cations present in raw water and the amount of salt used to regenerate the resin bed. Typically, 6 lb of NaCl/ft^3 (9 kg/m^3) of resin is used for regenerating strong acid cation resins.

Ion exchange dealkalisation usually employs WAC resins operating on the hydrogen cycle. Dealkalisation and alkaline hardness removal are synonymous terms. The high selectivity for divalent ion such as calcium and magnesium over monovalent sodium ions results in the preferential exchange of calcium and magnesium ions:

$$2R—COOH + M^{2+} \rightarrow (RCOO)_2M + 2H^+ \quad (7)$$

In the presence of hydrogen ions, the above exchange does not occur. However, if alkaline hardness (bicarbonate) is present, the exchanged hydrogen ions are immediately neutralised by the basic bicarbonate and carbonate anions to give carbon dioxide, which dissolves in water as weak carbonic acid:

$$H^+ + HCO_3^- \rightarrow H_2O + CO_2 \quad (8)$$

$$2H^+ + CO_3^{2-} \rightarrow H_2O + CO_2 \quad (9)$$

The exchange of calcium and magnesium ions continues until all basic anions are neutralised after which time no further exchange can occur. Hence, the extent of hardness removed is equivalent to the alkaline hardness of the water. The resins are regenerated with either dilute hydrochloric acid or dilute sulphuric acid over a period of 30 min. The weak acid resin is regenerated at virtually 100% efficiency. The weak base anion resins are also effective in removing strong mineral acid anions such as sulphates, chlorides, and nitrates. Hence, like IX softeners, they are sometimes used to treat RO feed water, thereby, reducing the potential of mineral scaling of RO membrane surface. The exhausted bed is regenerated with sodium hydroxide (caustic soda).

NF softening

Membrane softening by nanofiltration (NF) is a relatively new application, as discussed in Chapter 1. NF membranes ('loose RO') operate at a lower feed pressure than RO membranes, and have a high rejection (99%) of divalent hardness ions. It is a more attractive alternative to lime softening and IX softening because it is not only a reliable process, but it does not require chemicals for regeneration, and minimises waste. NF separation is a continuous process and is independent of the plant capacity (flow rate) and feed water hardness. It reduces both the hardness and the total dissolved solids to a much greater degree than IX and lime softening.[5]

2.2.3 Granular media filtration

Granular media filtration (GMF) is a process for removing suspended or colloidal particles, for example, removing suspended solids remaining after sedimentation–clarification. It reduces turbidity and improves clarity by removing various-sized particles, from coarse sediment down to 10.0 μm.[10,12] Filtration protects the IX resin beds and RO/NF membrane elements downstream from particulate fouling. GMF has different size exclusion

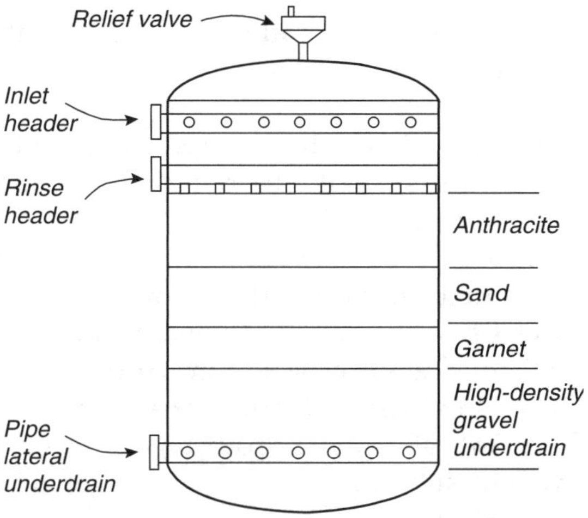

Figure 2.4 A granular media filter with multimedia layers. Source: USFilter

ratings, from 10 to 100 μm, depending upon the size of particles to be removed. In general, the removal of suspended and colloidal particles can be done by GMF, dead-end MF and cross-flow MF when the water Silt Density Index (SDI) is about 5. For raw waters containing high concentration of colloidal matter (SDI > 5), coagulation and flocculation is required before media filtration.

Granular multimedia filters feature layered beds of anthracite coal (0.8–1.2 mm size), sand (0.5–0.8 mm), finely crushed garnet (0.4–0.6 mm), and magnetite (0.3–0.4 mm) or other materials, as shown in Figure 2.4. The top layer of the bed consists of the lightest and most coarsely graded material, e.g. anthracite coal, whereas the heaviest and most finely graded material, e.g. garnet or magnetite is the bottom layer. The intermediate layer is silica sand. The specific gravity of anthracite is one-half that of silica sand. A typical bed is 1 m in depth. The principle is 'filtration in depth' – larger particles are removed at the top layers, and smaller ones are removed deeper in the filter media, i.e. the entire bed acts as a filter rather than the top few centimetres.

During service flow water typically flows downwards from top to bottom under pressure. Typical service superficial flow velocities (flow rate/bed cross-sectional area) are: 7–12 m/h for single-media gravity and pressure filters, 14–20 m/h for multimedia gravity and pressure filters, and 12–24 m/h for upflow filters. Filter operation must avoid channeling and 'leakage' of suspended solids; otherwise the RO membranes may get fouled. Space above the bed (50–100% freeboard) provided to allow for bed expansion during backwashing.

Since suspended solids are collected on the media, regular cleaning (backwashing) is required. The rejected particles form a layer on the surface of the media and contribute to blockage of the pores in the filter medium resulting in an increase in the pressure drop (ΔP). Typically, when ΔP reaches 1 bar, the filter is backwashed. During backwashing water flow direction is reversed; it enters the bed through the bottom and flows upwards. This fluidises the bed, which along with the reversed flow, dislodges the silt and carries it to waste. Typical backwash flow rate range is 24–36 m/h depending on the temperature, high enough to expand the media bed by at least 50%. The backwash cycle lasts 10–15 min and is followed by a rinse cycle that is accomplished by passing water in a down-flow direction for 5–10 min.

Coagulants are injected in the feed waterline upstream of granular media filters to remove suspended matter of colloidal nature when the particulate matter is either too small or electrostatically repelled from the media. Polymeric coagulants are added in low dosages (<10 ppm) to remove particles down to 0.5 μm particle size as compared to 10 μm without the coagulant. Anionic coagulants have been shown to reduce the turbidity of raw water to less than 0.5 NTU at a nominal dosage of 1–2 ppm.

2.2.4 Activated carbon filtration

Activated carbon filtration (ACF) is one of the most effective methods for removing non-polar organic compounds (low, medium, and high molecular weight), precipitated iron and chlorine from water.[10,12,13] Activated carbon is a microcrystalline, non-graphitic form of carbon, e.g. bituminous form of coal that has been processed to develop internal porosity. Activated carbon has the highest volume of adsorptive porosity of any material known to man. Because of its large surface area (1 l of granules has a surface area of 50 ha), activated carbon has a great ability to adsorb organic molecules of liquids or vapours. During feed processing, an organic molecule in the feed enters a surface pore and diffuses to a micropore where it is adsorbed on the carbon surface by physical attraction due to van der Waals forces. Up to 95% of organics can be reduced. The organic loading is 2–6 kg per 50 kg of carbon. Aromatic compounds are removed more effectively than aliphatic compounds, and non-polar organic compounds are better removed than polar compounds. In addition, some trace inorganics can be removed when chelated with organic compounds.

Activated carbons have specific properties depending on the material source and the mode of operation. Generally speaking, carbons from bituminous coal have a smaller pore size, a larger surface area, and a higher bulk density as compared to lignite carbon. A typical surface area range is from 850 to 3,000 m^2/g. Bituminous coal also has a lower 'peroxide number' than lignite. Peroxide number is an indicator of catalytic activity of carbon; the lower the number, the higher is the catalytic activity. Property standards used for specifying carbons for a specific application are defined below:

- The Phenol Number is used as an index of carbon's ability to remove taste-and-odour compounds.
- The Iodine Number relates to the ability of activated carbon to adsorb low molecular weight substances (micropores have an effective radius of less than 2.0 μm).
- The Molasses Number relates to the carbon's ability to adsorb high molecular weight substances (pore size ranges from 1 to 50 μm).

Activated carbon is also used for removing chlorine. Chlorine manifests itself mostly as hypochlorous acid (HOCl) and hypochlorite ion (OCl^-). Hypochlorous acid is the primary disinfectant, and hypochlorite ions are less effective. Hypochlorous acid is removed by reduction with carbon as per the following reaction

$$C + 2HOCl \rightarrow CO_2 + 2HCl \qquad (10)$$

Chlorine is adsorbed by carbon in the top 5–10 cm of the bed. The bed depth is typically 1 m. Since chlorine is consumed within the top few cm of the carbon bed, the media is susceptible to bacterial growth when the water temperature is greater than 12°C. A pH range of 6.5–7.5 and water temperature between 12 and 27°C provide the best environment for chlorine removal by activated carbon.

During normal service, feed water flows downwards through the bed from top to bottom as in the case of GMF under pressure. The linear flow velocity is 5–10 m/h. The flow is reversed during backwash. The beds are backwashed at 7–12 m/h (linear velocity) based on service run time or when the pressure drop across the filter bed exceeds 1 bar. The backwash cycle is followed by a downward flow rinse cycle before the bed is returned to service. Backwashing only removes the material collected on the surface of the media, and not what is adsorbed in the pores. Once the pores are filled with organics, carbon can be either regenerated or reactivated. The modes of regeneration are thermal, steam, solvent extraction, acid or base treatment, and chemical oxidation. In water treatment, thermal regeneration is usually done but weight losses and a loss of capacity result from regeneration.

2.2.5 Membrane filtration

Membrane filtration encompasses microfiltration (MF) and ultrafiltration (UF) only. Cross-flow MF and UF processing is a recent application for potable water production: (a) as pretreatment in RO/NF membrane plants, (b) in water reclamation, and (c) for municipal water treatment. Membrane filtration is used for treating high-turbidity water, surface waters high in total organic carbon (TOC) and for tertiary water treatment municipal plants.[6,8,17] Membrane filtration provides a more stable and superior water quality than coagulation–sedimentation and media filtration discussed above. The separation aspects and various applications of MF and UF are discussed in Chapter 1.

Dead-end vs. cross-flow filtration

Traditional MF separation is of the dead-end type where the feed solution flows perpendicular to the membrane surface. Unlike cross-flow or tangential flow filtration, there is no reject stream, only a feed stream and a permeate stream, as shown in Figure 2.5. The permeate drags all the solids in the feed

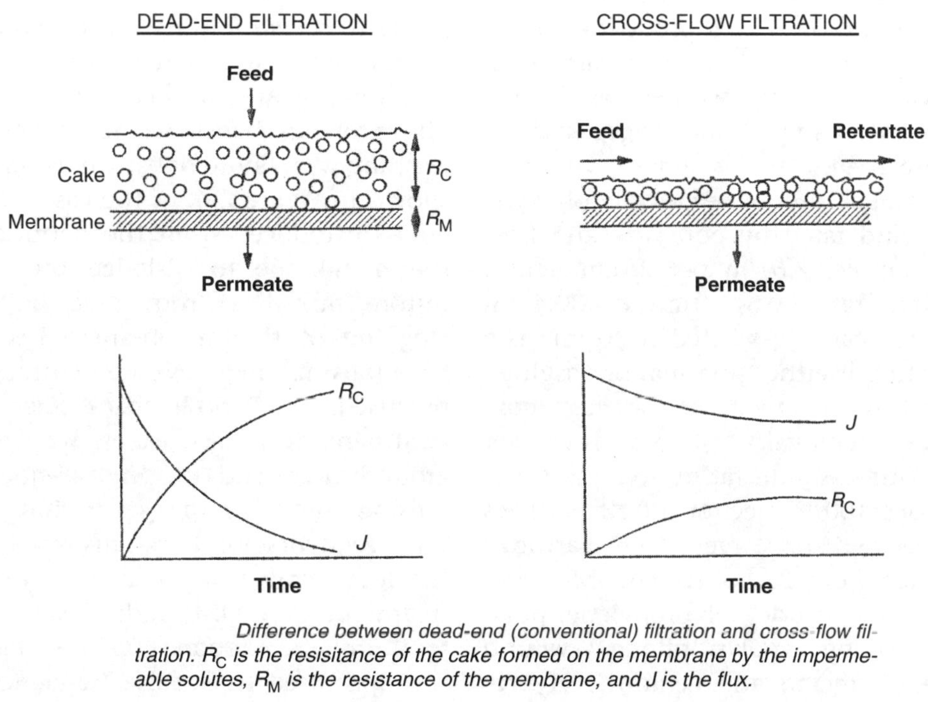

Figure 2.5 Dead-end vs. cross-flow filtration. Source: Cheryan, Copyright® 1998 from Ultrafiltration and Microfiltration Handbook by M. Cheryan. Reproduced by permission of Routledge/Taylor & Francis Group, LLC

solution to the membrane surface resulting in solute adsorption and deposition on the membrane surface and inside the pores. The separation mechanism is not a simple sieve mechanism where the particles with sizes smaller than the pore size flow freely through the pore while the larger particles are rejected.[18,19] Instead the particles to be separated are adsorbed onto the surface of the pore, resulting in a significant reduction in the size of the pore. Particles also deposit on top of the membrane, forming a cake-like secondary filter layer. Hence, the particles that can be separated by MF are often much smaller than the pore size.

In cross-flow filtration, the feed solution flows tangentially across the membrane surface. The cross-flow or tangential velocity continuously removes particles from the membrane surface by shear forces especially when operating in a turbulent regime. The net effect is reduced fouling. The advantage of dead-end filtration is high product recovery and simple operation. The filter cartridges, however, cannot be backwashed or cleaned because of internal pore blockage; instead, they are discarded.

The cartridges are either of depth type or pleated, and rated by pore size and flow rate, for example, 7.6 l/m per 25 cm length for cartridges rated 0.3–1.0 μm or 30.4 l/m per 25 cm for cartridges rated 3–30 μm. The pore size rating is either nominal or absolute. Nominal ratings are related to average pore size, e.g. 95% removal of all particles larger than 5 μm. An absolute rating, on the other hand, implies 100% rejection of all particles of size specified or larger. The cartridge lengths range from 25 to 100 cm. Materials of construction include polypropylene, polycarbonate, nylon, polysulphone, wound cotton, paper, metal, and ceramics. Hydrophobic filters (usually 0.2 μm pore size) are used in vents of high-purity water storage tanks to protect the water from atmospheric contamination. A cartridge filter vessel may contain 1–100 cartridges per vessel depending on the feed water flow rate and pressure drop.

Tangential flow microfiltration

Cross-flow MF membranes have a nominal pore size of 0.1–1.0 μm. They are, thus, effective in removing the largest sub-micron particles from low turbidity waters, and for removing low suspended solids containing bacteria or chemically flocculated solids at low pressure (~1 bar). Effluent turbidity levels less than 0.2 NTU are achieved consistently. Hollow fibre (nominal pore size 0.2 μm) MF membranes are being increasingly used for removing bacteria (*Giardia* and *Cryptosporidium*) in municipal water treatment plants and other matter larger than 0.05 μm in size.[6,7] Pre-filtration typically consists of a 200-μm fine screen.

MF like other membrane processes is prone to severe fouling during potable water processing.[20] Several techniques for controlling fouling are summarised in Table 1.15. The problem has been addressed by backwashing with liquid or back-pulsing with air as shown in Figure 2.6. In the case of the latter, air is introduced from the permeate side of the membrane to dislodge the solids and restore flux lost from cake build-up and clogging of the membranes. Two types of back-pulsing regimes are currently being practised: (a) Periodic back-pulsing for the treatment of surface water, wastewater, and groundwater, and (b) High-frequency back-pulsing used in bio-separation processes. The frequency of back-pulsing depends on the feed water solids concentration. For the treatment of high-turbidity water and high total organic carbon (TOC) surface waters, the optimised back-pulse frequency is every 15 min for a duration of 30 s.[19,21] Back-pulsing can only be used with high-pressure-resistant hollow-fibre membranes.

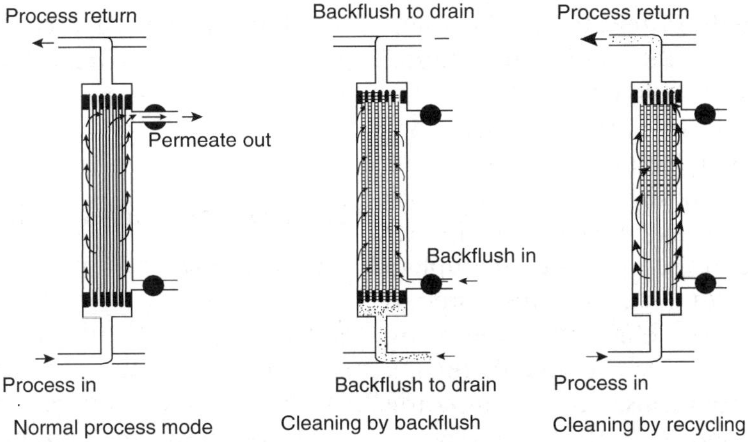

Figure 2.6 Modes of operation for hollow-fibre ultrafiltration and microfiltration modules. Source: Lonsdale, J. Memb. Sci. 10 (1982), 81–181.

Ultrafiltration

UF membranes (nominal pore size of 0.002–0.1 μm) have been traditionally used for removing high molecular weight organic compounds such as proteins, and colloids and oils, as discussed in Chapter 1. Since the early 1990s, however, UF hollow fibre (HF) membrane systems have been used in a variety of municipal water treatment applications, including treatment of surface water for production of safe drinking water and advanced tertiary treatment of municipal wastewater for recovery and re-use.[6,17] UF membranes remove very small size contaminants in feed water, including essentially all of the suspended solids, colloidal particles, and large size microorganisms such as *Cryptosporidium* and *Giardia*. UF generally produces product water with less than 0.1 NTU turbidity and less than 3.0 SDI (SDI values higher than 5 indicate high fouling potential for RO membrane spiral-wound elements). Hence, UF product water is ideal feed water for spiral-wound RO/NF membrane systems. High quality and consistent feed water improve RO/NF system operation including higher sustained membrane flux, lower operating pressure, increased product recovery, and less frequent membrane cleaning.

These UF membrane systems provide a simple and cost-effective alternative to conventional water treatment operations. HF membranes can operate at higher solid levels than SW membranes because they have open feed channels, and can be periodically backflushed with filtered product water to control the build up of solids on the membrane surface. Combinations of sodium hypochlorite (bleach), caustic, and acid are sometimes added to the backflush solution to 'enhance' the backflush and achieve a more effective removal of solids from the membrane surface. Since UF membranes can be sterilised with saturated steam at 121°C, they are especially suited for producing water that is free of bacteria and pyrogen as in the case of water for injection (WFI) in pharmaceuticals production.

Membrane bioreactors

A process that combines a biological stage with a membrane element is called a membrane bioreactor (MBR) process. In this process, biological degradation of organic pollutants is carried with microorganisms in the bioreactor, and MF or UF membranes are used to separate microorganisms from treated wastewater.[5,22,23] The use of membranes to remove solids from treated wastewater is the main difference between MBRs and conventional secondary treatment plants. MBR systems have a higher removal efficiency than conventional treatment plants, e.g. the MBR allows a higher biomass concentration, higher COD removal (>90%), and higher separation of solid suspensions (complete retention of the biomass). The MBR process has shown to be more efficient in removing total biochemical oxygen demand (BOD), turbidity, and coliforms. By eliminating the problem of poorly settling flocs when using a MBR system, there is more biological degradation resulting in higher treatment efficiency.

There are two types of MBRs: (a) External MBR (EMBR) in which the membrane modules are placed outside the bioreactor, and (b) Submerged MBR (SMBR) where the membrane module is placed inside the bioreactor.[22] The SMBR is more economical and energy efficient: (a) there is no recycle pump since aeration generates a cross-flow across the membrane surface, and (b) the operating conditions are milder because of lower values of transmembrane pressure (TMP) and tangential velocity. The TMP values are 1–4 bar for EMBR and 0.5 bar for SMBR systems. The permeate flux for SMBR, however, is lower; 15–50 l/h.m^2 vs. 50–120 l/h.m^2. Polymeric hollow fibre membranes with a pore size of 0.1–0.4 μm are the main membranes used in SMBR systems while tubular inorganic membranes are generally used in EMBR units. The EMBRs are preferred to SMBRs when treating concentrated effluents or concentrated biomass to avoid membrane fouling; higher shear rate is achieved because of higher recycle flow rate.

A key area of process deficiency is fouling by microorganisms as a result of microbial products, concentration, and size of particles. Different strategies have been considered for controlling fouling including backwashing. In addition, there is a concern whether oxygen can become the limiting factor during aerobic biological activity.

2.2.6 Deaeration–decarbonation

Air stripping is used to remove oxygen, carbon dioxide, ammonia, and volatile organic compounds from water. Decarbonation removes alkalinity – carbonate, bicarbonate, carbon dioxide – by acidification and stripping the resulting carbon dioxide. Carbon dioxide is removed from RO product water to reduce the load on IX deioniser resins downstream. This also increases the pH of process water thereby reducing the corrosive properties of water. Stripping of carbon dioxide increases the water pH, and thus reduces the corrosive properties of water. Forced air decarbonation is always used in conjunction with either RO or IX.

Ammonia is easily stripped since it is volatile. When the pH of water is raised above pH 9.3, the equilibrium point, ammonium ions in water convert to ammonia as per equation (11) given below as the equilibrium shifts to the left. The normal pH range for stripping ammonia is 10.8–11.5,

$$NH_3 + H_2O \leftrightarrow NH_4^+ + OH^- \qquad (11)$$

In the case of stripping volatile organic compounds (VOCs) such as halogenated hydrocarbons that are carcinogenic, the process relies on the tendency of moderately soluble organic compounds to vapourise based on Henry's law.

Air stripping requires packed towers for maximum operational efficiency.[13] Water is pumped to the top of a tower packed with media, as shown in Figure 2.7. The water is evenly distributed across the media. As it flows down under gravity it forms a film layer on the packing surfaces. Air is blown upwards from the bottom contacting the large surface

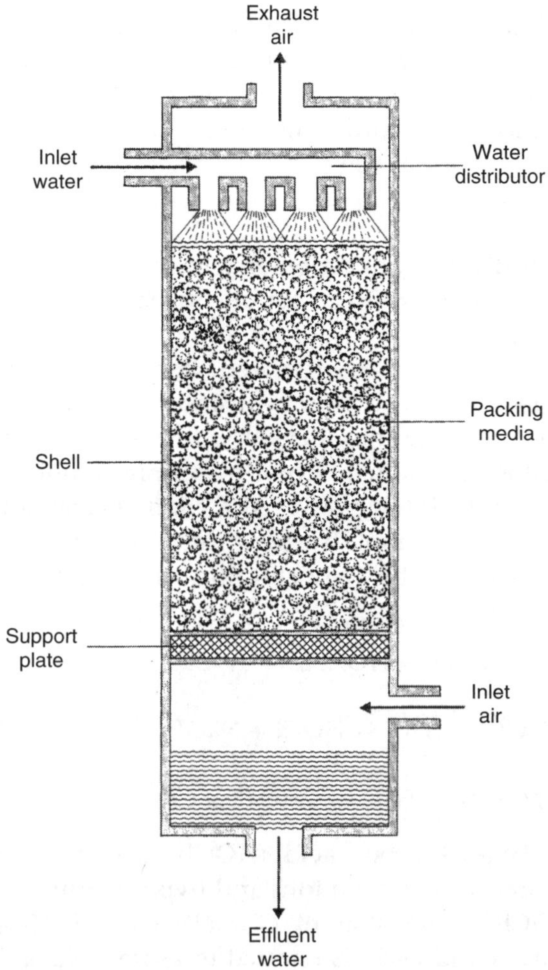

Figure 2.7 A typical packed tower. Source: Eckenfelder, Reproduced with permission from McGraw-Hill

areas. The blown air enhances the removal of the volatile species by mass transfer. Degasification is usually not economical if water alkalinity is less than 100 mg/l or 20% of total anions, or if the water flow rate is less than 5–6 m^3/h. The ratio of gas to liquid is in the range of 100 to 1 (volume basis).

2.2.7 Chemical oxidation

Chemical oxidation is used to control inorganics (manganese, ferrous iron, sulphur, sulphite, and cyanide irons) and organics (phenols, amines, humic acids, toxic compounds, bacteria, and algae) by converting them to unharmful products.[10] The oxidising agents used include air, oxygen, ozone, hydrogen peroxide, chlorine, chlorine/hypochlorite, and chlorine dioxide. Aeration followed by chlorination is very efficient at precipitating sulphur, which can then be removed via filtration. Contact times of three to four hours can reduce the hydrogen sulphide levels by up to 3 ppm via aeration alone. Hydrogen sulphide is generally found in well water, particularly shallow wells. Chlorination increases the kinetics of precipitation, and at the same time acts as a germicide to minimise the growth of sulphur bacteria, which are the source of hydrogen sulphide. Chlorine, potassium permanganate, and ozone are also effective in destroying many odourous compounds.

Manganese greensand oxidation-filtration

Some well waters, usually brackish waters, contain divalent ferrous iron, manganese, and sometimes sulphide in the absence of oxygen. When water is exposed to air or is chlorinated, these compounds are oxidised resulting in the formation of insoluble colloidal hydroxides and elemental sulphur as:[12,14]

$$4Fe(HCO_3)_2 + O_2 + 2H_2O \rightarrow 4Fe(OH)_3 + 8CO_2 \quad (12)$$

$$4Mn(HCO_3)_2 + O_2 + 2H_2O \rightarrow 4Mn(OH)_3 + 8CO_2 \quad (13)$$

$$2H_2S + O_2 \rightarrow 2S + 2H_2O \quad (14)$$

The manganese greensand filtration (MGF) process employs manganese greensand as the filter medium to remove soluble iron, manganese, and hydrogen sulphide by one-step oxidation-filtration. The filter bed consists of anthracite filter media over manganese zeolite (greensand). Greensand is processed from glauconite, which is an iron potassium aluminosilicate material of marine origin. The filter-bed material acts as a catalyst for the reaction of iron with oxygen. It can supply the necessary oxygen, which must be replenished with an oxidising regenerant solution such as potassium permanganate ($KMnO_4$).

The most common method of oxidative MGF is known as continuous regeneration. Permanganate solution (1–2%) is injected into the raw water feed waterline. Precipitation of iron and manganese occurs in the anthracite bed as per the reaction in equation (15). Heavier precipitates are filtered out in the anthracite bed, and the remaining fines and residual dissolved metal are removed by greensand. The water pH should be about 7.2.

$$Fe(HCO_3)_2 + KMnO_4 \rightarrow Fe(OH)_3\downarrow + MnO_2\downarrow \quad (15)$$

Iron concentrations up to 15 ppm can be effectively removed although concentrations greater than about three ppm require low service flow rates, which results in short runs between the backwash cycles. At low feed water iron concentrations, the flow velocity is 14–20 m/h for up to 36 h between the backwash cycles. Lower flow velocity 4 m/h and shorter service run lengths are advised when the feed water iron concentration is 3–15 ppm.

2.2.8 Disinfection

Disinfection is the selective destruction of pathogenic organisms (bacteria and viruses). Disinfection is not the same as sterilisation, which implies the destruction of all organisms. The disinfection treatment methods commonly used include (a) chlorination or ozonation chemical means, (b) photochemical means, e.g. ultraviolet disinfection, and (c) physical means such as membrane filtration. Parasites such as *Giardia lamblia* and *Cryptosporidium parvum* are resistant to conventional forms of disinfection, and are removed by other methods such as ozone and microfiltration.

Chlorination

The disinfectant action of chlorine results from its strong oxidising action on bacterial cell's chemical structure that destroys the enzymatic processes required for life.[12] The effectiveness of chlorine disinfection is a function of the product of contact time and chlorine residual. Chlorine gas is soluble in water (7,160 mg/l at 20°C and 1 bar), and hydrolyses rapidly to form hypochlorous acid or as the hypochlorites of sodium and calcium:

$$Cl_2 + H_2O \rightarrow HOCl + H^+ + Cl^- \quad (16)$$

$$NaOCl + H_2O \rightarrow HOCl + NaOH \quad (17)$$

$$Ca(OCl)_2 + H_2O \rightarrow 2HOCl + Ca(OH)_2 \quad (18)$$

Hypochlorous acid (HOCl) dissociates in water to hydrogen ions and hypochlorite ions (OCl^-). The sum of Cl_2, NaOCl, $Ca(OCl)_2$, HOCl and OCl^- is referred to as free available chlorine. The power of free chlorine residual decreases with increasing pH. Hypochlorous acid concentration at 20°C is 90% at pH 7, 50% at pH 7.6, and 10% at pH 8.6. Almost the reverse is true for hypochlorite ions. Hence automatic monitoring of residual chlorine and automatic feedback of injection rate is necessary to prevent overdosage or inadequate disinfection. Minimum chlorine residuals for bactericidal disinfection after 60 min of contact vary between 1 ppm at pH 6.0 and 1.8 ppm at pH 8.0. A hypochlorous residual of 0.5–1.0 ppm is effective within 30 min.

In gaseous form chlorine is extremely hazardous. Liquid chlorine is shipped in pressurised steel cylinders. One volume of liquid chlorine yields about 450 times vapour volume. Because of safety concerns liquid chlorine compounds such as sodium hypochlorite are used instead. Sodium hypochlorite (NaOCl) is handled in liquid form at concentrations between 5 and 15% available chlorine. Calcium hypochlorite ($Ca(OCl)_2$) contains about 70% available chlorine.

Ozonation

Ozone is a powerful disinfectant, and hence an effective disinfectant. It has similar bactericidal properties to chlorine and is at least equal to chlorine in it ability to perform virus inactivation.[12] Ozone is also an option for odour control in reclaimed wastewater treatment. Ozone is generated on-site from air or oxygen when a high voltage is imposed on a discharge gap. It is generated on-site since it is a relatively unstable gas with a half-life of about 10 min. When ozone is added to water, it rapidly reverts to oxygen so that no chemical residuals remain in the ozonated water. Chlorine is added for post-disinfection, if required. There are, however, some adverse health effects due to the formation of bromates and aldehydes as byproducts. The maximum allowable limit of bromate, a carcinogen, in drinking water is 25 ppb.

UV irradiation

Ultraviolet (UV) light represents a band of electromagnetic light in the 100–400 nm range. UV is a non-chemical disinfectant process that uses the least contact time (<5 s). Disinfection by UV radiation involves damaging the genetic material of organisms by energy in the form of light of wavelength 254 nanometres (nm) emitted from a low-pressure mercury vapour lamp.[12] This particular wavelength alters the genetic material (DNA) of bacteria, viruses, and other microorganisms. With their DNA altered, they are unable to reproduce and die within minutes. The dead bacteria produce TOC. UV disinfection avoids a major disadvantage of chlorination, the generation of harmful by-products such as trihalomethanes. However, some types of bacteria, which are commonly found in domestic water supplies, are known to be extremely resistant to UV radiation.

For microbial and ozone destruction, 254 nm UV energy is used, whereas shorter (and more powerful wavelength) electromagnetic radiation, 185 nm UV energy, is used to reduce organic compounds and chlorine destruction. The energy of a light beam is inversely proportional to the wavelength. Thus, 185 nm UV irradiation carries more energy and is more powerful than the 254 nm light. The 185-nm energy oxidises total oxidisable carbon (TOC) to form carbon dioxide and water. The UV irradiation required for effective destruction is referred to as dosage and is expressed in microwatt seconds/sq. cm. Most microorganisms are damaged at a dosage level of 10,000–30,000 $\mu W\,s/cm^2$, whereas the dosage level required for reducing ozone, chlorine, and TOC is 90,000 $\mu W\,s/cm^2$.

UV systems generally consist of a reactor with a number of lamps that emit UV radiation. Each lamp is encased in a quartz tube. As the water flows thorough the lamps, it gets radiated. Polytetrafluoroethylene (PTFE) is also used, as PTFE is transparent to UV radiation. UV radiation dosage is determined by residence time in the reactor and the intensity of radiation. The effectiveness of a UV system depends on the hydraulic design of the reactor, absorbance of liquid, presence of particles, and transmittance of the lamps and the quartz tubes. Intimate contact between the liquid and the lamps needs to be maintained at all points in the reactor since radiation diminishes with distance. Contact time for radiation exposure is maximised by maintaining plug flow in the reactor and by creating turbulence to cause transverse dispersion. Most UV lamps intensity drops to 60% in a year or after 8,000 h of use.

2.2.9 Electrodialysis

Electrodialysis (ED) was introduced in Chapter 1. The ED process is the reverse of RO separation in that the ions are removed and purified water is left behind. ED is sometimes used as a pretreatment for RO systems by removing the bulk of dissolved ions from high TDS waters, or for treating the RO reject stream by concentrating the dissolved ions to 80,000 ppm prior to evaporation and disposal in zero-liquid discharge systems for treating industrial wastewaters.

ED is a low-pressure process that uses ion selective membranes to desalinate water. Under the driving force of a direct current (DC) electric potential that is applied perpendicular to the flow, positive ions travel toward the cathode and the negative ions toward the anode, as shown in Figure 1.13. However, since it can remove only ionised compounds, silica and most colloids are not removed. ED is also susceptible to fouling like other membrane processes. The fouling problem is overcome by using an EDR process in which the applied DC polarity is reversed every 15–20 min. This removes the insoluble compounds that deposit on the membrane surfaces. EDR thus initiates automatic self-cleaning.[10]

Depending on water temperature, solids content, flow rate, and applied voltage, one ED membrane stack produces about 30–60% desalination. To achieve a practical level of 75–95% or higher salt removal, the membrane stacks are staged, e.g. two EDR stages for 75% salt removal, four stages for 94% salt removal, and six stages for 98% removal.[10] The water recovery with EDR is up to 95%. Since the ED process does not remove silica and most colloids, EDR has been used with UF and RO in hybrid systems for producing deionised water at nuclear power plants. In this triple-membrane process, UF removes most of the suspended solids and large organic molecules followed by EDR for removing most of the dissolved ions, and RO for removing the remaining contaminants especially silica and low molecular weight organics.

2.2.10 Ion exchange

The primary function of the ion exchange (IX) operation is to remove ions from a dilute solution and concentrate these into a relatively small volume. An IX reaction may be defined as the reversible interchange of ions between a solid phase (the ion exchanger) and a solution phase, the ion exchanger being insoluble in the medium in which the exchange is carried out.[15] IX resins remove undesirable ions by replacing them with an equivalent number of desirable ones. For example, a cation resin (R) selectively removes calcium ions by the following reaction:

$$\mathrm{R{-}2Na + Ca^{2+} \rightarrow R{-}Ca + 2Na^{+}} \qquad (19)$$

The reaction is reversible; when the resin becomes saturated with calcium it can be regenerated with a concentrated solution of sodium ions (sodium chloride). This is known as sodium-cycle cation exchange.

IX resins

Most IX resins are based on a styrene–divinylbenzene copolymer, which is appropriately treated to graft on ionic or functional groups. For example, sulphonation produces a cation resin, while amination produces an anion resin.[15]

Strongly acidic cation (SAC) resins

These resins have a sulphonic acid ($-SO_3$) functional group, as shown in Figure 2.8. The ones in the H-form ($-SO_3H$) are capable of removing all cations from water. For most deionisation applications, SAC resin with an 8% divinylbenzene (DVB) content are used. Generally, they show maximum selectivity for trivalent ions, followed by divalent and

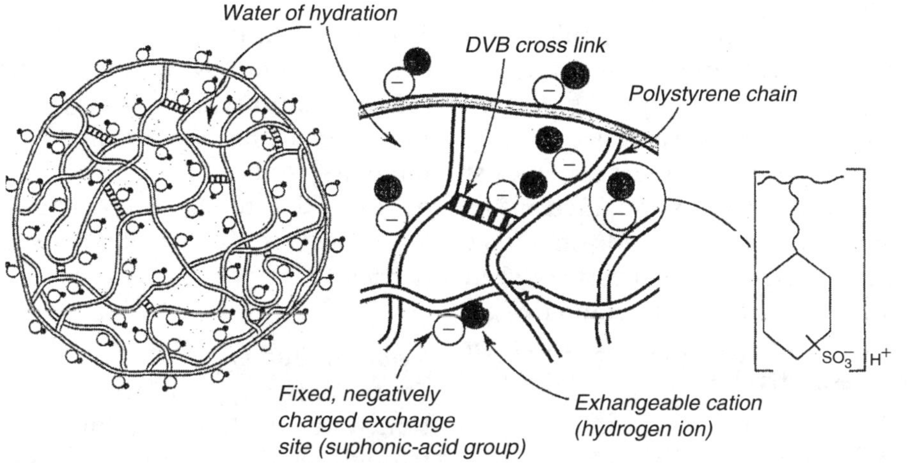

Figure 2.8 Schematic of a cationic ion exchange resin bead. Strongly acidic cationic exchange resin with fixed, negatively charged sites (sulphonic acid groups) and mobile hydrogen (positive) ions attached for the IX resin in the hydrogen form

monovalent ions. They are regenerated with strong acids, usually HCl or H_2SO_4, and require 200–300%, of the theoretical stoichiometric quantity. When used in the hydrogen cycle, the effluent is acidic. The maximum operating temperature of the resin is 135°C.

Weak acid cation (WAC) resins

This class of cation exchange resin is typified by the carboxylic acid (—COOH) functional group, which in its acid form is only very weakly ionised, e.g. methacrylic divinylbenzene, $—RC(CH_3)COOH$. The degree to which dissociation occurs is highly pH dependent, increasing with increasing external pH. The resins are used for removing alkaline hardness as discussed under 'softening.'

Strongly basic anion (SBA) resins

There are two types of SBA resins: Type I and Type II. The Type I resin has the highest basicity, and, therefore gives the optimum effluent quality. Type II has a lower basicity and, therefore, requires less caustic during the regeneration cycle. In general, Type II resin is used where silica effluent quality is not critical. Both resins are usually based on styrene DVB; however, recent resins now include acrylic materials. The functional group in Type I is quaternary benzyltrimethyl ammonium chloride, $RCH_2\,N(CH_3)_3^+\,Cl^-$, and the functional group in Type II class is, $RCH_2N(CH_3)_2\ (C_2H_4OH)^+\ Cl^-$. These resins are capable of removing all anions from water. Selectivity is maximum for divalent ions. SBA resins are regenerated with strong alkali such as caustic soda. The maximum operating temperature of SBA resin is 60°C (OH^- form) and 77°C (Cl^- form).

Weak base anion (WBA) resins

The WBA resins come in two forms: free base form, $CH_2N(CH_3)_2$, and chloride form $—CH_2\,N(CH_3)_2^+Cl^-$. WBA operate essentially as acid absorbers. They are more efficient than strong base anion resins for removing free mineral acids (FMAs) such as HCl, H_2SO_4, and HNO_3, but are not effective at removing carbon dioxide or silica.

Resins selection

Other than the selection of appropriate resin type (cation or anion) and its strength (strong or weak) there are no firm rules, and experience outweighs all other considerations. Guidelines concerning the options related to matrix, structure, and particle size grading are provided by resin manufacturers. Cost is also a major factor. Matrix-modified resins (acrylic) and structure-modified materials (macroporous) can increase resin cost between 20 and 150% compared with standard gel styrenic products when comparing anion or cation resins, respectively.

Physical characteristics

The size and range (0.3–2 mm) of resin beads is an important consideration vis-à-vis performance. Uniform beads have been developed to produce stronger resins that resist attrition and eliminate cross-contamination of the resins during backwashing in mixed-bed systems. Uniform size is important since it imparts:[10,15,16]

- A higher operating capacity,
- A lower ionic leakage,
- Cleaner separation of resins during mixed-bed regeneration, and
- Faster rinsing.

Diffusion of ions into and out of a large bead, limits its performance. Hence, the diffusional path length should be as short as possible to achieve:

- Increased utilisation of the functional sites located within the bead, and
- Higher efficiency with which regenerating solution chemicals get into and out of the bead.

Limitations

The basic limitation of an IX resin to remove and exchange ions and concentrate them within its structure is determined by the 'selectivity' and the 'capacity' of the resin. The ions being exchanged compete for the exchange sites with the resins having a different affinity for each type of ion. Note that since IX resins are primarily designed for removing soluble ionic species, they cannot be expected to absorb macro- or colloidal ionic species except at the surface of the particle.

Demineralisation

There are two types of resins used in deionisation water treatment applications. These are: strongly acidic cation exchange resins, and strongly basic anion exchange resins. If the regenerating solution were a strong acid, such as hydrochloric or sulphuric acid, instead of a salt, the metallic ions would be replaced with hydrogen, and the resin would then remove cations, replacing them with hydrogen ions. This is called hydrogen-cycle cation exchange. Thus,

$$R{-}2H + Ca^{2+} \rightarrow R{-}Ca + 2H^{+} \quad (20)$$

Anions are exchanged in a similar fashion. Anion resins usually operate in either the hydroxide cycle (regenerated with a strong alkali) or the chloride cycle (regenerated with sodium chloride):

$$R{-}2OH + SO_4^{2-} \rightarrow R{-}SO_4 + 2OH^{-} \quad (21)$$

By using hydrogen-cation and hydroxide-anion resins in series, all cations are replaced with H^+ ions and all anions with OH^- ions; the result is pure water. This process is called deionisation or demineralisation.

Dual-bed demineralisation

In a two-stage or dual-bed IX demineralisation process, raw water is first passed through a strong acid cation exchange resin bed in the hydrogen form. The effluent from the cation column is then passed through a strongly basic anion exchange resin bed in the hydroxide form. Across the cation exchanger, all

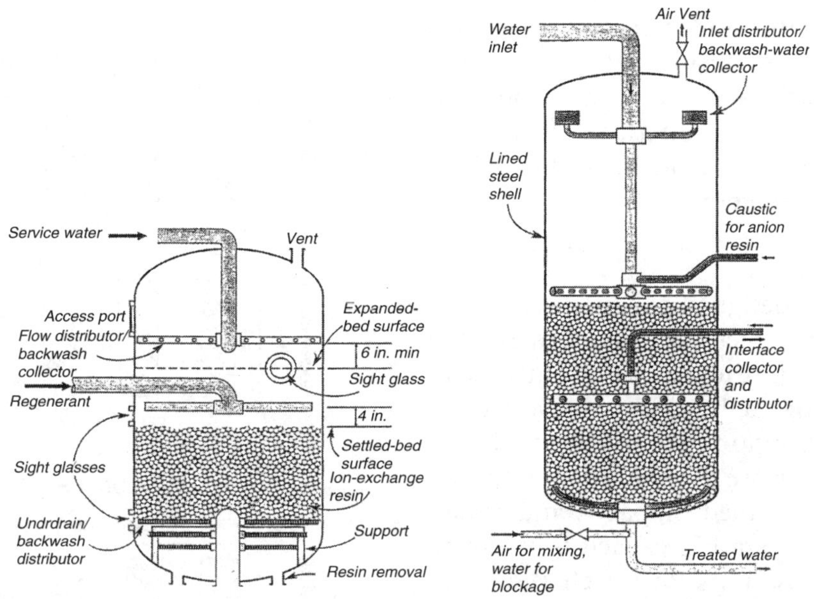

Figure 2.9 A single-resin demineraliser vessel (left) and a mixed-bed demineraliser vessel (right). The mixed-resin bed separates into lighter anion (top) and heavier cation (bottom) resin beds during backwashing. Acid is used to regenerate the cation resin, and caustic soda is used for regenerating the anion resin. Source: USFilter

cations exchange for hydrogen ions to give a dilute acidic effluent ('free mineral acids' or FMAs) made up of acid sulphates, nitrates, and chlorides together with dissolved carbon dioxide. Upon passing through the anion exchanger, neutralisation of the FMAs occurs through the exchange of all anions for hydroxide to give deionised water. A typical IX vessel is shown in Figure 2.9 (left). Dual-bed demineralisation sample design calculations are illustrated in Chapter 5.

Mixed-bed ion exchange

A mixed bed (MB) contains a uniform mixture of strong acid and strong base (Type I) resins in the H^+ and OH^- forms, respectively. Demineralisation occurs through simultaneous exchange of cations and anions in a single vessel. In effect, this is equivalent to having an infinite series of two-bed demineralisers in series. Every anion bead reacts instantly with the acid produced by a neighboring cation bead, removing the acid as it forms and driving the reaction to complete demineralisation. A pictorial drawing of a MB vessel is shown in Figure 2.9 (right).

The development of the MB–IX system is considered to be the most important contribution to the application of IX technology since the development of IX resins. The key to the successful operation of the mixed bed is the ability to separate the two resins after the service cycle so that they can be regenerated separately. Uniform beads have been developed to produce stronger resins that resist attrition and eliminate cross-contamination in separation of the resins from one another during backwashing. Not only is cross-contamination of the resins undesirable from the obvious loss of capacity, but it can also reduce the quality of water. One way to reduce cross-contamination occurring at the interface distributor is by the use of an inert resin. The inert resin has a specific gravity intermediate between those of the anion and cation resins that, hydraulically, classifies between the anion and cation resins when backwashed. The quantity of the inert resin (10–15% of total resin volume) should be enough to cover the interface distributor, thereby, effectively reducing the cross-contamination problem.

Normal operation
An IX vessel is designed to distribute influent water evenly over the resin bed so that it passes through the bed uniformly. This ensures that the bed remains in a packed condition. Disruption of the bed during service would increase the amount of impurity allowed through (called slippage), and reduce the resin bed's operating capacity. In down-flow service, those ions for which the resins have the strongest affinity or selectivity are held at the top of the bed, i.e. those with the lowest selectivity are displaced from the exchange sites by other ions and move down the bed. When the exchange front eventually reaches the bottom of the bed, as shown in Figure 2.10, the vessel is taken off-line for regenerating the resins. The first sign of column breakthrough or ionic leakage is indicated by the most loosely held ions, silica ($HSiO_2^-$) for anion resins and sodium (Na^+) for cation resins.

The operation of a MB is similar to a single-bed IX operation. The regeneration cycle shown in Figure 2.11 is, however, complex. Backwashing, in addition to cleaning as discussed above, helps to classify the resin: the lighter anion beads rise to the top of the bed, the heavier cation beads dropping to the bottom. An inert resin bed with intermediate density forms a layer between the two resins helping to separate them during regeneration and reduces likely cross-contamination of resins. Acid solution flows through the bottom and upward through the cation resin bed and out through the interface collector. Caustic soda solution is introduced above the bed level and flows down through the anion resin bed and out through the interface collector. The resins are then rinsed as described above and, mixed with air or nitrogen before returning to service. The regeneration process is described in detail in Chapter 4.

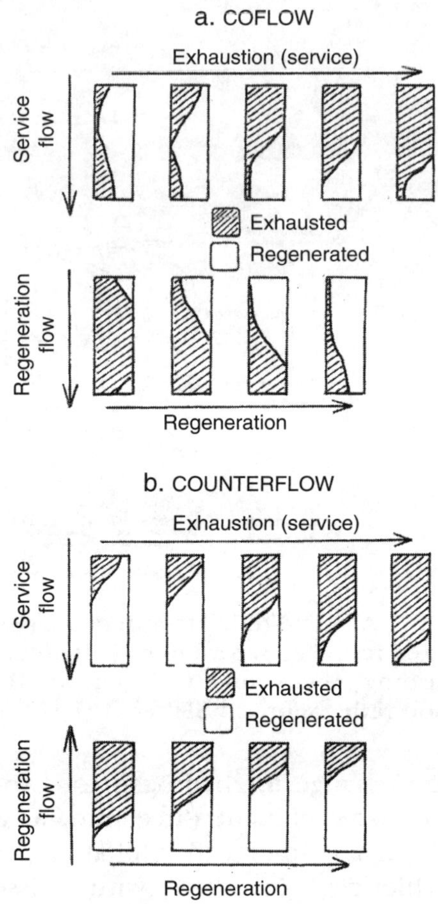

Figure 2.10 Resin bed loading distribution profiles. Ion leakage is indicated when the exhausted bed touches the bottom. Source: Harland

Regeneration
Step I – Backwashing. This is accomplished by passing water upward through the bed for 10–20 min at a velocity of 7–15 m/h sufficient to expand the bed by 50–100%. Backwashing removes particulate matter, relieves any bed compression, and allows trapped gases to escape.

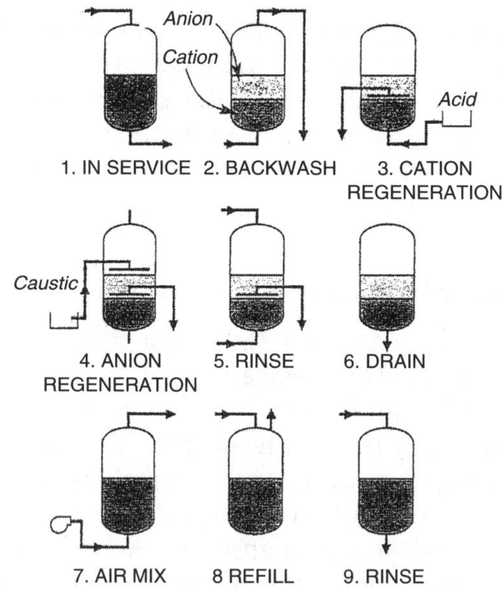

Figure 2.11 Regeneration process of mixed-bed resins

Step II – Chemical regeneration. Regeneration displaces the ions that were exchanged during the service run, returning the resin to the original ionic form. This is accomplished by using solutions of sufficient strength (e.g. 8–10% NaCl for softener cation resin) to drive the equilibrium established during operation in the reverse direction. Cation exchanger resins are regenerated with strong acids such as H_2SO_4 and HCl. Regeneration with H_2SO_4 is usually done at two different concentrations; initially at 2–3% acid followed by 5–6% dilute acid. Anion exchanger resins are regenerated with 4–5% dilute caustic soda solution.

The regenerating solution flow rate is low enough (2–4 $m^3/h.m^3$ of resin) to allow the chemical to diffuse into the resin and to allow the largest impurity ions to migrate out of the resin. Regeneration efficiency can sometimes be improved, significantly, by heating the regenerating solution. Heating caustic soda to 49°C aids in removing polymerised silica from SBA resins.

Step III – Rinsing. Excess regenerating chemicals from the resin bed are flushed by a two-step rinsing process. The first step is a displacement (slow) rinse, in which the flow rate is the same as that used for adding the regenerant solutions (typically replacement of two-bed volumes). This ensures that the chemicals are fully utilised. The second step is a fast rinse, typically 10 min. The high flow rate of water during rinsing flushes out any traces of the regenerant solution.

2.2.11 Non-DI water ion exchange applications

Besides DI water treatment and oxygen removal, there are several other applications where IX is used. Some of these applications are reviewed briefly.

Organics removal

Organic matter in surface water is supposed to behave as a colloid, and is often complexed with silica and heavy metal atoms such as iron, aluminium, and manganese. In order to minimise the risk of organic fouling, SBA resins operating on a co-flow chloride cycle are often employed. It is not essential that the resin has a high ion-exchange capacity. Rather, it is more desirable that they possess greater porosity even at the expense of capacity. Typically, 50–70% of organics are removable.

Nitrates removal

Biological denitrification and IX are the processes used for removing nitrates from contaminated groundwaters when present in carcinogenic concentrations (>50 mg/l). Ion exchange is the easiest and most economical. The process uses single SBA resin in either co-flow or counter-flow arrangement. To ensure reliable operation nitrate-selective resins have been developed. Regeneration is carried out using NaCl solutions.

Industrial wastewater treatment

SAC and SBA resins are typically used to treat wastewater effluents from metal finishing processes such as plating and anodising, pulp and paper manufacture, chemical leaching, zinc smelting, metal pickling, and photographic processing plants. For example, in metal pickling where hydrochloric acid is used in the steel galvanising process, cation is removed as a complex anion. Strong base anion resins in the chloride form readily take up the chloro-complex ions ($Fe^{3+}Cl_4^-$ and $Zn^{2+}Cl_4^-$), thereby, rejuvenating hydrochloric acid. Similarly, SBA resins and sometimes chelating resins are used for recovering heavy metals from various process streams.

Carbohydrate refining

IX processes in carbohydrate treatment are used for the purification of juices and syrups from cane sugar, beet sugar, and corn starch hydrolysates. The main operations are decalcification (softening), deashing (demineralising), and decolourising (removal of organic colour bodies). These processes improve the yield and quality of the final recrystallised sugar or concentrated syrup. One interesting application is the inversion of sucrose to invert sugar (fructose + glucose) using SAC resins in the hydrogen form. Since raw sugar processes yields a fairly viscous syrup, macroporous resins are often used in sugar refining. Decolourising is usually done with macroporous WBA resins.

Pharmaceutical processing

IX is widely used to concentrate and recover antibiotics and vitamins from fermentation broths. For example, streptomycin molecule, which is moderately basic is very favourably exchanged by the sodium form of macroporous WAC resins. General categories of biological compounds that are recovered and purified by IX include antibiotics, vitamins, nucleotides, amino acids, proteins, enzymes, and viruses.

2.2.12 Membrane degasification

Membrane contactor (MC) is a new type of phase-contacting device for use in gas transfer and liquid–liquid extraction processes, as discussed in Chapter 1. MCs are used for stripping (degassing) dissolved gases including oxygen and carbon dioxide from deionised water and many high-purity water industrial applications where deaeration towers are not suitable because of possible contamination. The CO_2 level is reduced to 1–5 ppm. Because of their compact size, MC systems are easily integrated with separation process such as RO or IX providing a highly efficient hybrid system. They can also be used in aerobic wastewater treatment without bubble formation. The process flow sheet of a MC unit with typical operating conditions for removing O_2 is shown in Figure 2.12.

Because of the large surface area of the fibres, a MC module can provide a large number of separation stages in a relatively short length of the module; the size of an MC is more than an order of magnitude smaller than a vacuum tower degasifier with comparable capacity. In general, it is the large surface area per module and not the enhanced mass transfer that makes the process more attractive than conventional contactors. In addition to being very compact in size, MC can be operated over a wide range of flow rates and outlet oxygen specifications. Because of its modular design, a membrane degasifier system can be expanded by adding MC modules in either series or in parallel. Adding more in series allows for lower oxygen concentration in water\outlet values, while

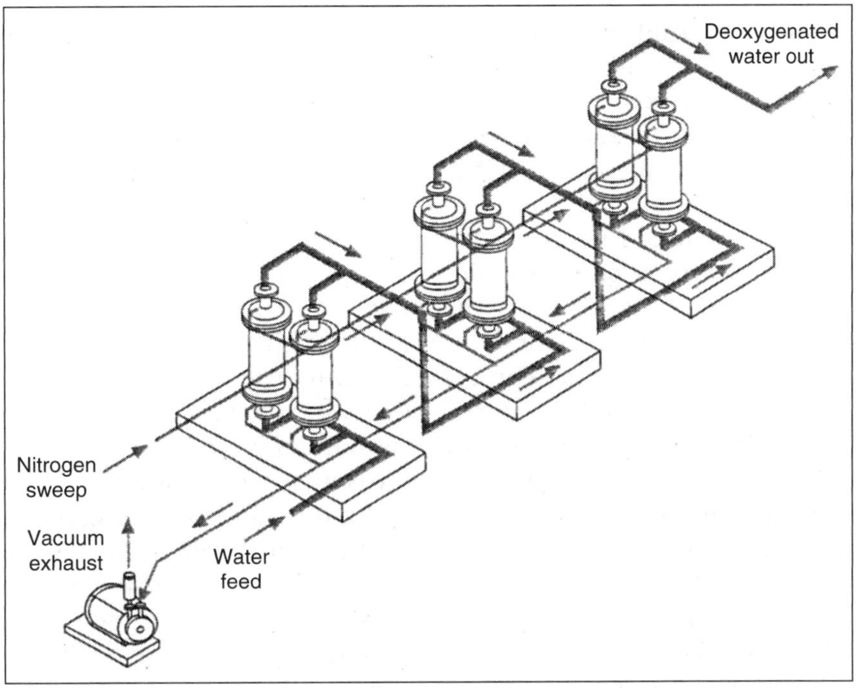

Figure 2.12 Membrane contactors (MC) system for degasification. Typical operating conditions: water flow rate 55 m^3/h at 25°C; feed and outlet O_2 concentration is 5 ppm and <10 ppb; MC 25 cm diameter × 70 cm long; 3 series × 2 parallel array; vacuum 700 mm Hg; N_2 sweep flow rate 1900 cm^2/min. Source: Hoechst-Celanese

additional units in parallel allows for higher flow rates.[24]

For a given MC, the performance can be described as a function of liquid flow rate. The performance of an MC is based on the percentage of oxygen removal (OR) defined as:

$$\text{OR, \%} = \frac{\text{Inlet concentration} - \text{Outlet concentration}}{\text{Inlet concentration}} \times 100$$

2.2.13 Electrodeionisation

Electrodeionisation (EDI) is a hybrid membrane-ion exchange process for producing high-purity water that combines ED/EDR and IX resin. The EDI module consists of multiple cell pairs stacked end to end, as shown in Figure 1.14. Direct current (DC) is the driving force for removing ions from the feed stream while continuously regenerating the IX resins. The electrochemical regeneration of the resin produces high-quality water consistently, without the need for traditional chemical regeneration, as described in Chapter 1.

Feed water enters the product compartments in parallel, and flows from top to bottom. As the water travels through the product compartments, it becomes increasingly pure. The product stream flow rate may be increased or decreased within the range given in the specifications to meet the changing usage requirements. Increasing the product water flow rate above the specified range decreases the product water quality. Up to 18.0 MΩ-cm water can be produced using RO permeate of conductivity <1.0 μS/cm as feed water.[25]

The feed water temperature, hardness, alkalinity, free chlorine, and pH must meet the quality requirements given in Table 2.3. RO pretreatment provides all the necessary pretreatment for optimum operation. If the carbon dioxide concentration is greater than 5.0 ppm, the feed water should be degasified using membrane contactors to prevent build-up of CO_2 in the system, which lowers the product water recovery and efficiency. In critical applications where bacteria control is important, a bacteria destruct UV light is installed to reduce the bacteria load in EDI feed water. EDI modules are sanitised with chemicals or hot water at 80°C.

EDI units are designed to operate with little or no downtime. An EDI system that is fed with RO permeate normally operates at a recovery of 80–95%. The maximum allowable recovery must be determined before adjusting the reject flow rate. If the reject flow rate is too low, the system recovery will be too high, and may result in scaling of the module.

Table 2.3 EDI process feed water requirements*

Item	Process condition
Pressure	
Minimum	2 bar g
Maximum	7 bar g
Temperature	
Minimum	10°C
Maximum	45°C
pH	4–10
Free chlorine	<0.02 mg/l (intermittent)
Iron, manganese, sulphide	<0.01 mg/l
Silica	<2.5 mg/l at 90% recovery
Total hardness	Depends on alkalinity and recovery; <1.0 mg/l at 90% recovery.

*The total hardness, CO_2, and alkalinity of feed water determine the maximum recovery (yield) at which the EDI unit may be safely operated without damage to the module. Source: USFilter/IonPure

2.3 Membrane fouling, scaling, and controls

One of the limitations of membrane processes or liquid separations is severe loss of productivity due to concentration polarisation and fouling, as discussed in Chapter 1. The drop in flux can be as much as 80% in a few minutes or may take days. The loss of productivity is substantial vis-à-vis membrane permeability measured with water.[18–20,26] Several novel techniques for controlling fouling and enhancing flux are given in Tables 1.13 and 1.15.

Membrane pretreatment processes discussed earlier are designed to minimise the potential problems of scaling resulting from the precipitation of the slightly soluble ions because these foul the membranes relatively quickly. Scaling is largely due to the deposition of colloidal materials in close proximity to the membrane surface that often results in fouling. One problem with scaling in membrane systems is that the process introduces a large amount of potential foulants in the system, for example colloidal fouling occurs if total suspended solids (TSS) are too high. Cakes of colloids formed on the membrane surface result in blinding the membranes reducing the flux drastically in most cases, as shown in Figure 2.13. Thus the efficiency of a membrane system, especially the life of a RO or a NF membrane (RO and NF membranes are less rugged than UF and MF membranes) depends on effective treatment of the membrane system feed water.

General rules of thumb for preventing or minimising fouling and scaling in RO/NF membrane systems are: turbidity less than 1.0 NTU, SDI less than 4.0, and LSI less than zero. Pre-treatment methods discussed earlier and also below are summarised in Tables 2.2 and 2.4.

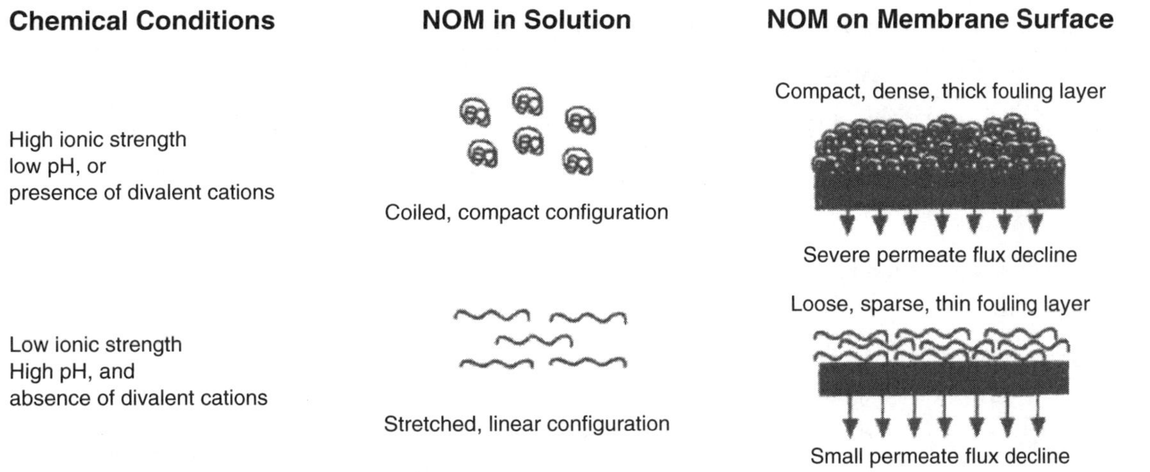

Figure 2.13 Effect of solution chemistry on the deposition of natural organic matter (NOM) on a membrane surface. Source: Schafer et al.

Generally speaking, the fouling and scaling problems are spread as follows: (a) organic fouling (50%), (b) colloidal fouling (30%), and (c) mineral scaling (20%). The effects of these on membrane performance are given in Table 2.5. Organic fouling consists of biofouling, natural organic materials such as humic and fulvic acids, and organics added during pretreatment, e.g. coagulants and antiscalants. Colloidal fouling occurs in almost all membrane systems, and is the most serious of the fouling problems.

Measures to prevent or control fouling and scaling include using solution compatible membranes, i.e. hydrophilic or hydrophobic membranes; using low fouling membranes, e.g. charge compatible membranes; reducing product water recovery; optimising pretreatment strategies such as antiscalants, acidification, water softening and fine filtration, and pre-filtration in the case of RO and NF systems.[19,27] Process parameters – temperature, flow/shear, pressure, feed pH, and concentration – influence fouling, for example, at temperatures below 30°C, the solubility of calcium and phosphate decreases resulting in fouling. Similarly, since protein molecules denature at high temperatures, fouling of UF membranes increases with temperature (30–60°C) due to greater adsorption.[18,19] Proteins are least soluble at pH 4–5 resulting in maximum fouling in this range.

VSEP fouling control. Advanced membrane processes such as the vibratory shear-enhanced process (VSEP®) have proven to be very effective in reducing fouling.[28] The process uses vibrational energy that creates shear wave at the membrane surface, repelling solids and foulants, as shown in Figure 1.24. The VSEP membrane module uses RO, NF, UF, or MF membranes depending on the application.

Start-up conditions. In the case of high flux UF and MF membranes, it has been shown that if the system is started slowly, fouling is reduced, for example a slow start resulted in 10–25% higher long-term flux than abrupt start-up when processing raw sugar solutions.[19] Maintaining a high permeate backpressure, ramping up the pump motor speed and reaching the final trans-membrane pressure (TMP), minutes or hours after start-up, allow operation below the critical flux for that system.

Table 2.4 Summary of conventional pretreatment methods

Problem	Pretreatment	Purpose	Limitations
Ca/Mg Bicarbonate	IX softening	Removes: Replaces Ca/Mg by Na which has a soluble bicarbonate	High TDS causes slip max. 800 mg/l; too expensive at greater than 9,000 m^3/day.
	Lime softening	Removes: Precipitates as $CaCO_3$ and $Mg(OH)_2$	Not suitable for less than 5,000 m^3/day.
	Acid addition	Removes: Replaces the bicarbonate with the more soluble chloride or sulphate	Difficulty and cost of obtaining acid
Ca Sulphate scale	IX softening	See above	
Silica scale	Raising temperature	Stabilises: Increases solubility	Cost of heat
	Lime softening	Removes: Brings down some of the silica with the $CaCO_3$ and $Mg(OH)_2$	Not suitable for less than 5,000 m^3/day
Iron precipitation	Aeration (oxidation) and filtration	Removes: Precipitates the iron as ferric hydroxide which is easily removed by filter	Presence of other oxidisable material, such as H_2S
	Exclude oxidising agents, e.g. air or Cl_2	Stabilises: Keeps the iron in the ferrous state in which it is soluble	Not good for intermittent use or where operating and maintenance staff are of poor quality
	Acid addition	Stabilises: Keeps the iron in solution	pH needs to go to approximately 5, so acid is expensive
Colloids	Coagulation and filtration	Removes: Causes the colloids to form larger particles which can be filtered out	
	IX softening	Stabilises: Discourages coagulation as the solution is concentrated and the colloids are rejected by the RO membrane	Not suitable for high TDS waters where slip through the softener causes more than 5 ppm total hardness in the softened water
Bacteria	Disinfection and filtration	Removes: Bacteria are killed and then removed by sand filtration	
Hydrogen sulphide	De-gas and add Cl_2	Removes: Most of the H_2S that comes off as gas and the remainder oxidised to sulphate	Hard waters require acid dosing before degasifier to prevent scaling of the degasifier packing
	Exclude oxidising agents, e.g. air, Cl_2	Stabilises: Stays in solution	Presence of bacteria or poor operating/maintenance staff
Chlorine or other strong oxidising agent	Dose sodium meta-bisulphite, sodium bisulphite, or sodium sulphite	Removes: Chemically destroys chlorine	
	Activated carbon filtration	Removes: By absorbing the oxidising agent	Filter often becomes a breeding ground for bacteria; regeneration or replacement needed periodically

Source: Amjad, Reverse Osmosis, 1993

Table 2.5 Effect of fouling and scaling on membrane performance

Type of fouling	Permeate flow[1]	Feed pressure	NDP[2]	Solute passage
Organic	Loss	Increases	-	-
Colloidal	Loss	Increases	-	Increases
Scaling	Loss	-	Increases	Increases
Biological	Decreases	Increases[3]	Increases	Increases

[1]Normalised permeate flow
[2]Net differential pressure across the membrane array
[3]Due to restrictions in the fouled feed channel spacer of SW membrane elements

2.3.1 Scaling and fouling mechanism

Each ionic compound has its own solubility limit, which is the maximum amount of the compound that can remain in solution. This is defined as the solubility product constant, K_{sp}. It is a function of temperature, pressure, and pH; slight variations in any of these properties can shift the solubility point and cause scaling. K_{sp} values are given in Chapter 5. For this reason, membrane processes are not operated under conditions of solubility limits (unless an antiscalant is in the feed) because in this zone – metastable region – precipitation can occur when favourable conditions exist.[14,27,28]

Phase changes occur at solid (membrane)–liquid interfaces where precipitation occurs easily if there are nucleation sites. Precipitation occurs at a finite rate depending on the number of nucleation sites, the degree of saturation, temperature, pressure, and time. Once the process of precipitation starts, the rate is controlled by the size of the solid–liquid interfacial area. If the particles are attached to a membrane surface, they grow in only one direction because the membrane limits access to the adjacent surface. Thus particles grow as a sheet until they form a layer of precipitate on the membrane surface, and the membrane gets fouled. The solute concentration is maximum at the membrane surface and decreases into the bulk liquid above the membrane surface; the concentration profile is the reverse of a fluid flow profile in a channel or a tube, as shown in Figure 1.21.

Crystallisation and scaling initially occurs at sites with the greatest amount of available free energy. Most membrane surfaces are relatively low-energy surfaces unlike the bulk solution, which has many imperfect solid surfaces such as suspended solids, which are ideal for crystal growth. In the bulk solution, there are nucleation sites such as colloidal particles in the feed solution, or are formed by chemical precipitation. Fouling occurs when the number of nucleation sites at the membrane surface is large in comparison to the number of sites away from the membrane surface.[28] Hence, it makes a difference whether the nucleation sites are at the membrane surface or in the bulk solution.

2.3.2 Membrane scaling

Scaling occurs when sparingly soluble salts get concentrated beyond their solubility limits in the reject stream of the membrane element. During the scaling process, colloids of insoluble mineral salts are formed. Minerals that precipitate and form scale are predominantly divalent metal ions such as calcium, iron, magnesium, barium, and silicon because they are almost insoluble in the presence of sulphate, phosphate, and carbonate ions. As these molecules precipitate they form crystals that start growing at nucleation sites especially at a liquid–solid interface.[27,28] Monovalent metals such as sodium and potassium, on the other hand, are almost completely soluble. Sparingly soluble salts in the order of decreasing scale formation are:[14]

$CaCO_3 > CaSO_4 > Silica > SrCO_3 > BaSO_4$
$> SrSO_4 > CaF_2 > CaSiO_4 > MgSiO_3$
$> MgSiO_3 > Ca_3(PO_4)_2 > Fe(OH)_2$

The most common scale found in RO/NF systems is calcium carbonate because it precipitates quickly once concentrated beyond its solubility limit, and also because most natural waters are almost saturated with respect to calcium carbonate. Calcium carbonate, carbon dioxide, calcium and bicarbonate ions are in equilibrium according to:

$$Ca^{2+} + 2HCO_3^- \leftrightarrow CaCO_3 + CO_2 + H_2O \quad (22)$$

According to equation (22) calcium carbonate precipitation is favoured by increasing calcium or bicarbonate concentration, decreasing carbon dioxide, reducing total dissolved solids, and increasing temperature or pH. Carbonate – $CaCO_3$, $SrCO_3$, $BaCO_3$ – scaling is prevented by lowering the pH, adding a scale inhibitor, softening (discussed earlier), reducing product water recovery, preventive cleaning, or a combination of all these techniques.[14] Scaling compounds such as hydroxides of aluminium, iron, and manganese are typically removed during RO pretreatment. Because of the low solubilities of most scales, they are very difficult, if not impossible, to remove from RO membranes. For example, barium sulphate solids can form at barium concentrations as low as 0.05 mg/l.

Process and feed water requirements for minimising mineral scaling in membrane systems are given in Table 2.6. The most susceptible elements of an RO/NF system to scaling are the last stages of the membrane array where the reject stream is most concentrated (see Table 2.9). Why does the RO concentrate or reject water come into play? This is discussed below but simply, as the water in the feed/reject stream gets concentrated with the removal of the permeate, sparingly soluble salts can reach their solubility limits and precipitate to form a scale on the membrane surface. Concentration factor (CF) is defined as:

$$CF = 1/(1 - \%recovery) \quad (23)$$

where recovery is (permeate flow rate)/(feed flow rate). In other words when the product water recovery is 75%, the CF is four, and the salts get concentrated by a factor of four. A more reliable indicator of scaling used by membrane manufacturers is called the Beta factor (β). It quantifies the effect of concentration polarisation (CP),

$$CF_M = CP/(1 - \%recovery) \quad (24)$$

where CF_M is the modified concentration factor. CP depends on the turbulence of the bulk stream in the feed channel above the membrane surface, and varies from 1.13 to 1.2 for RO/NF membranes, meaning that the concentration of solutes at the membrane surface is 13–20% higher than in the bulk reject stream. In short, the actual salt concentration at the membrane surface must be accounted for when designing a RO/NF membrane system.

2.3.3 Membrane fouling

One of the most important issues in the development of membrane filtration has

Table 2.6 Feed water requirements to minimise scaling

Parameter	Value
LSI (TDS < 10,000 mg/l)	< 1.0 w/ antiscalant
	< 0 w/o antiscalant
SDSI (TDS > 10,000 mg/l)	< 0.5 w/ antiscalant
	< 0 w/o antiscalant
Barium	< 0.05 mg/l
Strontium	< 0.1 mg/l
Silica, reactive	60–150 mg/l

been fouling. The main reasons for membrane fouling listed in Table 2.5 are:

- Organic molecules adsorption (organic fouling, e.g. proteins, humic substances)
- Particulate deposition (colloidal fouling, e.g. clay, iron, and alumina silicate)
- Microbial adhesion (biofouling, e.g. bacteria)

Thus, fouling is a result of membrane properties, and specific interactions between the membrane and solutes in the feed, as discussed in Chapter 1.[18,19,27] Membrane foulants can be both inorganic and organic components especially the latter. Several mechanisms of colloidal and organic fouling are explicitly illustrated in Figure 2.14.

In the case of traditional UF and MF operations (i.e. non-water treatment applications), the problems can be addressed by using or modifying membranes that are compatible with the liquid being processed, for example, positively charged UF membranes are used for recovering anodic paint in electrocoating whereas negatively charged membranes are used for recovering cathodic paint; otherwise fouling is severe, as shown in Figure 2.15. Back-pulsing is used with hollow-fibre UF and MF membranes to control fouling, as discussed earlier in this chapter. Pretreating the feed stream is also an option, for example, in the case of milk processing by UF using polysulphone membranes, fouling occurs due to colloidal calcium phosphate, which is associated with casein micelle.[19] Addition of EDTA or sodium citrate reduced fouling due to a reduction in the calcium–caseinate complex.

In the case of membrane filtration (UF/MF) applications in water treatment, there is a rapid and often irreversible loss of flux due to interactions between the membrane and the components of natural raw water. The most important foulant is small colloidal matter (3–20 nm in diameter). Material in the colloidal size range such as proteins and macromolecules are known to be the worst foulants. In addition, most of the foulants have a narrow size distribution when dealing with natural surface waters. Scaling by inorganic deposits also results in fouling because different types of fouling often occur simultaneously and influence each other.

General guidelines to minimise fouling are given in Table 2.7. As the SDI increases, the potential for fouling increases. Most manufacturers require that the feed water SDI_{15} (the SDI determined by a 15-min test described in Chapter 5) be less than five, while some require an SDI_{15} less than four. Turbidity specifications must also be satisfied to meet warranty conditions. Note that SDI and turbidity are only generally related – the higher the turbidity, the higher the SDI. However, water with low turbidity (less than 0.2 NTU) can still exhibit a SDI that is greater than five.

Iron and manganese are included as suspended solids since under most operating conditions, iron and manganese that are soluble in the RO feed water oxidise and deposit on the surface of the membranes.[14] Furthermore, oxidation of the metal can catalyse the oxidation of the membrane, leading to membrane failure. The levels of iron, manganese, and nickel in feed water must be less than 0.1 mg/l. Hydrogen sulphide is considered as a suspended solid because it is readily oxidised to form colloidal sulphur; it reacts instantaneously with chlorine to precipitate sulphur at the pH of typical feed water sources. Hydrogen sulphide is found almost exclusively in well water make-up sources. Reaction with dissolved oxygen in water precipitates sulphur almost as rapidly as exposure to chlorine. Colloidal sulphur is difficult to remove from the membrane surface.

Various techniques can be used to reduce the loading of suspended solids, organics,

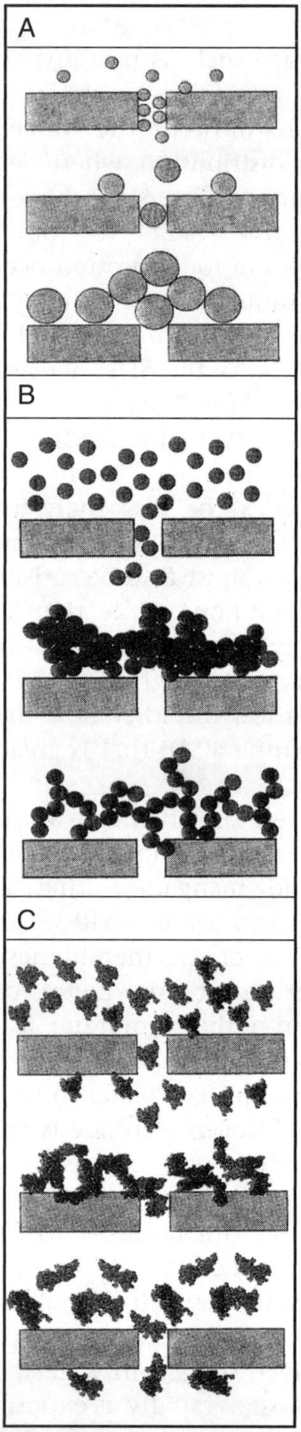

Pore versus Surface Fouling

Pore Adsorption ($d_{solute} << d_{pore}$): Colloids or solutes adsorb on the membrane walls, effective pore size is restricted and flux declines.

Pore Plugging ($d_{solute} \approx d_{pore}$): Colloid or solutes of a similar size to pore diameter block pores completely, reduction in membrane porosity and severe flux decline

Cake Formation ($d_{solute} >> d_{pore}$): Colloid or solutes larger than the pores are retained due to sieving effects and form a cake on the membrane surface, depending on pore to particle size ratio flux decline occurs (permeability of the cake layer as well as the cake thickness are important)

Impact of Colloid or Solute Stability

Stable Colloids smaller than the pore size are not retained by membrane, unless adsorbed by the membrane material

Tight Aggregates are formed by slow coagulation, are retained and form a cake on the membrane. The aggregate structure may collapse depending on forces on the aggregate and the aggregate stability. Flux through the tight aggregates is usually low unless the aggregates deposit as a porous cake of large particles.

Loose Aggregates are formed by rapid coagulation and are also retained. Such aggregates form a cake on the membrane. The aggregate structure may collapse depending on forces on the aggregate and the stability of the aggregate. Flux through the open aggregates is high if the structure is maintained during filtration.

Solute-Solute Interaction

Colloids < pores and stabilised with organics (for example) are not retained by the membrane, unless adsorbed by the membrane material or destabilised with high salt concentrations.

Aggregates with organics adsorbed after aggregation (for example) are fully retained by the membrane, but may penetrate into the upper layer of the membrane. This could also be organics destabilised with multivalent cations.

Colloids which are partially aggregated and destabilised such as a variety of solutes that interact with each other in heterogeneous ways in the presence of salts, collids and dissolved organics, form small and diverse aggregates which may block pores.

Figure 2.14 Colloidal–organic matter complex membrane fouling mechanism. Source: Schafer et al.

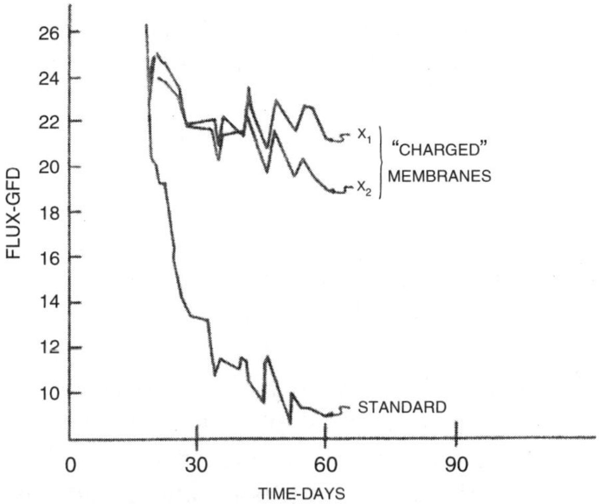

Figure 2.15 Ultrafiltration of cathodic paint using charged membranes (X_1/X_2) and a standard hollow-fibre membrane element. Source: Romicon

Table 2.7 Feed water requirements to minimise fouling

Parameter	Value
SDI_{15}	≤4
Turbidity	<1*
Iron**	<0.05 mg/l
Manganese	<0.5 mg/l
Hydrogen sulphide	<0.1 mg/l
Organics (TOC)	<10 mg/l

* Some membrane manufacturers recommend that turbidity be <0.2 NTU
** At pH>7.0 and 5–10 mg/l dissolved oxygen; at lower pH and lower oxygen levels, slightly higher iron levels can be tolerated

and microbes in feed water. These include physical processes such as dead-end MF to filter the foulants from feed water, and chemical treatments that enhance the filterability of the solids such as the addition of lime coagulants (Table 2.4). Foulants and their control strategies are addressed in Table 2.8. Since any remaining solids and organics are generally removed in the first membrane modules in RO and NF systems, these

Table 2.8 Treatment methods for controlling fouling

Foulant	Fouling control
General	Hydrodynamics/shear, operation below critical flux, chemical cleaning
Inorganic (scaling)	Operate below solubility limit, pretreatment, reduce pH to 4–6 (acid addition), low recovery, additives (antiscalants).
	Some metals can be oxidised with oxygen
Organics	Pretreatment using biological processes, activated carbon, ion exchange (e.g. MIEX), ozone, enhanced coagulation
Colloids (<0.5 μm)	Pretreatment using coagulation and filtration, MF, UF
Biological solids	Pretreatment using disinfection (e.g. chlorination/dechlorination), filtration, coagulation, MF, UF

Source: Schafer et al.

materials typically foul the first stages of RO/NF system (Table 2.9). Once deposited on the membranes, foulants attract additional solids, thereby accelerating any fouling problem that might already exist. In addition to solids, microbes, and organics, soluble heavy metals (such as iron) can foul RO membranes when oxidised within the membrane modules. Oxidation can occur in any stage of an RO system when the pH and dissolved oxygen concentration are suitable. Microbes, if left

Table 2.9 Fouling/scaling location

Type of foulant	Most susceptible location*
Scaling/silica	Last element of last stage
Metal oxides	First element of first stage
Colloids	First element of first stage
Organic	First element of first stage
Biofouling (rapid)	First element of first stage
Biofouling (slow)	Throughout the whole installation

* RO/NF membrane array
Source: Schafer et al.

untreated, can reproduce and spread, thereby fouling the entire RO system.

Organic fouling

Humic acid and fulvic acid represent up to 80% of the total organic carbon – dissolved organics – of natural waters. Humic substances are weak acidic electrolytes with carboxylic- and phenolic-OH groups with a micelle-like structure and with a molecular weight between 500 (fulvic) and 100,000 (humic). The chemical structure of humic acid is shown in Figure 5.1.[15] They become hydrophobic as pH decreases, and thus foul hydrophobic membranes more.[20] Humic fouling in the case of natural waters is aggravated by the presence of calcium (Ca^{2+}) by forming a bridge between the membrane and the negatively charged membrane surface, and/or between negatively charged carboxyl groups of the humic acid. Other organic components that foul membranes include microbial slime, polyhydroxy aromatics, and polysaccharides. Feed water organic levels measured as total organic carbon (TOC) should be low to prevent fouling with organic molecules as well as to minimise the potential for microbial fouling since organics provide nutrients that support microbial growth.

Colloidal fouling

Most common colloids found in natural waters are clays, silica, iron and aluminium hydroxides, and organic debris. In the case of industrial process streams, colloids may be paint pigments, proteins, bacterial and yeast cells, and high molecular weight alcohols. Colloidal fouling in membrane system is caused by the convective deposition of colloids on the membrane surface, and higher the permeate flux, higher is the rate of colloidal fouling. Membrane systems and operating conditions are designed to reduce the risk of colloidal fouling since it is the most severe.[29] These control measures include: feed water pre-filtration, reduced recovery, higher cross-flow velocity and frequent chemical cleanings.

Silica fouling

Silica is one of the most common elements, and is present in many natural waters. It is found in waters in the form of (a) reactive silica, (b) colloidal silica, and (c) suspended particles (sand). The reactive silica is called silicon dioxide, and in this form is generally not ionised at normal pH levels of water.[30] The colloidal species is either polymerised silicon with multiple units of silicon dioxide, or silicon that has formed loose bonds with organic compounds or other complex inorganic compounds such as calcium oxide and aluminium silicate. Well waters contain the most silica, from 50 to 100 mg/l. It is mostly reactive silica, a result of dissolving from rock whereas surface waters contain more colloidal silica even though surface waters contain more reactive silica than colloidal silica. In the case of surface waters, silica chemistry is complex due to biological activity.

Silica has a low solubility of 120 mg/l at pH of 7.0 and 25°C, and silica-scale-like barium sulphate scale is difficult to dissolve during cleaning and must be avoided. Silica scaling usually occurs in the last stage of a membrane array (Table 2.9). Silica fouling is usually associated with iron and alumina (e.g. clays, mullite, and feldspars); metal hydroxides adsorb silica to form silica–metal complexes such as aluminium silicates that readily foul the membranes. Iron fouling can be prevented by (a) antifoulants, or (b) removing iron by oxidation as by greensand filtration. Alumina fouling is prevented by (a) pH control and coagulation prior to multimedia filtration, or (b) alumina compatible antiscalants. If silicates are less than 5 ppm, iron should be less than 0.5 ppm and manganese less than 0.2 ppm, and when the silicates are 30–50 ppm, iron should be less than

0.05 ppm and manganese less than 0.02 ppm. In addition, hardness should be removed and alumina should be less than 0.05 ppm. A dispersant such as high molecular weight polyacrylate scale inhibitor is helpful in silica scale control by slowing agglomeration of silica scale particles.

The solubility of silica is a function of pH and temperature, as shown in Figure 5.12. Lime softening and lime plus soda ash softening are most effective in removing silica. Other options are process related: (a) run the RO system at lower recovery, (b) increase the feed water temperature – silica solubility increase with temperature, and (c) use silica inhibitors. Colloidal silica is difficult to remove by IX because it is not ionised, and can foul the resins when the levels are high. Colloidal silica flocculates easily in boundary layers resulting in severe fouling. It can be removed with UF membranes with a MWCO of up to 100,000 Da. Since the solubility of the silica increases below a pH of about 7.0 and above a pH of about 7.8, the actual solubility of silica in the concentrate stream is further affected by the pH of reject water.

Biofouling

Biological fouling by microorganisms has an adverse effect on membrane performance (Table 2.5). A biofilm is difficult to remove because it protects the microorganisms from shear forces and disinfection chemicals. Microorganisms – bacteria, algae, fungi, viruses, and higher organisms – can be regarded as colloidal matter (typical bacteria size is 1–3 μm). Biofouling can be complicated due to mutual interactions between different types of fouling, for example if iron oxide or a biofilm accumulates on the membrane surface it prevents the migration of other ions (calcium, sulphate) back into the bulk solution, as discussed in Chapter 1, resulting in a supersaturated condition. Thus gypsum (calcium sulphate) scale is formed beneath the primary foulant, and is not easy to remove as compared to the iron scale or the biofilm. The potential for biological fouling is higher with surface water than well water.

Biofouling has been investigated extensively.[31,32] Key characteristics are summarised below:

- Appropriate sampling of biofilms is necessary to identify biofouling. A biofilm has the following characterisitics: (a) high content of water and organic matter (70–95%), (b) high numbers of colony forming units and cells, (c) high contents of carbohydrates and proteins, (d) high content of adenosine triphophate (ATP), and (e) low content of inorganic matter.
- RO membranes reject bacteria absolutely. However, microorganisms can be found in the permeate due to (a) leakage at O-rings, (b) microscopic imperfections in the membrane surface, (c) microbial contamination of feed water, and (d) growth of microorganisms from contaminated piping.
- Biofilm mode of growth enables microorganisms to survive and multiply even in extremely low nutrient habitats (5–100 ppb TOC).
- Generally biofouling is a slow process.
- The cumulative effects of membrane biofouling are: (a) increased cleaning and maintenance costs, (b) a noticeable deterioration of product water quality, (c) significantly reduced membrane life.
- Biofouling can lead to secondary mechanical deformation ('telescoping') of SW membrane elements.
- Biofouling of membrane surfaces is invariably accompanied by some degree of mineral deposition.
- Biofilm can be considered as a dense gel layer, and dissolved minerals tend to accumulate in this layer and increase CP.

- Biofouling causes flow losses due to constriction of the flow channel, increased roughness of the surface, and increased drag because of their viscoelastic properties.
- The first step in biofilm formation prior to microbial adhesion is the irreversible adsorption of macromolecules, which leads to a 'conditioning film' (humic substances, lipopolysaccharides, microbial products). This conditioning film alters the effect of the membrane; the electrostatic charge and the critical surface tension may change.
- Primary adhesion is a function of several factors: (a) microorganisms, (b) membrane surface, (c) fluid properties, and (d) bacteria growth phase. Many bacteria have a slight negative charge, and have to overcome the repulsion barrier when they attach to slightly negatively charged membrane surfaces. Cell hydrophobic property, however, does not seem to be crucial in adhesion to polysulphone (PS) hydrophobic membranes.
- Primary adhesion occurs within a very short time; the highest rate is during the first one hour followed by a slow plateau after 4–6 h. Dead cells adhere as fast as living cells. This means that destruction of bacteria is not enough, dead cells must be removed.
- Higher cross-flow velocity provides higher shear forces, and thus, a thinner biofilm. However, the shear forces of the turbulent flow do not 'penetrate' the viscous sub-layer and affect the bacterial monolayer. Fluid velocities in SW modules have very little effect on impeding the initial rate of microbial (or colloidal) foulant deposition although they can reduce the thickness of the fouling layer.
- Surface roughness has a significant effect on biofouling as it: (a) increases the convective transport near the surface, (b) protects small particles from surface shear, and (c) increases the surface area of attachment.
- Polycations used in pretreatment enhance biofouling because of the electrostatic attraction to the slightly negative overall charge of microbial cells.
- A common cause of biofouling is the overdosing of flocculants (used in removing suspended solids during pretreatment) which provide a suitable habitat for microbial growth.
- Chlorination of seawater has been observed to induce biofouling by degrading humic acid into biologically assimilable material, which supports the growth of a biofilm.

There are several risk factors that are also applicable to non-biofouling membrane operating conditions. These should be borne in mind when operating any membrane system:

- Feed water characteristics – Temperature $>25°C$, high amounts of organic and inorganic nutrients, large number of cells ($>10^4$ colony forming units per ml), and high SDI.
- Operational characteristics – Infrequent monitoring of performance characteristics, use of microbial contaminated pretreatment chemicals, lower cross-flow velocities, and long storage periods,
- System design – Extended piping runs, dead-legs, and disinfected holding tanks.

In most cases, biofouling is assessed based on performance characteristics: (a) flux decline, (b) decrease in salt rejection, or (c) increase in feed-brine pressure drop, as given in Table 2.5. These indirect methods, however, are not absolute indicators of biofouling, and cannot be used for any correlation. Common techniques for controlling biofouling include:

- Removal of the biofilm is essential for effective sanitisation because a biologically deactivated biofilm can cause biofouling.
- Chlorine is a biocide. However, it does not sterilise water, and after chlorination the surviving bacteria can grow especially between 25 and 35°C. Water has to be dechlorinated when using polyamide (PA) membranes.
- Activated carbon used for removing organics is often the source for colonising microorganisms.
- Cartridge filtration using 5.0 μm pore size cartridges are too large for removing bacteria.
- 254 nm UV irradiation is used for disinfecting water. Many types of bacteria are, however, resistant to UV light. Once a biofilm is formed UV light is not able to deactivate or remove the cells. UV treatment is limited to relatively clean water because of interference by suspended solids in water.
- Regular ozone treatment is effective for sterility, and is used in high-purity water systems.
- Hydrogen peroxide requires long contact times (a few hours), high concentrations (>3%), or high temperatures (>40°C). It is an oxidant and, therefore, not compatible with PA membranes.
- Detergents are effective in inhibiting biofouling in many cases. Alkaline cleaners are quite effective in controlling microbial slime. Biz® bleach is an effective cleaner but is an oxidant.
- Sodium bisulphite is effective in controlling biofouling and colloidal fouling. Typically, 500–1000 mg/l is dosed for 30 min every 24 h to control aerobic bacteria. The permeate contains 1–4% bisulphite during this shock treatment.
- Hot water sanitisation is an effective control strategy.
- Preventive sanitisation is much more effective than corrective disinfection because a single attached bacteria is easier to kill and remove than a biofilm. Membrane life is shortened by extensive sanitisation.

2.3.4 In-line control strategies

The water pretreatment UOPs discussed in Section 2.2 are the basic essentials for providing acceptable water to RO/NF membrane plants. Additional treatment may be required to further minimise the potential for control fouling and scaling to ensure that the RO/NF membrane plants operate from three to six months without shutdowns for cleaning unlike MF and UF plants that are usually subject to daily or weekly cleaning. For example, synthetic antifoulants such as Hypersperse™ are used to prevent carbonate and sulphate scaling, and to reduce the rate of colloidal fouling.[29]

In-line coagulation

Polymeric coagulants are sometimes added in low dosages (<10 ppm) to remove particles down to 0.5 μm particle size as compared to 10 μm without the coagulant upstream of a cartridge filter. Even very low amounts of cationic polymeric coagulants in RO/NF feed water can, however, foul negatively charged membranes. In addition, some organic coagulants are not compatible with acrylic-based antiscalants used in RO/NF systems. Typical coagulant dosage ranges from 0.5 to 20 ppm. Inorganic coagulants generally require a higher dose than polymeric coagulants. The best technique for determining the proper dosage of coagulant is to feed the product in-line and then measure the SDI in the filter effluent as a function of coagulation feed rate.

In-line dechlorination
Oxidising agents such as chlorine damage PA membranes by breaking down the polymer backbone, as discussed in Chapter 5. The maximum allowable chlorine limit for PA membranes varies between 0 and 1000 ppm h. It is, therefore, mandatory that RO feed water is dechlorinated when using PA membranes. A reducing agent, such as sodium metabisulphite ($Na_2S_2O_5$) is invariably used to scavenge any traces of chlorine in carbon filter effluent water. Sodium bisulphite ($NaHSO_3$) is formed when sodium metabisulphite is dissolved in water. $NaHSO_3$ then reduces hypochlorous acid as per the reaction:

$$NaHSO_3 + HOCl \rightarrow NaHSO_4 + HCl \quad (25)$$

In theory, 1.34 mg of sodium metabisulphite (100%) is required to remove 1.0 ppm of free chlorine. In practice, 3.0 mg is used. Similarly, it takes 1.46 ppm of sodium bisulphite, and 1.77 ppm of sodium sulphite (Na_2SO_3) to remove one ppm of free chlorine. A minimum contact time of 5 s is required. Sodium metabisulphite also helps in reducing dissolved oxygen in water.

In-line pH adjustment
Membrane scaling is linked to system recovery and cross-flow velocity; generally, for recovery up to 70%, and with or no iron present, acidification may be the only chemical pretreatment necessary to prevent scaling by calcium carbonate. When water recovery is 75–80%, additional processes such as scale inhibitors and/or IX softening instead of acidification is required. The solubility of calcium carbonate depends on the pH of feed water. The equilibrium can be shifted to the right to convert calcium carbonate to soluble calcium bicarbonate ($(HCO_3)_2$) by lowering the pH to 6.0 with sulphuric acid or hydrochloric acid as:

$$CaCO_3 + H \leftrightarrow Ca^{2+} + HCO_3^- \quad (26)$$

Acid reacts with bicarbonate alkali to produce carbon dioxide. RO permeate is often high in anions, and always acidic (pH <7.0) due to the presence of dissolved carbon dioxide (carbonic acid). Carbon dioxide concentration can be several hundred ppm when using acidified feed.[33] Dissolved carbon dioxide is removed by tower decarbonation or by membrane degasification to reduce the loading on anion resins when EDI or mixed-bed IX is used for producing DI water. Membrane degasification is preferred in high-purity water systems to prevent contamination. Alternately IX softening is used to remove calcium ions followed by raising the pH of the softened water to 8.3–8.5 by adding caustic soda. Raising the pH with sodium hydroxide converts the carbon dioxide to sodium bicarbonate, which is easily rejected by the membrane.

Acid addition is determined by the LSI or SDSI of RO reject water. To control calcium carbonate scaling by acid addition alone, the LSI or SDSI in the concentrate stream must be negative as indicated in Table 2.6. When an antiscalant (A/S) is used, the LSI can be 1.0.[14,33] LSI is applicable when the total dissolved solids (TDS) are less than 10,000 mg/l whereas SDSI is applied when the TDS is greater than 10,000 mg/l.

Sulphuric acid is commonly used but hydrochloric acid is preferred when the scaling potential is high due to $CaSO_4$, $SrSO_4$, and $BaSO_4$. Calcium sulphate is more soluble than $BaSO_4$ and $SrSO_4$. However, since calcium ion is present in natural water sources more abundantly than barium and strontium ions, $CaSO_4$ is a greater problem. Nevertheless, $BaSO_4$ and $SrSO_4$ scale is difficult to re-dissolve once precipitated. Hence, overdosing of sulphuric acid must be avoided.

Acidification, however, has several limitations:

- Low pH increases fouling by natural organic matter (NOM) such as humic acids.
- Low pH lowers permeate quality due to higher TDS of feed water, increase in silica and carbonic acid.
- Permeate TDS increases when using hydrochloric acid because added chloride has a lower rejection than sulphate. Hence, acid treatment is usually used for carbonates and phosphates scale prevention.

Antiscalant threshold treatment

Scale inhibitors or antiscalants (A/S) are generally organic compounds containing sulphonate, phosphonate, or carboxylic acid functional groups and chelating agents such as carbon, alum, and zeolites that sequester and neutralise a particular ion which may be formed. The majority of scale inhibitors can be classified as threshold inhibitors. In addition, chelating agents such as EDTA (tetra sodium salt of ethylene diamine tetra acetic acid) are used to control hardness (at $pH > 6.0$) and metallic ion deposits. Antiscalants prevent mineral scaling by getting absorbed on the scale, forming salt crystals thereby preventing the attraction of the supersaturated salt to the crystal surfaces. Since A/S inhibit the growth of crystal, it does not grow to a size or a concentration large enough to precipitate out of the suspension. Many scale inhibitors also contain dispersants that keep the precipitates suspended in solution.[14,33]

Antiscalants are preferred to IX softening when the feed water hardness is less than 100 ppm because of cost and ease of operation. Although antiscalants are very effective in preventing carbonate and sulphate scaling of membranes, they do not prevent scaling from occurring; rather they delay the formation of large crystals that form scales. The effectiveness of scale inhibition is approximately 30 min. A RO/NF system that uses A/S must be designed with automatic flush cycle after shutdown to prevent scaling by concentrated salts in the feed–reject channel above the membrane surface.

Antiscalants can be used alone, but are usually used with acid feeds. When acid is used with an A/S, the LSI value (of the reject stream) of 1.0 is acceptable (see Table 2.6) although some A/S manufacturers claim an LSI of 2.7 is acceptable when using their recommended product.[34] The A/S dosage ranges between 2 and 10 ppm, depending on the scale-forming potential of the RO feed water, product water recovery, and A/S manufacturer's recommendations. The main advantage of a higher LSI value is that the RO/NF system can be operated at higher recovery resulting in lower operating costs. From a process point of view, higher recovery results in higher salt concentration in the feed–reject channel that in turn results in sparingly soluble salts exceeding their solubility limits faster. Hence, high recovery is only an option when operating a second-pass RO unit or when the feed water is pure.

Sodium hexametaphosphate (SHMP) is a threshold agent derived from the dehydration of orthophosphoric acid or its sodium salt. It is used to inhibit the formation of calcium carbonate and metallic sulphate scale. It is most widely used because it offers good inhibition at a low cost. Depending on the concentration of calcium and sulphate, and depending on the CF (equation (23)), the dosage is in the range of 2–5 ppm. SHMP can prevent calcium sulphate precipitation up to 150% of the saturation limit. Organophosphonates are an improvement over SHMP since they are more resistant to hydrolysis, even though more expensive. They offer scale inhibition and dispersion ability similar to SHMP.[14,34]

Polyacrylic acids (PAA) are good at both scale inhibition and dispersion, and are more effective than SHMP. PAA with high molecular weight distribution show the best dispersion ability at the cost of scale inhibition ability. However, precipitation may occur with cationic polyelectrolytes or multivalent cations such as aluminium or iron resulting in fouling the membranes. Blend inhibitors are a combination of low (2,000–5,000 Da) and high molecular weight (6,000–25,000 Da) of PAA or a blend of low molecular weight PAA and organophosphonates giving excellent dispersive and inhibitor performance.

Antiscalant treatment can, however, contribute to fouling largely due to either under-dosing or overdosing; the former can result in scaling while the latter can lead to fouling.[33] Overdosing can also lead to biofouling and can result in complexes formed with hardness ions since assimilable organic carbon content is a food source for microorganisms.[33]

2.4 Membrane systems design

The design of a hybrid membrane water system entails a comprehensive design of the feed water treatment system, the design of the membrane array based on optimum product water recovery and solute rejection, membrane feed water pressure, process control, and post-treatment for higher purity product. Membrane system design, therefore, is an iterative procedure encompassing membrane selection (size of membrane element and type of membrane), membrane array (number of stages and passes, and number of elements) design and product quality calculations until the desired quality is obtained. In the case of RO and NF systems, membrane manufacturers have generated computer programs for designing their membrane-based systems. In the case of UF and MF systems, however, pilot data is required to design a membrane system.

2.4.1 RO/NF system basics

The RO and NF membrane processes were discussed in detail in Chapter 1. RO membranes are well suited for rejecting dissolved ions and most organics (some organics such as ethanol and acetone have very low rejections of 45–55%). The rate of water transport through a membrane depends on membrane properties (polymeric, chemical, morphological), water temperature, and the difference in applied pressure across the membrane, less is the difference in osmotic pressure between the concentrated and dilute solutions. Osmotic pressure is proportional to the solution concentration and temperature, and depends on the type of ionic species present. For solutions of predominantly sodium chloride at 25°C, a rule of thumb is that the osmotic pressure is 0.7 bar per 1000 mg/l concentration.

The choice of a membrane is usually determined by the composition of feed water and the final product water quality. Homogenous asymmetric cellulose acetate (CA) membranes (blends of cellulose acetate and cellulose triacetate) and thin-film composite (TFC) membranes are the most commonly used, as discussed in Chapter 1. The data in Table 2.10 show that CA membranes have a neutral charge, are more hydrophilic, less prone to fouling, and can tolerate low levels of chlorine. Because of these properties, CA membranes are often preferred in wastewater applications where the SDI is high, and in potable water purification where

Table 2.10 General characteristics of CA and PA membranes

Property	Regenerated CA	Polyamide[1]
Membrane type	Homogenous asymmetric	Homogenous asymmetric; Thin-film composite
Salt rejection	~95%	>99%
pH range	4–6	1–13
Feed pressure[2]	15–30 bar g	10–20 bar g
Hot water sanitisable (80°C)	No	Yes[3]
Surface charge	Neutral	Anionic
Chlorine (oxidants) tolerance	1–2 ppm	<0.1 ppm
Pretreatment	High	Very strict
Organics removal	Good	Effective
Biological growth	Problematic	Not a problem
Fouling tolerance	Good	Fair
Surface roughness	Smooth	Rough
Membrane module types	[4]SW, FP, HF, T	SW, FP, HF[5], T[5]

[1] RO and NF membranes only
[2] Not seawater desalination
[3] TFC membranes only
[4] Spiral wound, flat-plate, hollow-fibre, tubular
[5] Not applicable to TFC membranes

residual chlorine is required. TFC polyamide and polyetherurea membranes have higher rejection and flux, higher durability than CA membranes but have very limited tolerance to chlorine. TFC membranes have a higher rejection than CA membranes due to co-ion repulsion since the TFC membrane has a net negative charge at a pH greater than 5.0. Below a pH of 4.0, the membrane has a net positive charge. Hence, at low pH values the rejection decreases. In order to increase the dissolved ion rejection when CO_2 is present, caustic soda is often injected (unless the feed water is degasified) to raise the pH between 7.0 and 8.0. Under alkaline conditions, the CO_2 is converted to bicarbonate and carbonate ions by raising the pH to greater than 7.5. The bicarbonate ions are easily rejected by the RO membrane whereas carbon dioxide is not.

Membrane manufacturers are continuing to develop newer and more efficient membranes and modules, for example for (a) seawater, (b) brackish water, (c) high rejection, (d) high flux, (e) loose wrap, (f) hot water sanitisable (80°C), (g) low energy, and (h) nanofiltration. Standard SW modules are 8 in. diameter × 40 in. long with a membrane surface area of 440 ft^2 (40 m^2). Larger SW modules, 17 in. × 60 in., with a surface area of 2400 ft^2 (220 m^2) have been developed for seawater and brackish water desalination. These larger modules are more efficient and less costly.[35] SW modules are compact and inexpensive. The flow regime in SW elements is turbulent with superficial velocities ranging between 10 and 60 cm/s corresponding to Reynolds (Re) numbers of 100–1300.[19] Although these Re numbers indicate laminar flow, turbulence is due to the mesh spacers in the feed channel shown in Figure 2.16.

Spiral-wound loose wrap or full-fit modules (FFM) are used in many pharmaceutical systems because there are no brine seals to prevent the bypass of feed water. When a brine seal is utilised, a large pocket of water remains stagnant around the RO membrane. Since the water is not chlorinated in the case of PA membrane elements, stagnant water is prone to bacterial growth on the outer surface of the membrane. A FFM uses the feed pressure to expand the membrane's diameter. This expansion provides the control of feed water bypass while allowing a small flow of water to pass around the membranes outer surface. This small flow of water eliminates the stagnant pocket of water that the brine seal units retain, and can be rinsed in much shorter time. Hence the FFM offers better control of bacterial growth and is often used for pharmaceutical and beverage applications. The bypass flow may be as high as 20% of the feed flow to the membrane. This means that the actual cross-flow velocity is considerably lower and the permeate flow rate is produced with less feed water when compared with membranes that utilise brine seals (because some flow is bypassed and not available for filtration). The actual recovery is, therefore, considerably higher when compared with membranes that utilise brine seals.

An RO/NF membrane system consists of the following components: (a) a high pressure pump, (b) a membrane element/pressure vessel assembly (array), (c) instrumentation, and (d) a clean-in-place (CIP) unit. A typical RO/NF membrane assembly consists of either spiral-wound (SW) or hollow-fibre (HF) membrane elements housed in fibreglass-reinforced plastic (FRP) or stainless steel pressure vessels. A typical SW element is shown in Figure 2.16. Up to seven SW elements are connected in series in a single pressure vessel 8–10 m long, as shown in Figure 2.17. The desired system capacity and recovery are achieved by connecting pressure vessels in parallel, and by staging the reject stream in an array with a decreasing number of pressure vessels; the tapered configuration design is shown in Figure 2.18. The brine/reject volume gets reduced in each stage, so that the number of modules in successive stages is reduced usually in the ratio of 2:1 to maintain optimum reject flow velocities. The high feed channel flow velocity is desired in order to reduce concentration polarisation at the membrane surface. The number of elements required is calculated using the formula:

$$N_E = (\text{Permeate flow rate})/(\text{Permeate flux} \times \text{Active membrane area})$$

The amount of product water (permeate) recovered is generally dependent on: (a) the total area of membrane within each vessel; (b) membrane pressure supplied by the high pressure pump(s); (c) the reject flow rate; and (d) feed water quality. General rules of thumb for RO membrane product water recovery are as follows: (a) 70–75% for normal or brackish feed water, (b) 30–45% for seawater, and (c) 80–85% for purified feed water, e.g. first-pass RO permeate as second-pass RO feed.

During RO/NF membrane operation, water is forced into the membrane module pressure vessel by a high-pressure pump at pressures in the range of 10–30 bar g for municipal water or brackish water, and from 55 to 80 bar g for seawater. The desalted product (permeate) is removed from the opposite side of the membrane at low pressure. A flow-regulating valve on the reject side is used to create backpressure and increase recovery, as shown in Figure 2.19. The total pressure drop from the feed inlet to reject outlet is minimal (<2 bar g), which allows the high-pressure reject to be fed to successive RO stages to increase recovery or productivity.

The simplest membrane element assembly consists of one pressure vessel, containing one membrane element with a product water recovery (yield) of about 15%. In order to increase % recovery and still maintain an acceptable concentrate flow, a part of the concentrate stream is returned to the inlet side of the high-pressure pump. Concentrate recycling design is used with very small RO units and is discussed later in the chapter (see Figure 2.23). The main advantage is the compact size of the RO unit. The disadvantage is a larger feed pump to handle higher feed flow. Accordingly, the power consumption is relatively higher than that required in a multi-stage configuration. In addition, due to blending of the feed with the concentrate stream, the average feed salinity is increased. Therefore, both the feed pressure and the permeate salinity are higher.

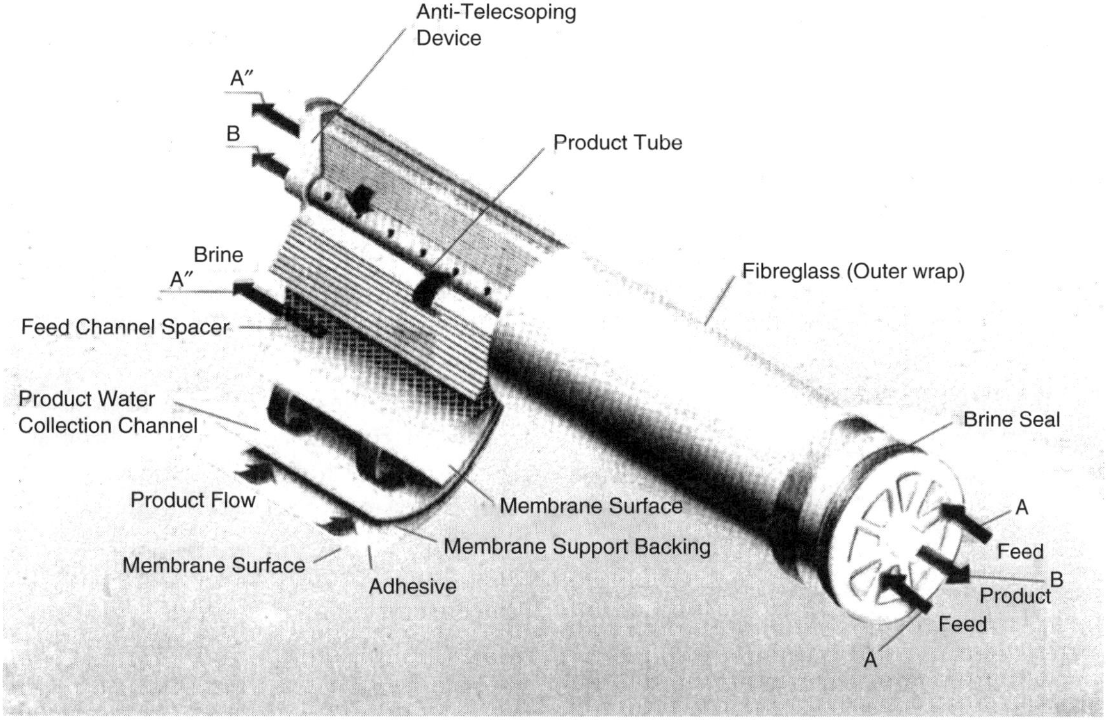

Figure 2.16 Cut-up diagram of a typical spiral-wound RO membrane element. Source: SAEHAN membrane brochure

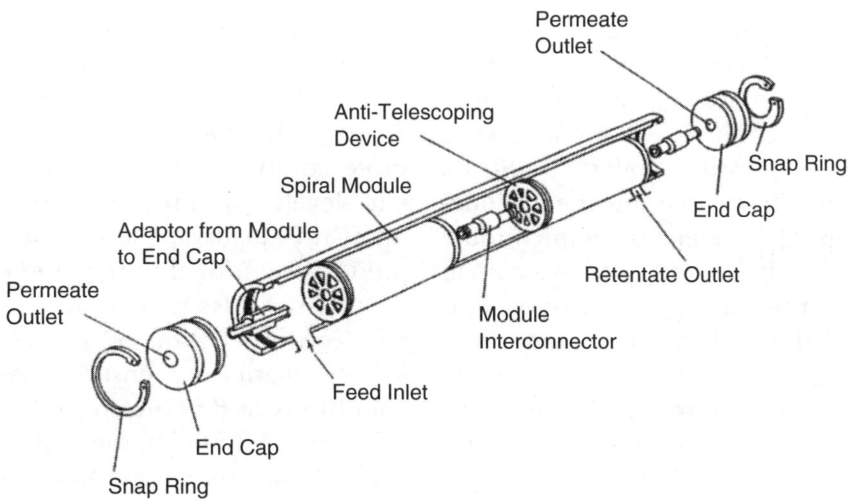

Figure 2.17 Multiple spiral-wound elements connected in series in a pressure vessel. Permeate tubes of each element are connected by module inter-connectors. Up to seven elements are connected for RO applications, and two to four elements in the case of UF and MF applications. Source: Cheryan, Copyright® 1998 from Ultrafiltration and Microfiltration Handbook by M. Cheryan. Reproduced by permission of Routledge/Taylor & Francis Group, LLC

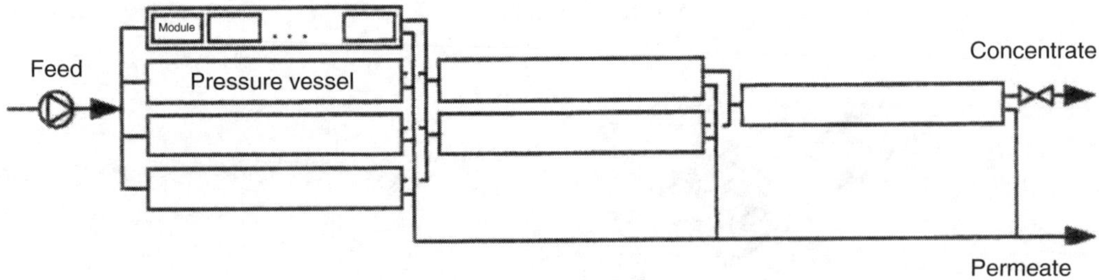

Figure 2.18 Typical RO and NF membrane array. The tapered configuration has four pressure vessels in stage l, two in stage 2 and one in stage 3. Source: Wittman and Thorsen

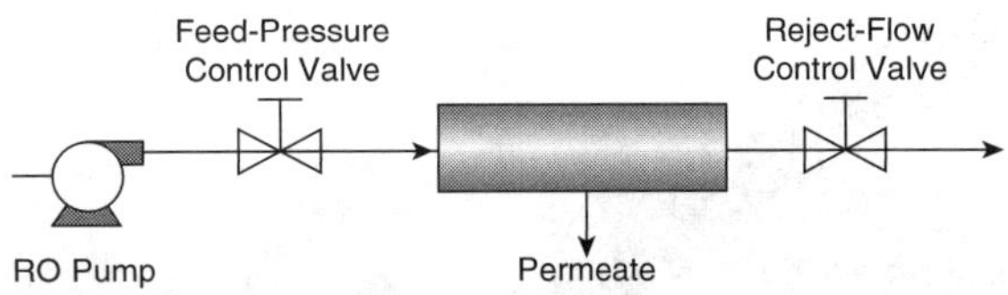

Figure 2.19 Schematic of a basic RO/NF control system diagram. Source: Singh and Tembrock, Reproduced with permission from Chemical Engineering Progress, Sept. 1999. Copyright 1999 AIChE

For some applications, the single-pass RO system may not be capable of producing permeate of a required quality. For example: (a) seawater RO systems, which operate on a very high salinity feed water, at high recovery and/or at high feed water temperature; and (b) brackish RO systems, which require very low permeate conductivity for supplying makeup water to pressure boilers. Two-pass RO systems are often used in the production of high-purity water for semiconductor, beverage, medical, and pharmaceutical production to optimise solute rejection and productivity. In this design, the primary (first-pass) unit permeate is fed to the secondary (second-pass) unit as shown in Figure 2.20a. Because of very low dissolved solids ion content of the first-pass RO permeate, the second-pass RO membranes used have a higher flux (e.g. $34\,l/m^2/h$) and high recovery (e.g. 85–90%). Further, to increase the overall recovery and conserve water, the reject stream from the second-pass is recycled back to the RO pump inlet. The dissolved ion concentration in the concentrate from the second-pass is usually lower than the concentration of the first-pass unit feed water. Therefore, blending the first-pass feed water with the second-pass concentrate reduces slightly the concentration of the feed, and increases the overall utilisation of the feed water.

2.4.2 Membrane system controls

Of all the major membrane processes, RO/NF separation is the most complex both in terms of operation and controls.[36] RO (and NF) membrane systems operate in a continuous mode with minimum recycle, and RO plants are by far the largest with a capacity of up to $76{,}000\,m^3$/day. Further, for hybrid membrane systems the process control becomes even more complex. RO/NF plants require different levels of process control depending upon the quality of feed water supplied and product water quality requirements.

Systems discussed provide monitoring and controls of major process variables in RO/NF plants to ensure that the design conditions of flux and rejection are achieved with minimal membrane fouling while ensuring fail-safe operation. New control systems are being designed that interlock the pressure drop across a membrane stage with the high-pressure RO pump or the membrane array inlet pressure control valve to maintain constant permeate flux. Similarly, new system

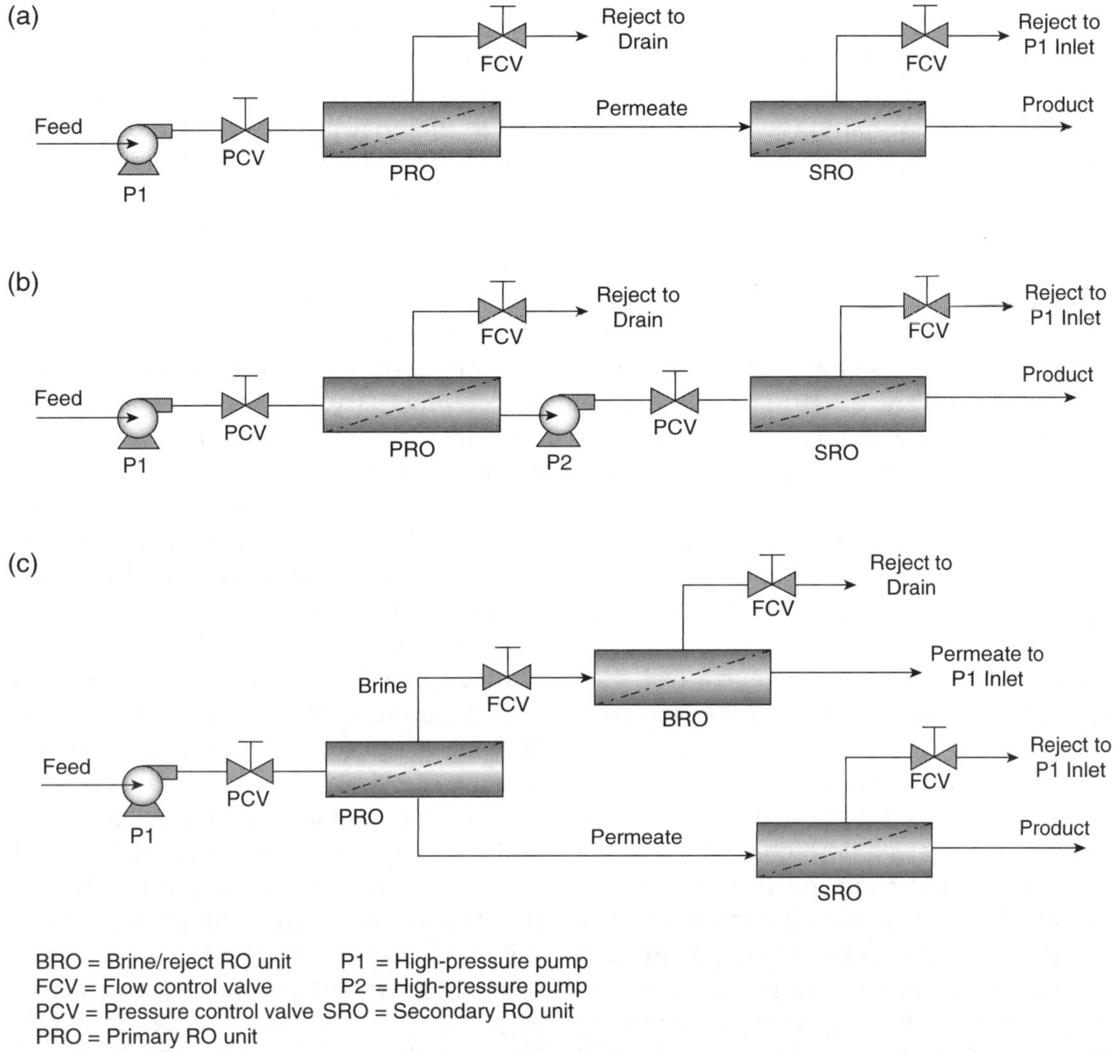

Figure 2.20 High recovery or high product quality two-pass RO systems. (a) two-pass RO without inter-stage pressure boosting; (b) two-pass RO with inter-stage pressure boosting; and (c) two-pass RO with brine RO reject recovery. Source: Singh and Tembrock, Reproduced with permission from Chemical Engineering Progress, Sept. 1999. Copyright 1999 AIChE

designs are envisaged where the pressure drop is monitored across each element in a pressure vessel rather than across the entire pressure vessel or stage. Since each pressure vessel has up to seven elements in series, closer monitoring of pressure drop would help in monitoring fouling and/or scaling, and enhance system efficiency. Nowadays an increasing number of RO systems use electric motors with variable speed drives, which enable adjustment of flow rate and pressure over a wide range with very little loss in efficiency. The variable speed drive, thus, reduces pressure losses.

Basic controls

The basic level of control routinely used in the chemical process industries involves sequencing operations such as manipulating valves or starting/stopping pumps, instrument verification, data acquisition, online maintenance, and fail-safe shutdown procedures. The next level of computer control involves process control of parameters such as flow, pressure, and temperature. RO/NF systems require both levels.[36]

The simplest form of RO/NF system contrail entails the adjustment of only two variables: membrane inlet pressure and reject flow rate, as shown in Figure 2.19. These functions affect the quantity and quality of RO permeate, and indirectly potential for membrane fouling. The reject flow control valve must be throttled in conjunction with the inlet pressure control valve to achieve the desired productivity or product water recovery, i.e. yield.

Several more complex valve control sequences are illustrated in Figure 2.20. In a two-pass RO system (Figure 2.20a), the first-pass RO inlet pressure and reject flow control valves are adjusted simultaneously with the second-pass RO reject flow control valve, to achieve the desired recovery. In the case of a two-pass RO system (Figure 2.20b), where the feed to the secondary RO unit is re-pressurised, the secondary membrane unit inlet pressure control valve is also adjusted. In order to reduce the size of the waste stream and increase the overall product water recovery, the reject stream from the primary RO unit is sometimes further processed in a brine RO unit. In such setups, the primary RO inlet pressure and reject flow control valves, the brine. RO reject flow control valve and the secondary RO reject flow control valve are all adjusted simultaneously to achieve the desired recovery (Figure 2.20c).

RO system performance (% recovery and % rejection) for a given membrane is a function of the operating conditions. Hence, the control system is designed to run the system at the membrane inlet pressure and recovery specified by the system manufacturer. Membrane performance, however, declines with time due to changes in the material properties of the polymer, so that control valve settings periodically require adjustments – for instance raising the membrane inlet pressure. In addition, the control system may be used to monitor RO/NF pump inlet and outlet pressures, membrane reject and permeate pressures, membrane array pressure drop (ΔP), feed water temperature, pH, conductivity, chloine level, and permeate conductivity. Continuous monitoring of RO/NF performance, alarms, shutdown regime, data logging, and performance trending are essential; for example, high ΔP across a membrane stage or an array is a good indicator of membrane fouling.

When specifying the high-pressure feed pump, membrane manufacturers assume that membrane flux will decline by about 20% in three years. The pump is, therefore, designed to provide feed pressure corresponding to the initial membrane performance and to compensate for expected flux decline. In the case of centrifugal pumps, the pump selected is oversised, and during operation the feed pressure is regulated by throttling (turning down the pump discharge valve).

Control systems

A programmable logic controller (PLC)-based membrane system utilises remote input/output (I/O) functionality. The PLC communicates with various remote I/O control enclosures in the membrane plant, and performs all sequencing and inter-locking functions. In a fully automated system, independent (non-PLC) controllers such as PID (proportional plus integral plus derivative) or other tuning allow the operator to modulate control valves, e.g. feed water temperature, membrane inlet pressure, and reject flow rate.[36] The PID controllers also give the operator easy access for changing set point values, and supervise alarm conditions without the need for an operator interface screen. Independent PID controllers can allow the operator to run the system without a working PLC; if a single independent PID controller fails, only the particular parameter it was controlling cannot be adjusted.

Process control valves

Membrane feed pressure control may be performed via two different techniques: proportional valve position or pump motor speed.[36] The valve solution is generally less expensive in initial capital but solid state control of pump motor speed reduces operating expense because electrical power savings can be realised over the life of the equipment. The long-term reliability of the motor speed solution is greater if done properly, but the process control is slightly better with the proportional control valve.

Control of the reject flow rate can only be achieved with the use of a proportional valve. Since the type of valve can affect the performance of the system, the valve should be capable of reducing the pressure (up to an order of magnitude bar, if necessary) while maintaining a smooth linear response throughout the required range of flows. In seawater RO applications, energy recovery from the high-pressure reject stream (>55 bar g) is an important economical requirement,[37] high-pressure drop control is, however, complex. Some systems may be designed with a fixed-pressure drop, using an orifice plate, upstream of the reject control valve. The orifice plate lowers the amount of pressure that must be dropped across the control valve. Although, this reduces the sensitivity of the valve, it enables the control-system output signal to operate over a greater range, thereby, increasing the accuracy of control.

The difficulty with controlling membrane feed pressure and reject flow rate simultaneously is that one affects the other. This gets more complicated with a double-pass system. If the membrane feed pressure is reduced, the reject flow rate also will decrease if all other parameters (feed water temperature, quality, and membrane condition) in the system remain constant. The danger, then, is that the whole system may oscillate if the PID controllers are not properly tuned.[36] Tuning one controller for fast operation and the other for relatively slow operation is often the solution. Fast operation generally is best applied to the membrane feed pressure controller because during start-up, this parameter must be adjusted first. Manual control of these valves is, therefore, often preferred.

Incorporating the controller start-up output condition is also necessary to prevent the controller from drifting towards 0 or 100% – i.e. fully closed or fully open, respectively – during idle periods; otherwise if the RO/NF pump starts when the membrane feed pressure or the reject flow control valve is closed, a sudden spike in initial pump discharge pressure can result in equipment damage. Conversely, if the membrane feed pressure valve is fully open during start up, the membrane elements are exposed to excessive pressure, which may cause compaction of the membranes in the leading elements.

Product quality controls

Feed water parameters that affect membrane life adversely are high pH, free chlorine, and high temperature are shown in Figure 2.21. Often, automatic on/off valves are provided to divert the feed or the permeate stream to drain when the stream quality is not in the desirable range, as shown in Figure 2.21. For example, if the feed water conductivity is much higher than the system is designed for, membrane scaling may result. A much higher than design permeate conductivity, on the other hand, indicates

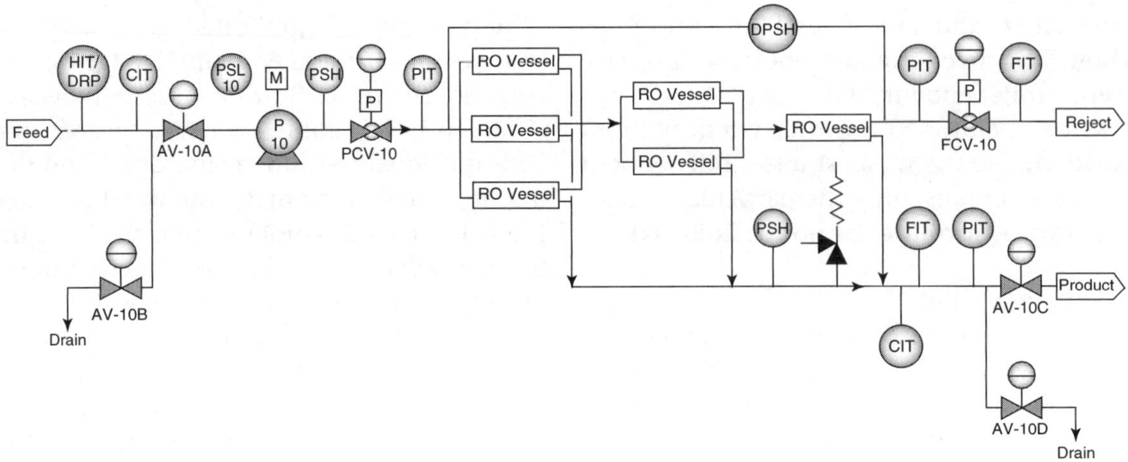

Figure 2.21 Typical first-pass RO unit process flow sheet. Source: Singh and Tembrock, Reproduced with permission from Chemical Engineering Progress, Sept. 1999. Copyright 1999 AIChE

membrane scaling, membrane element integrity, or membrane damage. Hence, a low rejection alarm is often provided. Similarly, when the permeate flow rate exceeds the design point (or the reject flow is less than the design value), the potential for membrane scaling is substantially increased. A high recovery alarm is also included to alert the operator to reset the reject flow control valve setting or shutdown the unit.

High pressure drop across a membrane array is a strong indicator of membrane fouling due to constriction in the reject/concentrate flow channel above the membrane, and calls for membrane cleaning. Usually RO/NF plant designs incorporate an automatic five minute flush cycle preferably with RO permeate before shutdown to protect the membranes from (a) scaling by sparingly soluble salts in the salt concentrated stagnant reject channel, and (b) biological fouling. Often the flush cycle is programmed in the PLC to run at a preset interval such as every four hours. During flushing the high-pressure RO pump is off. Flushing minimises the likelihood of scaling and fouling but is not a substitute for feed water pretreatment.

Safety features

RO control systems provide safety features for operating high-pressure RO feed water pumps. Pressures at pump inlet and discharge, membrane inlet, product output, and reject output are, therefore, monitored by pressure switches and transmitters, as shown in Figures 2.21 and 2.22. To ensure long-term safe operation of the equipment, pressure switches shut down the pump during low inlet pressure and high discharge pressure conditions. The high-pressure RO pump must be run in auto mode to ensure it shuts down automatically during an emergency. A relief valve is provided in the permeate line to protect membrane elements from damage due to any backpressure downstream. Further, in a fully automated RO system, the membrane inlet pressure and the reject flow control valves are of the air-to-open type. Air-to-close valves should not be used because it would result in excessive membrane pressure during start-up.

The following general process conditions are taken into consideration when designing a control system to ensure safe and

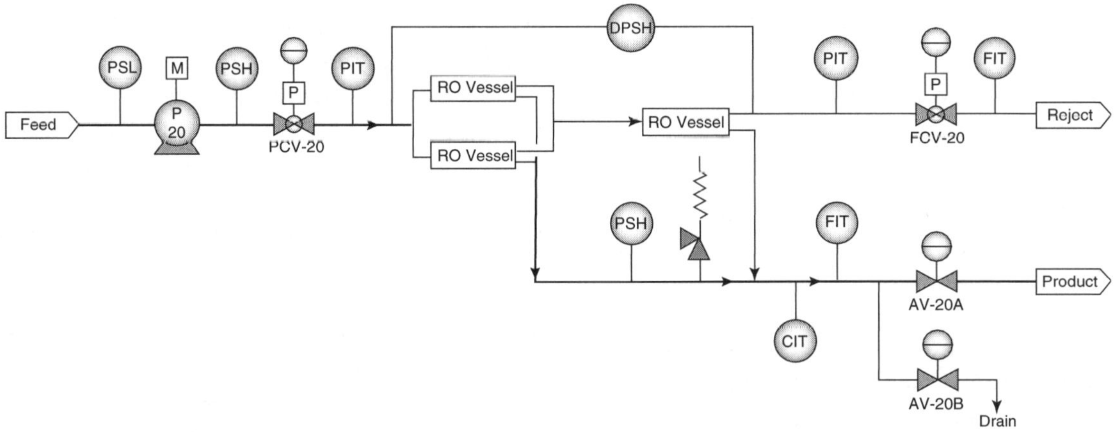

Figure 2.22 Typical second-pass RO unit process flow sheet. Source: Singh and Tembrock, Reproduced with permission from Chemical Engineering Progress, Sept. 1999. Copyright 1999 AIChE

reliable operation:[14]

- Product water backpressure greater than the system pressure at any time will damage the RO membranes.
- Operate the RO unit at the lowest system pressure that produces the design flow rate and salt rejection.
- The initial minimum salt rejection is based on chloride ions, and applies to each individual membrane element. The total % rejection may be lower depending on the array. Membrane deterioration may result in salt passage to double within three years.

2.4.3 RO/NF array design

The following basic membrane separation phenomena should be borne in mind when designing a membrane system:[14]

- The permeate flow rate (convective) is proportional to the net driving pressure (NDP) differential across the membrane.
- The salt flow rate (diffusive) is proportional to the concentration difference across the membrane, and is independent of applied pressure.
- Permeate TDS depends on the relative mass transfer rates of water and dissolved solutes through the membranes.
- The chemical and physical nature of the membrane determines the preferential transport of water over dissolved solutes.
- Higher the permeate flux, greater is the likelihood of higher concentration polarisation (CP). As CP increases, the osmotic pressure of the solution in the feed–reject channel increases, salt passage increases, and the risk of scaling and/or fouling increases.

Membrane manufacturers provide design guidelines based on the parameters given below. These guidelines are usually modified based on the type of feed water and pretreatment.[33,38]

- Maximum feed flow rate to any element in the pressure vessel
- Maximum reject flow rate from any element
- Maximum product water recovery for an element

- Maximum flow rate (or flux) for any element
- Maximum average flux for a system
- Maximum applied pressure.

The optimal design of a RO system incorporates certain rules of thumb based on the membrane selected for the particular application:

- Recovery per element is <19% for softened water or well water with an SDI <3
- Recovery per element is <16% for unsoftened water or surface water with an SDI = 3–5
- Net pressure drop across the array is <7 bar g
- Average flux for each element depending on the type of feed water*
- Percent variation in permeate flow rate is <10% between the first and last elements in the same pressure vessel
- Feed water flow rate to the first element of each stage is the same, <10%.

*14–20 l/m^2.h municipal wastewater (sewage); 17–24 l/m^2.h for treated river water; 20–27 l/m^2.h for lake water; 27–34 l/m^2.h for deep well water (low turbidity); and >34 l/m^2.h for RO permeate water when there is no fouling and/or scaling potential.[39,40]

In order to avoid excessive concentration polarisation at the membrane surface, permeate recovery per membrane element should not exceed 18%. In the case of brackish water RO systems, the average recovery per 40-in. long membrane element is usually about 9%. The overall recovery for a staged system with pressure vessels containing six elements is usually as follows:[39]

- One-stage array (1): 52–56%
- Two-stage array (2:1): 75–80%
- Three-stage array (4:2:1): 85–90%

The recovery in each element is controlled by the concentration of rejected species, especially, sparingly soluble salts of calcium and magnesium, and silica in the brine stream. When the product recovery is 50%, the salt concentration in the reject stream is doubled, whereas the salt concentration increases four-fold when the recovery is 75% due to the concentration factor (equation (23)). Hence, the RO system is operated below the design recovery point. In general, the product water recovery is maintained well below 15%, and the systems are usually designed for a recovery of 8–10% per element. The scaling and fouling potential is usually the highest in the last elements of the final stage (Table 2.9).

The above conditions are taken into consideration when modeling a RO/NF membrane system. The engineers use computer-generated performance projection software provided by membrane manufacturers to design an optimal membrane array design that maximises the operating conditions and minimises fouling and scaling conditions. A typical RO/NF program calculates permeate quality (conductivity, pH), feed pressure requirements, and the final concentrate stream solubility numbers such as LSI and SDI depending on: (a) permeate flow rate, (b) % recovery, (c) feed water composition, (d) feed water temperature, (e) type and number of membrane elements, (f) the rate of flux decline, and (g) the rate of salt passage increase.

The programme algorithm is an iterative calculation in which the computer first estimates a feed pressure to satisfy the desired recovery, and then calculates the performance of the first element of the system.[39,40] The concentrate from the first element becomes the feed to the second element, and a second calculation of membrane element performance is made, and so on from element to element through the complete array of the proposed design. The programme then sums the permeate flow from all elements, and compares this value to the target value. The programme adjusts the feed pressure based on this comparison, causing the solution to converge to the required feed pressure to achieve the required permeate recovery given the user-defined system parameters and until the programme has converged on a single unique solution. Calculations can be repeated with different design parameters, or membrane element array configurations. In case the programme does not converge, a warning is issued requesting a revised number entry. The programme also calculates the concentration polarisation coefficient called the β factor:

$$\beta = \text{Concentration at the membrane surface/Concentration in the bulkfluid.}$$

β is a function of the ratio of permeate flow from an element to the feed-brine average flow for that element. The optimum value is 1.13 for the last element of the last stage of a membrane array. To maintain this value of β when using 130 cm (40 in.)-long SW elements, the maximum recovery is usually limited to 15% for one element, 30% for two elements in series, and so on to 50% for six elements in series in a pressure vessel.

Computer-generated design of a single-pass, three-stage RO membrane (Hydranautics) array (4:3:3) with concentrate recycling is shown in Figure 2.23 and given in Table 2.11. The stream numbers in the figure are: 1 is raw water, 2 is blended water, 3 is membrane array feed, 4 is reject, 5 is reject/concentrate recycle, 6 is reject-to-drain, and 7 is permeate. The RO unit is designed to produce 27 gpm (6.13 m^3/h) permeate corresponding to an overall recovery of ~75% (27 gpm permeate/36 gpm raw water feed). The thin-film composite polyamide RO membranes reduce the total dissolved solids (TDS) content from 421 mg/l in blended feed water to 5 mg/l in product water at an average rejection of 98.8%. A portion of reject flows to drain at 9 gpm (2 m^3/h), and the remaining recycles to the RO pump inlet at 5.5 gpm (1.25 m^3/h). The design feed pressure is 145 psig (10 bar g). The β factor is within range, and the LSI is negative as required. The RO membranes are spiral-wound modules, 4 in. diameter × 40 in. long.

A typical RO skid is shown in Figure 2.24. It is a single-pass, two-stage (4:2 array) unit with pressure vessels containing six spiral-wound membrane elements (8 in. × 40 in.) in series in each vessel. There is room on the backside of the skid to double the number of vessels to make it into an 8:4 array with permeate flow rates approaching 70 m^3/h at 75%. The RO high-pressure pump is multi-stage, horizontal, submersible type. The skid includes a control panel, and instruments such as conductivity and flow monitors shown on the right-hand side. The end view of RO/NF membrane pressure vessels is shown in Figure 2.25.

The effect of colloidal fouling on membrane processes was discussed earlier in this chapter; it is directly proportional to the permeate flux and the solids content of the feed solution.[29] Since colloidal fouling has a strong negative effect on membrane performance, membrane systems are designed by limiting the permeate flux of each element based on the recommendations of the membrane manufacturers. One such plot for a TFC membrane with various natural water feeds is shown in Figure 2.26. The data show that as the quality of feed water improves, the recommended permeate flow rate also increases. For example, the membrane flux is linear with pressure in the case of nearly pure water or RO permeate, as discussed in Chapter 1.

2.4.5 UF/MF process design

The design and mode of operation of UF and MF systems are defined by the process needs and application. There are three main process configurations: (a) single-pass,

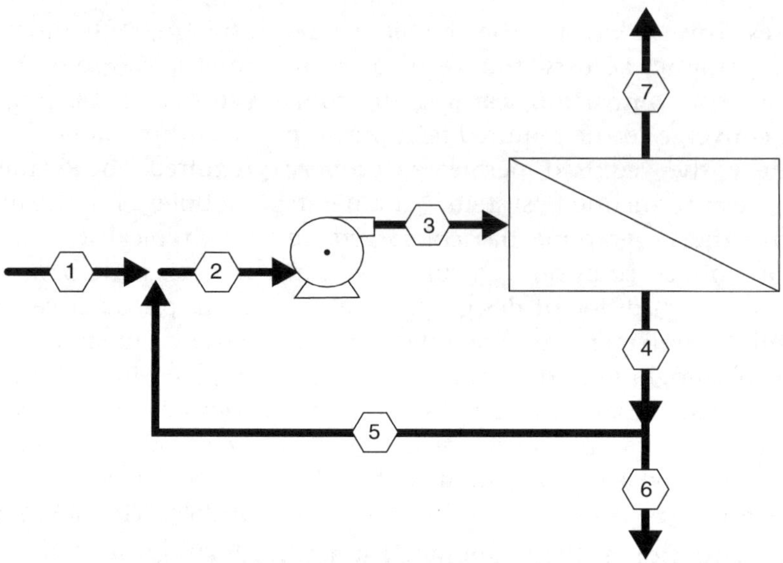

	1	2	3	4	5	6	7
Flow (gpm)	36.0	41.5	41.5	14.5	5.5	9.0	27.0
Pressure (psi)	0.0	0.0	115.6	72.2	0.0	0.0	0.0
TDS (ppm)	302.6	420.9	420.9	1195.9	1195.9	1195.9	4.7

Figure 2.23 Process flow diagram of a single-pass RO membrane design given in Table 2.11 with reject recycling. Stream 1 is feed, 2 is blended feed, 3 is membrane array pressurised feed, 4 is reject, 5 is reject/concentrate recycle, 6 is reject-to-drain, and 7 is permeate

(b) batch filtration, and (c) feed-and-bleed, as shown in Figure 2.27. Batch and feed-and-bleed designs involve extensive recycling to remove greater than 95% of the solution, and require heat exchangers in the recycle loop, to remove heat and protect biologicals and proteins from irreversible damage. UF and MF systems design is usually based on pilot plant data.[18,19]

Single-pass design

This is the simplest process configuration, and is almost only used in water purification systems such as high-purity water microelectronics plants where UF is used as the final polishing process. Product water recovery per single pass is >90%. This is possible because the feed water is of very high purity so that concentration polarisation is negligible. The objective is to remove to reduce the total particle count required in microelectronics production.

The main disadvantage of the single-pass system for non-high-purity water applications is that substantial membrane area is required to achieve the 95–99% permeate recovery. This in turn results in higher energy consumption and pumping costs for non-water applications – to maintain high velocities to minimise concentration polarisation and/or fouling as the feed becomes more viscous – and to overcome higher pressure drops to pump the viscous concentrate through the feed channels.

Batch filtration

As shown in Figure 2.27, the retentate recycles to the feed tank, and the concentration of the retained particles and solutes in the feed

Table 2.11 RO membrane system performance projection (Hydranautics)

Calculation created by	**RS**		
Project name	Single-pass RO design	Permeate flow	27.00 gpm
HP pump flow	41.5 gpm	Raw water flow	36.0 gpm
Recommended pump pressure	144.6 psi	Total system recovery	75.0%
Feed pressure	115.6 psi	Permeate recovery ratio	65.1%
Feed water temperature	25.0°C (77°F)	Concentrate recirculation	5.5 gpm
Feedz water pH	7.14 (0.00)	Element age	0.0 years
Acid dosage, ppm (100%)	0.0 H_2SO_4	Flux decline % per year	7.0
Acidified feed CO_2	11.1	Salt passage increase, % per year	10.0
Average flux rate	15.2 gfd	Feed type	Well water

Stage	Perm. Flow (gpm)	Flow/vessel Feed (gpm)	Flow/vessel Conc (gpm)	Flux (fgd)	Beta	Conc. and throt. pressure (psi)	Conc. and throt. pressure (psi)	Element type	Elem. no	Array
1–1	12.8	10.4	7.2	18.1	1.13	100.5	0.0	ESPA2-4040	12	4 × 3
1–2	7.8	9.6	6.9	14.8	1.11	83.8	0.0	ESPA2-4040	9	3 × 3
1–2	6.3	6.9	4.8	11.9	1.12	72.2	0.0	ESPA2-4040	9	3 × 3

	Raw water		Feed water		Permeate		Concentrate	
Ion	mg/l	$CaCO_3$	mg/l	$CaCO_3$	mg/l	$CaCO_3$	mg/l	$CaCO_3$
Ca	32.6	81.3	45.5	113.4	0.16	0.4	129.9	323.8
Mg	7.4	30.5	10.3	42.5	0.04	0.1	29.5	121.3
Na	47.8	103.9	66.3	144.2	1.11	2.4	187.8	408.3
K	0.0	0.0	0.0	0.0	0.00	0.0	0.0	0.0
NH_4	0.0	0.0	0.0	0.0	0.00	0.0	0.0	0.0
Ba	0.000	0.0	0.000	0.0	0.000	0.0	0.000	0.0
Sr	0.000	0.0	0.000	0.0	0.000	0.0	0.000	0.0
CO_3	0.1	0.2	0.1	0.2	0.00	0.0	0.4	0.7
HCO_3	69.3	56.8	96.5	79.1	0.80	0.7	274.7	225.1
SO_4	44.7	46.6	62.4	65.0	0.07	0.1	178.5	185.9
Cl	50.5	71.2	70.4	99.3	0.33	0.5	200.9	283.4
F	1.8	4.7	2.5	6.6	0.02	0.1	7.1	18.8
NO_3	44.7	36.0	61.6	49.7	2.11	1.7	172.4	139.1
SiO_2	3.7		5.2		0.03		14.7	
TDS	302.6		420.9		4.7		1195.9	
pH	7.0		7.1		5.3		7.6	

(Continued)

Table 2.11 Continued

	Raw water	Feed water	Concentrate
$CaSO_4/K_{sp} \times 100$	1%	1%	6%
$SrSO_4/K_{sp} \times 100$	0%	0%	0%
$BaSO_4/K_{sp} \times 100$	0%	0%	0%
SiO_2 saturation	3%	4%	11%
Langelier Saturaton Index	−1.21	−0.79	0.58
Stiff and Davis Saturation Index	−1.16	−0.74	0.60
Ionic strength	0.01	0.01	0.02
Osmotic pressure	2.5 psi	3.4 psi	9.8 psi

Figure 2.24 A typical RO membrane skid showing a 4:2 two-stage spiral-wound membrane array, high-pressure pump, instruments, and control panels. The high-pressure RO pump is horizontal, multi-staged submersible type. Source: USFilter

tank increases with time as the permeate is removed resulting in a continual decline in flux. The permeate recovery per pass is much smaller than in a single-pass system as is the membrane area, but the residence time is very long contributing to high pumping costs. The pumping costs can be reduced by incorporating a partial recycle stream, as shown in the figure. Batch filtration is used in small-scale and pilot-scale applications when the permeate is the product, for example, fruit juice and for effluent treatment where the concentrate is disposed off. It is not suitable for applications when the concentrate is the product because it can get damaged due to prolonged exposure by heat and shear as well as bacterial growth. Constant recirculation of the concentrate often results in denaturation of proteins and cells.

Figure 2.25 End view of a large RO/NF membrane system. Source: Ultrapure Water

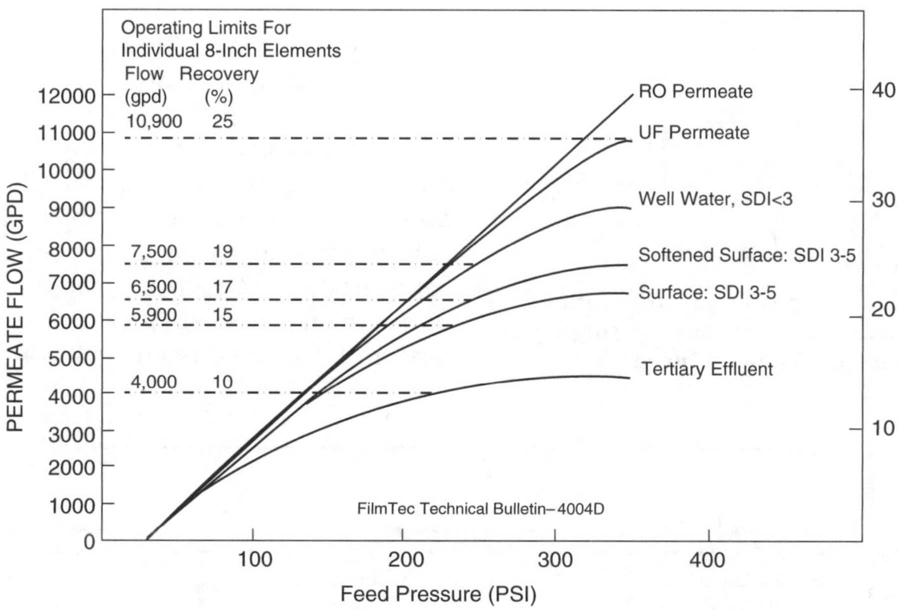

Figure 2.26 Membrane flux characteristics of a spiral-wound RO module for various feed waters. Source: Film-Tec membrane catalogue

Feed-and-bleed configuration

The feed-and-bleed (F&B) design is used in most large-scale commercial applications. As shown in Figure 2.27, only a part of the concentrate is recycled while the remainder is collected or fed to the next stage or processing step. F&B systems are most often configured as multi-stage units to reduce the required membrane area, as shown in Figure 2.28. In this case, the concentrate from the first stage flows to the second stage as feed and so on during series operation.

The area required to obtain a concentration factor of 10 in a F&B unit, for example is three times higher than required for batch processing and 60% higher than required for single-pass processing.[18] This is because the entire system operates at the final retentate concentration corresponding to the lowest permeate flux whereas in batch filtration, the initial concentration is relatively low, and flux relatively high.

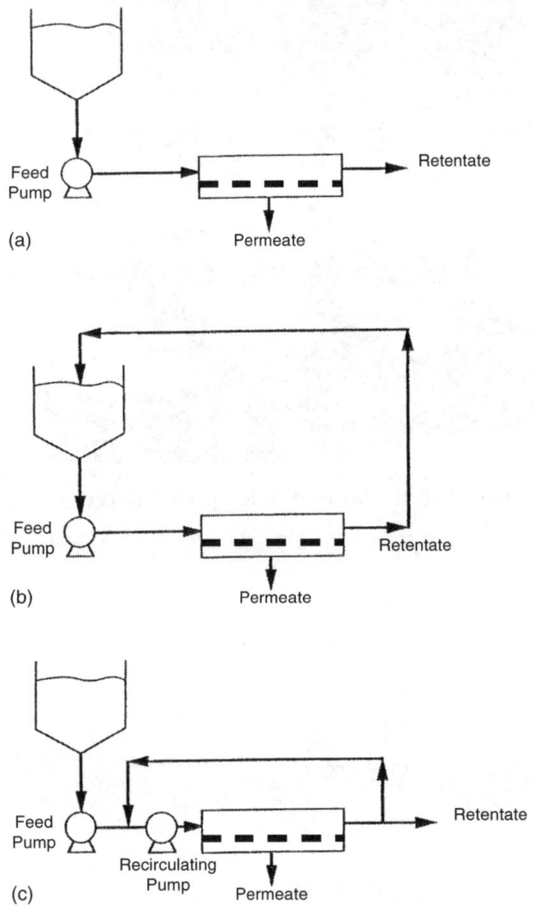

Figure 2.27 Typical process configurations of UF and MF membrane systems; (a) single-pass, (b) batch process, (c) feed-and-bleed

Multiple array design

The UF and MF membrane arrays are designed to operate in series as shown in Figure 2.28, in parallel as shown in Figure 2.29, or a combination of both. Most commercial processes use three to five stages to optimise required membrane area and pumping costs as shown in Figure 2.29. Flux vs. concentration curves for the system illustrated in Figure 2.29 are shown in Figure 2.30.

UF/MF performance design

The time required for a given UF/MF operation is controlled by the membrane area, volume of feed to be processed, and solute concentration. Of these, solute concentration is the most important factor as it controls flux and the membrane gel concentration,

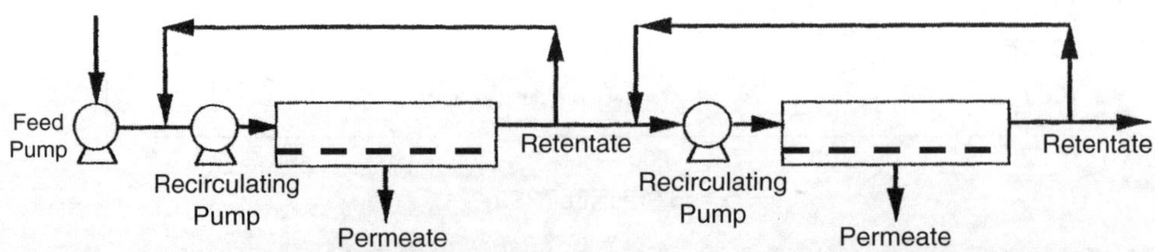

Figure 2.28 Schematic representation of a multi-stage UF membrane system

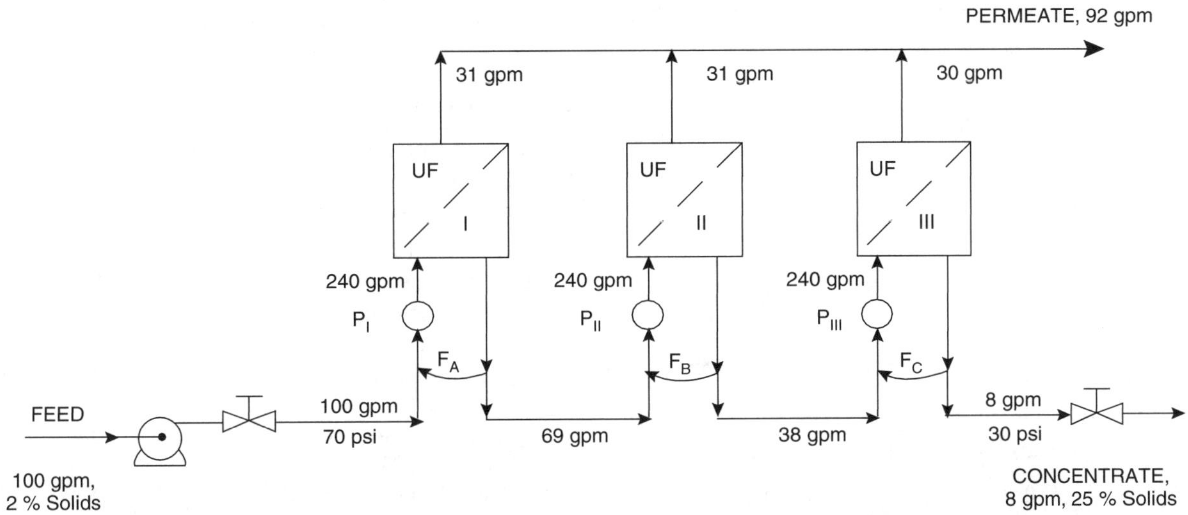

Figure 2.29 Design and operation schematic of a multi-stage, feed-and-bleed UF membrane system. Each membrane stage contains four UF modules in series. F_A, F_B and F_C represent recycle streams operating at 140, 171 and 202 GPM. P_I, P_{II} and P_{III} are recirculation pumps

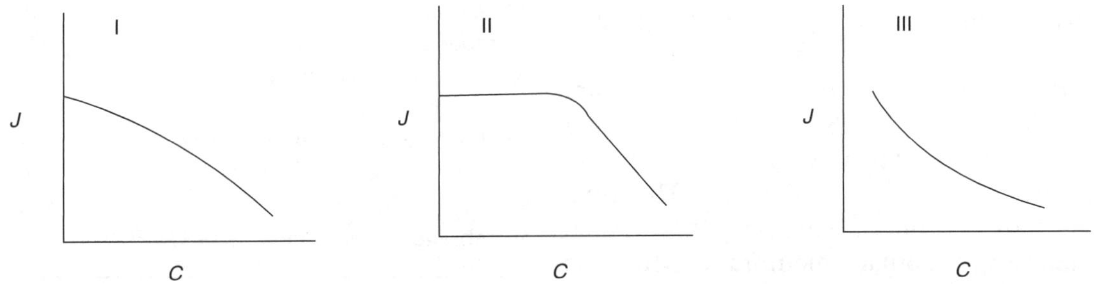

Figure 2.30 Typical flux (J) vs. feed concentration (C) curves for UF membrane system in Figure 2.29

which is determined by the well-known flux-concentration mass transfer limited gel-polarisation model introduced in Chapter 1 as:

$$J = k \ln C_g/C_b \qquad (27)$$

The semi-log relationship of flux vs. feed concentration for a batch filtration test run is shown in Figure 2.31[41] where C_g is the 'gel layer' concentration, and C_b is the bulk feed concentration (in the absence of a gel layer, $C_g = C_w$), as shown in Figure 1.7. The C_g value of 14% is considerably lower than that for macromolecules and colloids, but is consistent with the data for other surfactant systems. A low value of C_g restricts the maximum practical level of final concentration in similar UF applications. The mass transfer coefficient, k is the slope of the curve. The steeper the slope (higher Sherwood number, $Sh = k.L/D$ where k is the mass transfer coefficient, L is a characteristic dimension, and D is molecular diffusivity), the faster is the rate the solutes migrate to the membrane surface. The UF membrane used was a regenerated CA with a MWCO = 10,000 Da.

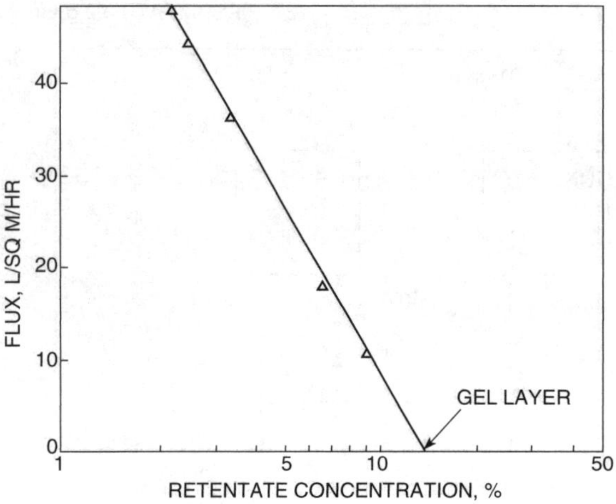

Figure 2.31 Batch UF membrane flux vs. concentration curve for a surfactant feed water. Cross-flow cell at 2.76 bar g; 1100 ml/min; 20°C; Amicon regenerated cellulose YM-10 membrane with MWCO of 10,000 Da. Source: Singh[41]

The experimental data was used to generate a preliminary design of a UF plant.[41] The design parameters are given in Table 2.12. The design is based on a PVDF tubular membrane (MWCO = 18,000 Da). A tubular membrane plant consumes 50% more power than either a HF or a SW plant. The final concentrate volume is, however, 50% smaller. In addition, tubular membranes have a longer life than HF membranes (5 vs. 3 years), and are five to six times cheaper than HF modules.

The curve shown in Figure 2.31 is the standard UF/MF plot and holds true for almost all cases. A slight variation of the plot for batch UF of a protein solution is shown in Figure 2.32.[26] In this plot, gel polarisation seems to be a two-step process involving build-up and clogging at low concentrations followed by compaction of the sub-layer. The test run used three membranes: DPO6, an isotropic microfilter with a pore size of 6,000 Å; PM-10, a polysulphone UF membrane with a MWCO = 10,000 Da; and PM-30, a polysulphone UF membrane with a MWCO = 30,000 Da. The pure water flux for these membranes was 0.66, 0.25, and 0.58 ml/cm^2 min. The isotropic microfilter fouled easily due to internal pore plugging, and had the lowest flux with the protein solution feed even though it had the highest water flux, as shown in Figure 2.33.

Table 2.12 Tubular UF plant design parameters

Item description	Design parameter
System flow rate (m^3d)	327
Permeate flow (m^3/h)	13.6
Reject recycle flow (m^3/h)	455
Product water flux ($l/m^2/h$)	51
Product water recovery (%)	96.5
Final concentrate volume (m^3)	11.4
Feed pressure (bar g)	3.45
Penreco oil/Igepal CO-660 conc in feed (%)	2.0/0.2
Oil/surfact injection rate (m^3/h) at 1.4 bar g	0.4
Final concentration in reject (%)	30.0
Membrane (Da)	PVDF, MWCO = 18,000
Membrane tube size:	
diameter (cm)	2.54
length (m)	3
Number of tubes	1,200
Tubes/bank	12
Number of tube banks	50
Number of parallel arrays	2
Membrane life (y)	5
Feed equalisation tank (m^3), FRP[a]	3.8
Feed/recycle tank (m^3), FRP[a]	38
Feed pump (m^3/h)	455
Membrane cleaning tank (m^3), LDPE[b]	1.9

[a] Fibre glass-reinforced plastic
[b] Low-density polyethylene
Source: Singh[41]

2.4.6 Membrane performance monitoring

Most membrane system problems occur due to a failure to record and analyse data after it is recorded. Data trending helps in

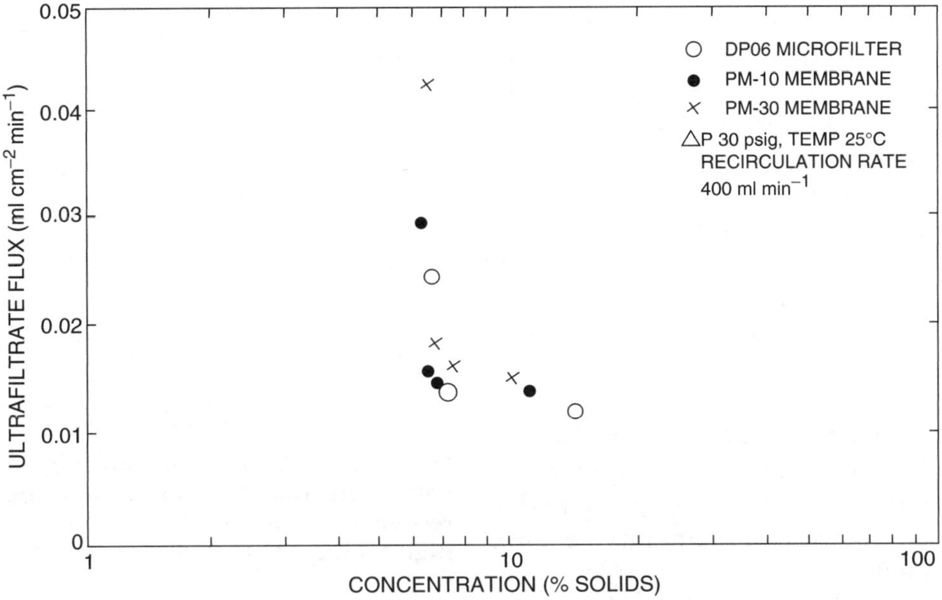

Figure 2.32 Batch cross-flow membrane filtration flux vs. concentration curve for a protein feed solution for various UF and MF membranes. Amicon polysulphone PM-10 and PM-30 membranes with MWCO of 10,000 and 30,000 Da. The 'gel layer' concentration is between 30 and 40%

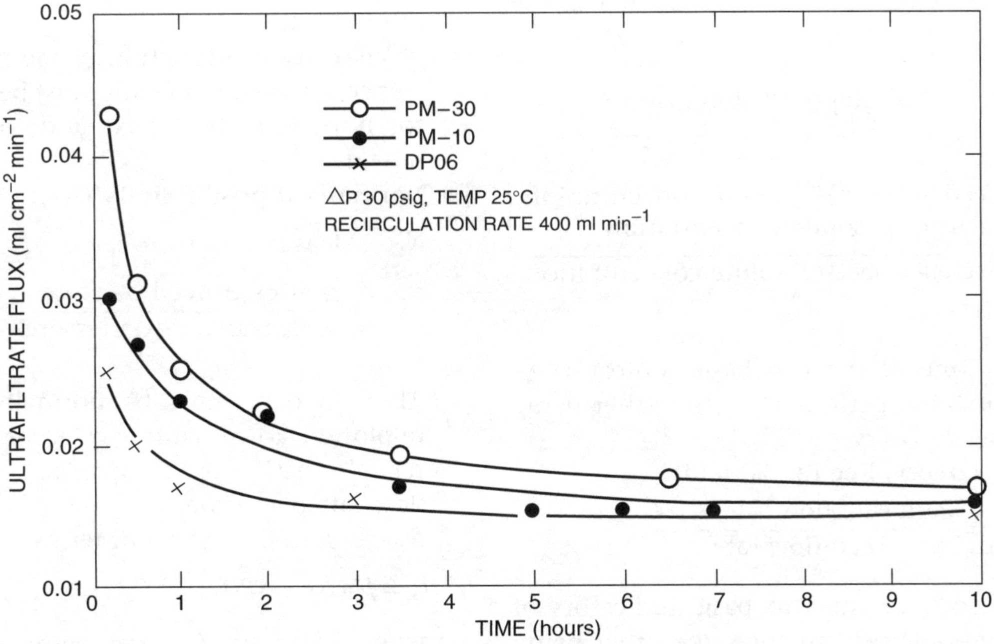

Figure 2.33 Typical cross-flow membrane filtration flux vs. time curve for the system in Figure 2.32

diagnosing the system. Without data trending it is difficult to analyse when the membrane unit should be shutdown for cleaning the membranes (e.g. see Figure 2.34). The performance data reflected in Table 2.5 provides markers for cleaning and sanitising the membranes, as discussed in Section 2.5. According to the membrane manufacturers, cleaning should be initiated when any of the following conditions are true although the initial flux is never restored.

- Permeate flow has dropped to 10–15% below the rated flow at normal pressure.
- Temperature-corrected feed pressure has increased 10–15% to maintain the rated permeate flow.
- Permeate quality has decreased 10–15%; salt passage has increased 10–15%.
- Applied pressure has increased about 10–15%.
- The differential pressure across an RO unit has increased noticeably (>15%).

The following relationships are used when evaluating data and monitoring membrane system performance:

1. **%Rejection, Rej (or %Retention):**

$$\text{Rej} = \frac{\left\{\begin{array}{l}[(\text{Feed} + \text{Reject})/2]\text{ solute concentration}\\ -\text{ Permeate solute concentration}\end{array}\right\}}{[(\text{Feed} + \text{Reject})/2]\text{ solute concentration}} \times 100$$

It is one of the two basic indicators of membrane performance, the other being water recovery.

2. **%Recovery, Rec (or %Yield):**

$$\text{Rec} = \frac{\text{Permeate flow rate} \times 100}{\text{Feed flow rate}}$$

It is one of the two basic indicators of membrane performance, the other being solute rejection.

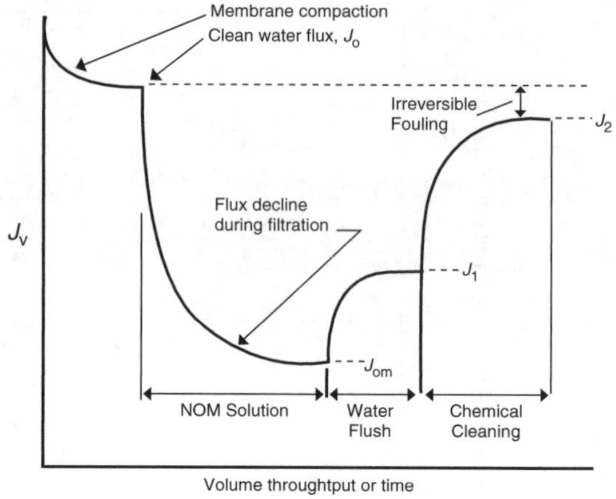

Figure 2.34 Typical membrane flux decline with time due to fouling, and post-cleaning recovery for a membrane system processing high fouling streams. Source: Schafer et al.

3. **Concentration factor (C_f):**

$C_f = 1/(1 - \text{Rec})$

It is a useful parameter but not often used.

4. **Temperature correction factor (T_{CF}):**

$T_{CF} = (1.024)\exp(t - 25)$

where, t is measured fluid temperature (°C).

Viscosity is a function of temperature. Hence, a correction factor must be applied to normalise the flux and permeate flow data.

5. **Normalised productivity (N_P):**

$$N_P = \text{Measured flow} \times T_{CF} \times \left(\frac{\text{Design feed pressure}}{\text{Measured feed pressure}}\right)$$

All field data must be normalised and re-plotted to provide an accurate and reliable basis for comparing permeate flow rate over time.

N_P can also be written in terms of flux as:

$$J_S = J_M(\Delta P_S/\Delta P_M) \times (1.024)\exp(T_S - T_M)$$

where J_S and J_M are permeate flux at standard and measured conditions;

ΔP_S and ΔP_M are net differential pressure (NDP) across the feed inlet and reject outlet at standard and measured conditions; and T_S and T_M are fluid temperature at standard and measured conditions, respectively.

2.5 Membrane cleaning and sanitisation

Membrane cleaning historically has not been the focus of major study vis-à-vis fouling even though it is a key element of long-term membrane life and operation. Membrane cleaning and sanitisation are required to restore membrane performance. Prompt and regular cleaning is necessary to prevent fouling from getting worse with long-term adverse effects. A typical flux vs. time curve for a UF membrane system in Figure 2.34 illustrates the effects of fouling and cleaning on flux. Flux decline slope values are used to judge the rate of colloidal fouling, and to determine when cleaning is required.[29] Figure 2.35 shows a severely fouled RO membrane that was not cleaned on time so that cleaning at that late stage became ineffective.

2.5.1 Membrane cleaning

Cleaning is the removal of foulants and scalants from the membrane element to restore membrane flux and rejection as far as possible (flux is never restored to the new membrane condition). The main chemical and physical cleaning methods used are:

1. Chemical detergents and cleaner solubilises or disperses the foulants. Chemical cleaning efficiency increases with temperature since the rate of chemical reaction doubles with every 10°C increase in temperature.
2. Hydraulic energy in the form of high flow velocity in the feed channel. Ideally, cleaning should be done at turbulent conditions, i.e. $Re > 2,100$, which is only possible with tubular membranes. Sometimes foam balls are used to scrape dirt from the surface of tubular membranes, as shown in Figure 2.36.
3. Backflushing is used to remove colloidal and particulate foulants from hollow fibre and inorganic membranes only. The technique is not very effective for removing some foulants such as macromolecules and small solutes.

Figure 2.35 A heavily fouled RO membrane element. Source: Amjad, Reverse Osmosis, 1993

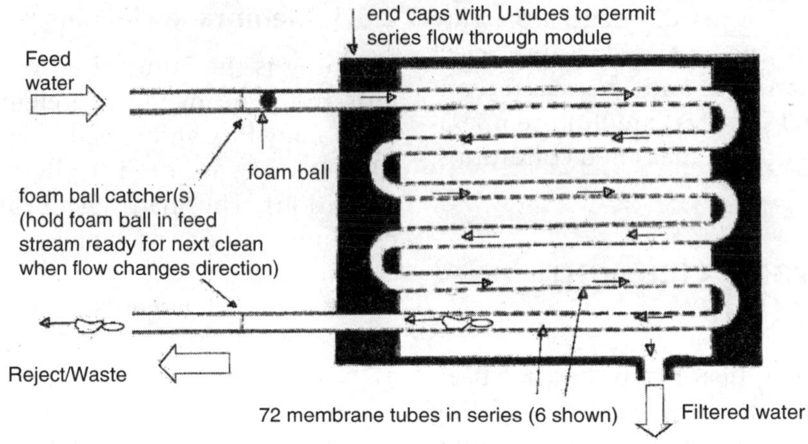

Figure 2.36 The operating principle of cleaning of tubular membranes with foam balls. Source: Wittman and Thorsen

4. Other physical methods include air sparging, vibrations, sonication, and CO_2 back permeation.

The frequency of cleaning can affect membrane longevity. Hence, the following general precautions during cleaning should be noted:

- Clean at the recommended temperature to optimise cleaning efficiency and membrane life.
- Use the minimal amount of chemical contact time to prolong membrane life.
- A 'gentle' pH range is 4–10, while the harshest is 1–12. Extremes of pH have an adverse effect on the life of RO and NF membranes. During recirculation, the temperature, must be less than 50°C at pH 2–10; <35°C at pH 1–11, and 30°C at pH 1–2 for PA membranes. Always monitor the cleaning solution pH.
- Improper selection of a cleaning chemical, or the sequence of chemical introduction, can result in worsening the fouling.
- The pressure drop across the membrane (TMP) should be minimal; high pressure drops aggravate fouling during cleaning. High pressure drop can be used after the membrane has been cleaned to remove matter from pores.

Cleaning chemicals

Chemical cleaning is based on chemical reactions – hydrolysis, saponification, solubilisation, dispersion, chelation, and/or peptisation – to breakdown bonds and cohesion forces between the membrane surface and the foulants.[27] Commonly used chemical cleaners for membrane systems are given in Tables 2.13 and 2.14. Acids being low pH cleaners are used first to remove precipitated salts or mineral scale. Phosphoric acid is the least aggressive (not recommended for ceramic membranes). It also has its own detergent action because of the phosphate groups. Nitric acid is very corrosive but effective. Citric acid is very effective for iron foulants. It combines acidity with detergent properties and chelating ability. Sulphuric acid and hydrochloric acid are not recommended when there are stainless steel components. Alkalis being high pH cleaners are used for removing organics and proteins; they act by solubilising them. Polyphosphates are added to help solubilise carbonates and emulsify fats.[19,27]

Table 2.13 Cleaning methods for RO and NF membrane systems

Foulant	Example	Chemical cleaner
Scale (pH adjustment and/or scale inhibitor)	$CaCO_3$, $CaSO_4$, $BaSO_4$, $SrSO_4$, SiO_2	Citric acid, 0.2% (wt.) HCl, 0.5% (wt.) phosphoric acid, or EDTA-based solution; Clean silicate-based foulants with ammonium bifluoride-based solutions
Colloidal clays/silt (filtration and/or charge stabilisation)	SiO_2, $Fe(OH)_3$, $Al(OH)_3$, $FeSiO_3$	EDTA- or BIZ-type detergents at high pH; Clean silicate-based foulants with ammonium-bifluoride-based solutions
Biological (sodium bisulphite addition or chlorination)	Iron-reducing bacteria, sulphur-reducing bacteria, mycobacteria, pseudomonas	EDTA- or BIZ-type detergents at high pH; Shock disinfection with hydrogen peroxide, peracetic acid
Organic (filtration)	Polyelectrolytes, oil, grease	Detergents/surfactants, isopropanol

Source: Amjad, Reverse Osmosis, 1993

Table 2.14 Cleaning techniques for UF and MF membranes

Foulant	Cleaning reagent	Process condition	Reagent action
Fats and oils, proteins, polysaccharides, bacteria	0.5N NaOH + 200 ppm chlorine	30–60 min at 25–55°C	Hydrolysis and oxidation
DNA, mineral salts	0.1M–0.5M acid (acetic, citric, nitric)	30–60 min at 25–55°C	Solubilisation
Fats, oils, proteins, biopolymers	0.1% sodium dodecyl sulphate, 0.1% TritonX-100	30 min overnight, 25–55°C	Wetting, emulsifying, suspending, dispersing
Cell fragments, fats, oils, proteins	Enzyme detergents	30 min overnight, 30–40°C	Catalytic breakdown (proteolysis)
DNA	0.5% DNAase	30 min overnight, 30–40°C	Enzyme hydrolysis
Fats, oils, and grease	2–50% ethanol	30–60 min at 25–55°C	Solubilisation

Caustic solutions act by breaking the bonds between the membrane surface and the foulant.

Chlorine in the form of household bleach (4 ml of 55 bleach in 1 l of 0.5 N caustic soda results in 200 ppm of chlorine) is an effective cleaner (not to be used for PA membranes) especially when the pores are fouled in the case of UF and MF membranes. The recommended dosage is 50 ppm for CA membranes, 20–400 ppm for polyether sulphone membranes and 1,000–2,000 ppm for inorganic membranes. In the case of polymeric membranes, chlorine action results in opening the pores that allows the foulant to be flushed out at high pressure.

Sometimes a number of different cleaning chemicals are used in a specific sequence for optimum cleaning, for example a high pH cleaner to remove foulants such as oils or biological matter followed by a low pH cleaner. Some cleaning solutions have detergents (surface-active agents) added to aid in the removal of heavy biological and organic debris, while others contain a chelating agent like EDTA added to aid in the removal of colloidal material, organic and biological material, and sulphate scale. Enzymes are effective for removing proteins and other severe foulants that form strong bonds with the membrane surface due to charge interactions. Organic solvents such as ethanol solutions are sometimes used to remove hydrophobic foulants including fats, oils, and grease. Cleaners require 30–90 min for effective action; enzyme cleaners require 60–90 min whereas optimum time for chlorine-containing cleaners (not applicable to PA membranes) is 60 min.[19] Prolonged cleaning after the optimum time, however, can foul the membranes again.

The cleaning action of chemicals is illustrated by a few examples. Detergents are used when the foulant is oil or grease. The addition of a detergent lowers the contact angle, θ at the solid–oil–water boundary, as shown in Figure 2.37.[42] If $\theta = 0$, the oil will detach from the membrane surface spontaneously, and will be washed away with water during cleaning. If the contact angle is in the range, $90^\circ < \theta < 180^\circ$, only part of the oil can be detached by water under hydraulic pressure while some remains attached. Hence, a different mechanism such as solubilisation is needed to remove the residual oil, for example increasing the temperature to 45°C increases the detergent efficiency markedly. Similarly the process of removing particulate foulants from the membrane surface by the action of a detergent and hydraulic pressure are depicted in Figure 2.38.

The effectiveness of cleaning a UF membrane fouled by organics – proteins, lipids, and carbohydrates – when processing aqueous soy flour extract is shown in Figure 2.39; as shown in Figure 2.34, the flux declined sharply due to fouling. The data in Figure 2.39 shows the effects of step-wise cleaning protocols: (a) rinsing; (b) 0.5% (wt.) sodium hydroxide cleaning; (c) 0.75% (wt.) protease detergent cleaning; and (d) sanitising with sodium hypochlorite (150 mg/l).[27] There is a dramatic increase in flux recovery after treatment with the enzyme-based detergent most likely because the enzymes are action specific (catalysts), and the enzyme used was directed

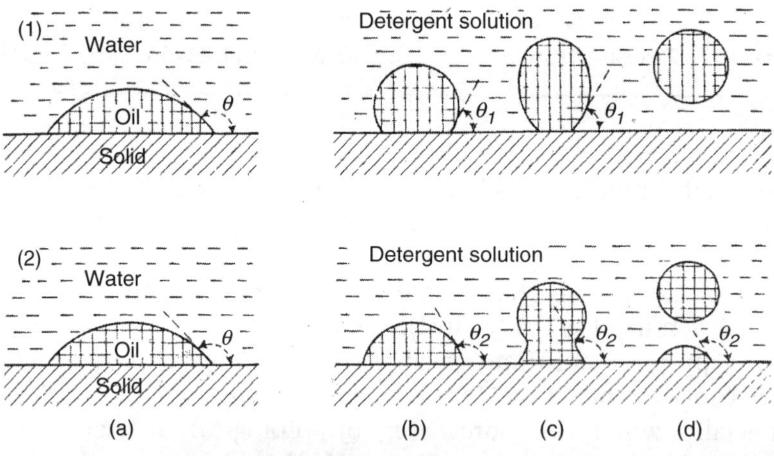

Detachment of oily dirt from a solid surface. The sequences (left to right) show: (a) the substrate/dirt system in contact with pure water, (b) lowering of contact angle caused by detergent [(1) $\theta < 90^\circ$, (2) $\theta > 90^\circ$], and (d) mechanical (hydraulic) detachment of oil droplets

Figure 2.37 Detergent cleaning action for removing oily dirt from a solid surface. Source: Shaw

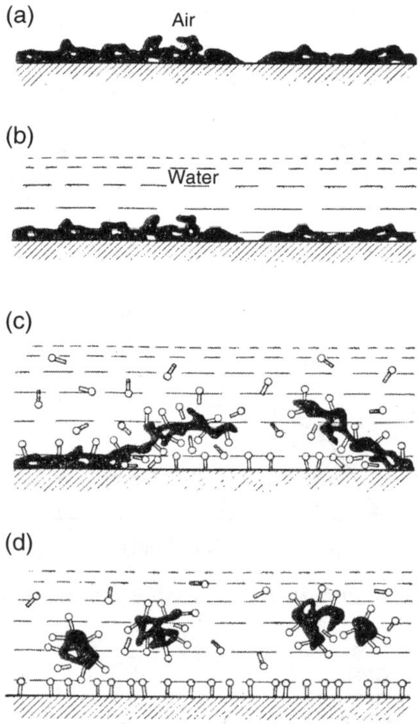

Removal of dirt from a solid surface by detergent and mechanical action. (a) Surface covered with greasy dirt. (b) Water by itself fails to dislodge the dirt mainly because of its high surface tension and inefficient wetting action. (c) Detergent is added to the water. The hydrophobic parts of the detergent molecules line up both on the dirt and on the solid surface, thus reducing the adhesion of the dirt to the solid. The dirt may now be dislodged by mechanical action: (d) Dirt is held suspended in the solution because the detergent molecules form an adsorbed layer on the cleaned surface and around the dirt particles

Figure 2.38 Removal of dirt from a solid surface by detergent and mechanical action. Source: Shaw

to the specific foulant. Enzymes are often used when the cleaning conditions are around neutral pH, when there is a likelihood of biofouling (Table 2.14). The effectiveness of sodium hypochlorite was most likely due to its ability to cause swelling of the membrane.[43,44]

2.5.2 In situ element cleaning procedures

The membrane elements are cleaned in the pressure vessels by re-circulating the cleaning solution across the high-pressure side of the membrane at low pressure and relatively high flow rates. A typical clean-in-place (CIP) unit schematic is shown in Figure 2.40. The time required to clean a stage or a pair of stages is 2–3 h for small systems and 4–8 h for large systems. A general procedure for cleaning RO/NF membrane elements is described below:

1. Start with a low-pressure flush (<5 bar g) by pumping clean water from the cleaning

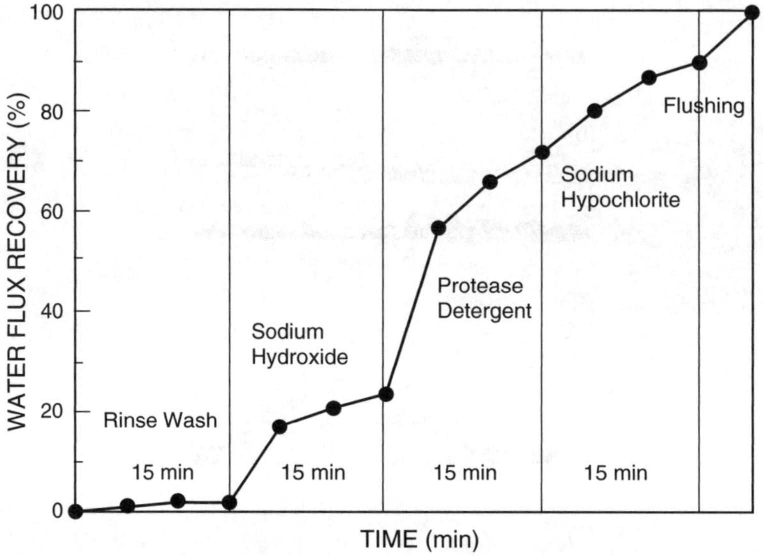

Figure 2.39 Membrane flux recovery with successive cleaning steps. Source: Schafer et al.

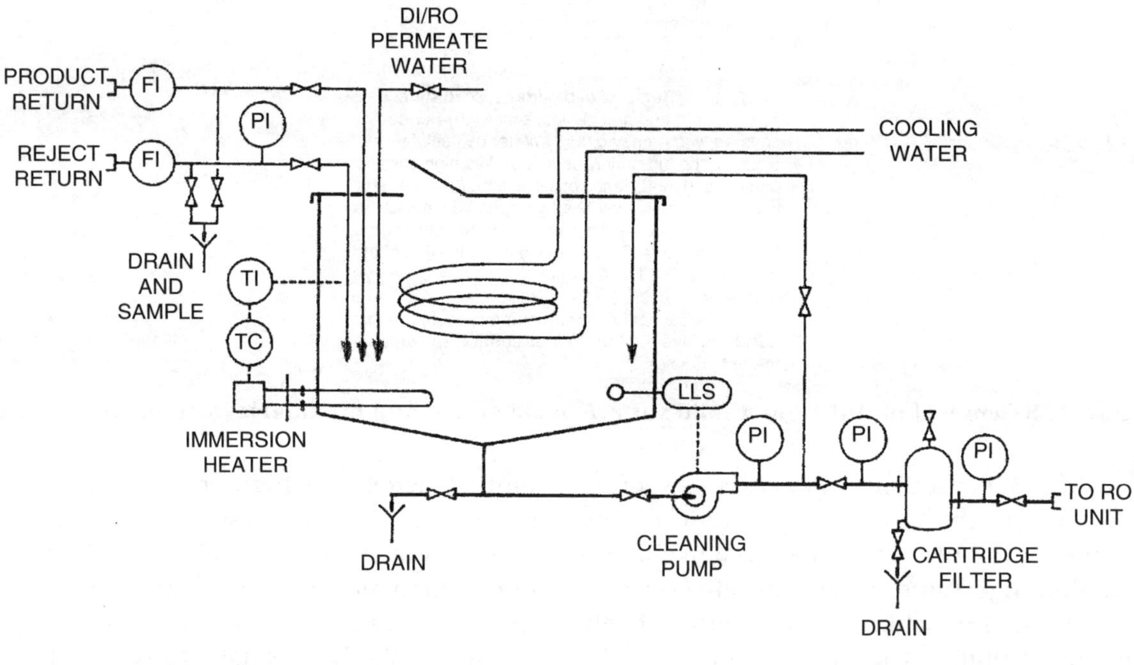

Figure 2.40 Schematic diagram of a typical membrane cleaning system (CIP). Source: USFilter

tank (or equivalent source) through the pressure vessels to drain for several minutes. Flush water should be RO permeate or DI water, and free of hardness, transition metals, and also chlorine-free when using PA membranes.

2. Mix a fresh batch of the selected cleaning solution in the cleaning tank.

The dilution water should be clean water; RO permeate or DI water.

3. Circulate the cleaning solution through the pressure vessels for 30–60 min at 8–10 m^3/h for 2 or 3 stages using 20 cm diameter SW elements. High flow rate flushes away the foulants. The maximum pressure drop should be 1 bar per element or 3 bar per pressure vessel. The solution temperature must be greater than 15°C.
4. Initially, send the displaced water to drain so as not to dilute the cleaning chemical. Divert up to 20% of the most highly fouled solution to drain before returning the cleaning solution back to the CIP tank. Initially for the first 5 min, throttle the flow rate to two-thirds of the maximum design flow rate slowly, and then increase the flow rate to the maximum design flow rate. If required, readjust the pH to the target when it changes more than 0.5 pH units.
5. An optional soak and recirculation sequence can be used, if required. The soak time can be from 1 to 8 h depending on the type of foulant and the membrane. Ensure proper temperature and pH are maintained. Also note that this procedure increases the chemical exposure time of the membrane.
6. Upon completion of the chemical cleaning steps, a low-pressure cleaning rinse with clean RO or DI water (20°C min) is required to remove all traces of chemical from the CIP skid and the membrane unit including piping. Drain and flush the CIP tank. Refill the CIP tank with clean water for the rinse step. Rinse the pressure vessels by pumping all the rinse water from the CIP tank through the pressure vessels to drain. A second cleaning can be started at this point, if required.
7. Once the membrane system has been rinsed of cleaning chemicals thoroughly, a final low-pressure clean-up flush can be performed using pretreated feed water. The permeate divert to drain line remains open. Feed pressure should be less than 4 bar g. This final flush continues until the flush water flows clean and is free of any foam or residues of cleaning agents. This usually takes 15–60 min. Sample the flush water going to the drain for detergent removal and lack of foaming by using a clear flask and shaking it. Measure the conductivity to test for removal of cleaning chemicals; the flushed water should be within 10–20% of the feed water conductivity. A pH meter can also be used to compare the flushed water with the feed water pH.
8. Once all the stages of a membrane unit are cleaned, and the chemicals flushed out, the membrane system can be restarted and placed into a service rinse. The permeate should be diverted to drain until it meets the process water quality requirements (e.g. conductivity, pH). It can take several hours and even a few days for the permeate quality to stabilise, especially after high pH cleaning.

2.5.2 Chemical sanitisation

Sanitisation means disinfection for reducing microbial contamination down to an acceptable level whereas sterilisation implies the inactivation of all microorganisms. Cleaning always precedes sanitisation. When a membrane system (membrane elements, pressure vessels, and piping) has been infected by bacteria or mould, the system requires chemical sanitisation. The CIP system is used for sanitising as well. Disinfection was discussed in Section 2.2. The most common chemical disinfectants are chlorine, hydrogen peroxide, and peracetic acid:

- Chlorine is mostly used in the form of hypochlorites, as discussed in Section 2.2.
- Hydrogen peroxide is not compatible with PA membranes. It is relatively non-corrosive and slow acting, and decomposes into oxygen and water.
- Per(oxy) acetic acid is a mixture of hydrogen peroxide and acetic acid. It is effective against bacteria, viruses, mould, and spores. It releases oxygen that affects microorganisms. Oxygen release is an added benefit as it sanitises the membrane pores. It is fast acting at low temperatures, is non-corrosive to stainless steel, and leaves no harmful residues. It should not be used with PA membranes, unless recommended by the membrane manufacturer.
- Sodium metabisulphite is effective against bacteria, yeasts, and moulds. It is non-corrosive and leaves no harmful residues. Since it is a reducing agent, it is also used for storing PA membranes in a 1% (wt.) solution.

General safety precautions

- Continuous exposure to per(oxy) acetic acid at the above levels may eventually damage the membranes. Hence, it should be used only periodically. The biocide efficacy of peracetic acid is much higher than that of hydrogen peroxide, but as most peracetic acid solutions also contain hydrogen peroxide, care must be exercised not to exceed the 0.2% concentration total for both compounds.
- When hydrogen peroxide is to be used, the pH of the solution must be less than 4.0, and preferably equal to 3.0. This will ensure optimal biocide result and longer membrane life. If the pH is high, the chemical attack on the membrane will be much faster. When a mixture with peracetic acid is used, pH adjustment is usually not required.
- If an alkaline cleaning has preceded sanitisation, the alkalinity must be carefully rinsed out of the membrane vessels, and the pH checked before hydrogen peroxide or hydrogen peroxide/peracetic acid solution is applied. This rinse must include the permeate side. If the alkalinity is not rinsed out, the membranes may get oxidised.
- The sanitising solution temperature must not exceed 25°C. Several hours of sanitising at higher temperatures results in a dramatic increase in the salt passage.
- Iron or other transitional metals must not be present because they will catalyse membrane degradation in the presence of hydrogen peroxide due to oxidative degradation resulting in an increase in salt passage.

2.5.3 Hot water sanitisation

Hot water sanitisation (HWS) or pasteurisation is commonly used to kill microorganisms. RO membranes reject bacteria and other microbes. However, due to defects in the membrane surface, the rejection is not 100% or absolute so that microorganisms can pass through the membrane into the permeate side of the membrane where they can multiply. Hence, sanitisation of a membrane system is essential for the production of US Pharmacopeia (USP) water and water for injection to meet the very low bacterial limits.[45–47] HWS of RO membrane systems is a relatively recent development made possible by the development of TFC membranes and membrane equipment capable of handling hot water as high as 85°C for brief period of time;

membrane elements are manufactured with special adhesives, permeate tubes, and connectors to withstand elevated temperatures. These membranes also make it possible to sanitise and protect the RO membranes from biofouling especially when the use of biocides is not acceptable. Typical membrane manufacturers' specifications of a hot water sanitisable TFC membranes are given in Table 2.15.

Table 2.15 Hot water sanitisable TFC–SW–RO operating membrane data

Item	Specification
Physical size	20 cm diameter × 100 cm long
Type of membrane	Thin film composite
Membrane element configuration	Full-fit
Membrane surface area	$36\,m^2$
Rejection at standard conditions*	96–98%
Membrane flow rate (m^3/day)[2]	22–29
Recovery per element	15%
Feed temperature (service)	≤45°C
Feed pressure (service)	≤40 bar g
Feed SDI rating	≤5
Feed turbidity	≤1 NTU
Feed free chlorine	<0.1 mg/l
Feed pH (continuous)	2–11
Feed pH (cleaning – 30 min.)	1–12
Sanitisation temp. (max.)	≤80°C
Sanitisation feed pressure (max.)	3 bar g
Sanitisation duration	60 min
Sanitisation cycles	≤150
Sanitisation feed flow rate (max.)	$10\,m^3/h$
Sanitisation permeate flow rate	$0.2\,m^3/h$ per element
TMP (max.)	1.7 bar

*Standard conditions are 2000 mg/l NaCl, 16 bar g, 25°C, pH 8, and 15% recovery
Source: USFilter/IonPure

The HWS equipment includes a CIP skid with a heat exchanger to heat water and a pump to circulate hot water through the membrane system loop. Sometimes a heat exchanger is added to aid in cooling during the cool down period. The HWS procedure recommended by membrane manufacturers involves the following steps:

1. Clean the membranes with a low-pH cleaner to remove mineral scale and foulants.
2. Heat the water to 80°C at of 3–5°C per minute at low flow rate ($<10\,m^3/h$) and pressure (<3 bar g).
3. Circulate the water in the CIP-RO skid loop at 80°C for up to 1 h. Maintain cross-flow velocity so that the pressure drop is less than 0.15 bar per membrane element.
4. At the end of the re-circulating period, start cooling the membrane system down to 25°C at 3–5°C per min.

HWS is considered a superior process to chemical sanitisation, and does not require prolonged rinsing as in the case of biocide sanitisation. However, two points need to be noted. First, HWS is performed more frequently than chemical sanitisation. Second, high temperature reorients the polymer chains permanently.[48] At high temperatures, the membrane structure becomes tighter with increased rejection and decreased flux due to changes in the membrane morphology as a result of changes in the glass transition temperature[48] because polymers shrink when heated.[49,50] The first shrinkage results in almost a 50% reduction in flux according to the membrane manufacturers, but after the initial shrinkage subsequent HWS cycles are unaffected. To compensate for the anticipated loss, HWS systems are designed with more surface area than ambient temperature membrane elements.[46]

2.6 Concluding remarks

Water is the world's most valuable and wasted resource. With growing demand outstripping supply in many regions of the world, the situation is likely to get exasperated by the year 2020 as shown in Figure 2.41. Estimated use of the world's water in 1996 was as follows: approximately 63% for agriculture, 25% for industry, 7% for cities, and 5% for system losses.[51]

Less than three percent of the world's water has a salinity content that can be considered safe for human consumption. According to the World Health Organisation (WHO), the total dissolved solids (TDS) should be less than 1,000 mg/l (ppm) in drinking water based on taste considerations. By comparison seawater has an average TDS of 35,000 ppm and brackish water TDS varies between 1,000 and 10,000 ppm. It is estimated that less than 1.0% of the world's water supply is easily accessible and has acceptable water quality. According to the National Council Research Report,[7] the principal water problem in the early twenty-first century will be inadequate and uncertain supplies. Since water availability includes both water quality and quantity, reduction in water quality due to water pollution will further diminish the available water supply. World's water needs, however, can be met by better water management, e.g. by reducing losses using micro-irrigation techniques most of the 66% water lost during irrigation can be recovered,[51] by using advanced water treatment technologies for treating polluted water and for reclaiming water, and by conservation.

Hybrid membrane water treatment has a great potential for meeting the challenges of anticipated water shortages around the globe by treating raw water to produce high-quality potable water for applications as diverse as drinking water, ultrahigh purity and reclaimed water for crop irrigation. Appropriate membrane technologies for targeted applications can be summarised as follows: high salinity water (RO), industrial product water (NF/RO/EDI), municipal water (MF/UF/RO), recycled water (UF/RO), reclaimed water (MF/RO), and industrial wastewater (UF/RO). RO is well established for desalinating seawater and brackish water. UF has been traditionally used for treating industrial wastewaters. Recently introduced submerged and cross-flow MF

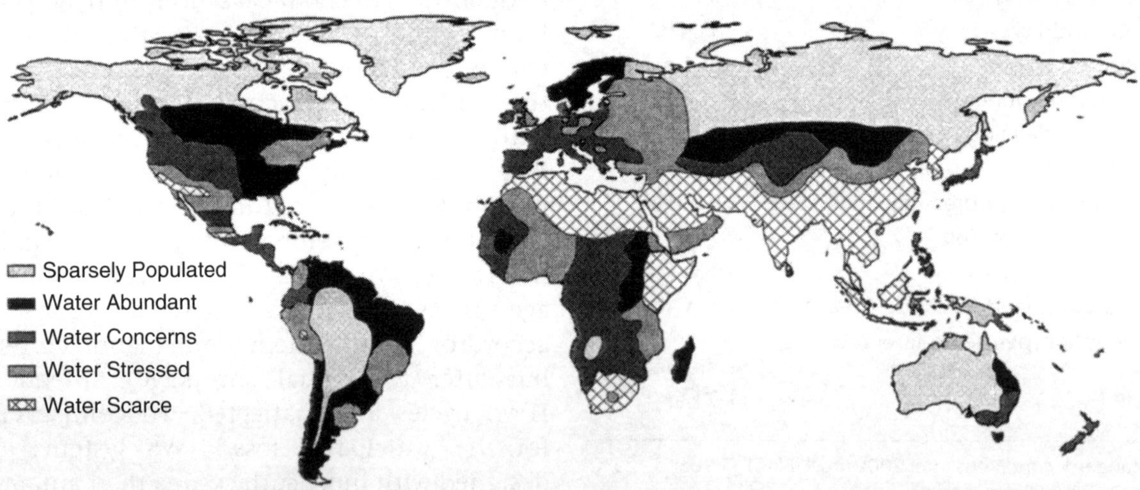

Figure 2.41 Projected worldwide water scarcity through 2020. Source: USFilter

along with MBR are being increasingly used for reclaiming wastewater and for municipal water treatment. Some of these applications are discussed in detail in Chapter 3. Further cost reduction and performance enhancement are, however, the key goals that need to be addressed in the near-term.[5–7]

References

1 Water, a survey, The Economist, 19–25 July (2003), 42.
2 Water scarcity: a looming crisis? BBC News, 20 October (2004).
3 Parliament of World's religions, NISHAAN, 4 (2004), 31–35.
4 CHEMRAWN XV Conference, Paris. Chem. & Eng. News, 82 (29) (2004), 25–30.
5 A.I. Schafer, A.G. Fane, T.D. Waite, eds., Nanofiltration, Elsevier: Oxford, UK (2004).
6 C.T. Cleveland, Big advantages in membrane filtration, J. AWWA, 91 (6) 1999, 24–32.
7 National Research Council, Review of the Desalination and Water Purification Technology Roadmap, National Academic Press: Washington, D.C., USA (2004).
8 B. Durham, W.K. Koh, G. Thompson, B. Biltoft, Membrane filtration – An effective pretreatment to RO in water reclamation. Experience in the municipal and industrial sectors, Waste/Wastewater Management Conference, 20–21 November (2002).
9 A. Bennet, Recent developments in RO technology reduces whole life project costs. Filt. & Sep., September 2003, 20–22.
10 S.D. Strauss, Water treatment, Power, June (1993), 1–52.
11 D. Bhattacharyya, Membranes in environmental applications, Env. Prog., 20 (1), A12.
12 J.W. Clark, W. Viessman, M.J. Hammer, Water Supply and Control, Harper & Row Publishers: New York, NY (1977).
13 W.W. Eckenfelder, Industrial Water Pollution Control, McGraw-Hill: New York, NY (1989).
14 S.V. Cabibbo, D.B. Guy, A.C. Ammerlaan, A. Ko, R. Singh, Reverse Osmosis Technical Manual, NTIS Publication: Springfield, Virginia, USA (1979).
15 C.E. Harland, Ion Exchange, Royal Society of Chemistry: Herts, UK (1994).
16 D.L. Owens, Practical Principles of Ion Exchange Water Treatment, Tall Oaks Pub: Littleton, CO (1995).
17 Antonio von, Gottberg, J. McArdie, B. Antrim, D. Halperm, Hollow fiber UF pretreatment for RO brackish water and sweater desalination systems, Ultrapure Water, 22 (3) (2005), 23–29.
18 L.J. Zeman, A.L. Zydney, Microfiltration and Ultrafiltration, Marcel-Dekker: New York, NY (1996).
19 M. Cheryan, Ultrafiltration and Microfiltration Handbook, CRC Press: Boca Raton, FL, (1998).
20 K.J. Howe, M.M. Clark, Fouling of microfiltration and ultrafiltration membranes by natural waters, Env. Sci. & Tech. 36 (16) (2002), 3571–3576.
21 S. Wang, The use of fluid instabilities to control MF/UF membrane fouling, Membrane Quarterly, NAMS, 20 (1) (2005), 7–11.
22 B. Marrot, A.P. Barrios-Martinez, N. Roche, Industrial water treatment in a membrane bioreactor: a review, Environ. Progress 23 (1) (2004), 59–68.
23 G.T. Daigger, B.E. Rittman, S. Adham, G. Andreottiola, Are membrane bioreactors ready for widespread application? Env. Sci. & Tech. 39(19) (2005), 399A–406A.
24 J.L. Humphrey, G.E. Keller II, Chapter 5, in Separation Process Technology, McGraw-Hill: New York, NY (1997).
25 CDI-LX Systems Technical Bulletin. USFilter/IonPure.
26 R. Singh, Surface properties in membrane filtration, Chem. Eng. Prog. 85(6) (1989), 59–64.
27 Schafer, A.I., et al., Fouling in nanofiltration, in Nanofiltration (eds. A.I. Schafer, A.G. Fane, T.D. Waite), Elsevier: Oxford, UK (2004), 169–240.
28 G. Johnson, B. Culkin, M. Monroe, Vibratory shear guards against mineral scale membrane fouling. Filt. & Sep., January/February (2003), 24–29.
29 D.L. Comstock, Reduction of colloidal fouling in membrane systems, Industrial Water Treatment 2 (4) (1991), 39–42.
30 L.F. Comb, Silica chemistry and reverse osmosis, Ultrapure Water 13 (1) (1996), 41–43.
31 H.-C. Flemming, Mechanistic aspects of reverse osmosis membrane biofouling and prevention, in Reverse Osmosis (ed. Z. Amjad), Van

Nostrand Reinhold: New York, NY (1993), 163–209.

32 H.F. Ridgway, Microbial adhesion and biofouling of reverse osmosis membranes, in Reverse Osmosis Technology (ed. B.S. Parekh), Marcel-Dekker: New York, NY 429–481.

33 R. Bradley, Design considerations for reverse osmosis systems, in Reverse Osmosis (ed. Z. Amjad), Van Nostrand Reinhold: New York, NY (1993), 104–138.

34 J. Visconti, USFilter, Chandler, AZ, personal communication, May (2005).

35 W. Eykamp, Section 22 Alternative separation processes, in Perry's Chemical Engineers' Handbook (eds. D.H. Perry, D. Green), 7th Edition, McGaw-Hill: New York, NY (1999).

36 R. Singh, J. Tembrock, Effectively control reverse osmosis systems, Chemical Engineering Progress, 95 (9) (1999), 57–64.

37 D.B. Guy, R. Singh, Some alternate methods of energy recovery from reverse osmosis plants, WSIA J., 1982, 9(2), 37–51.

38 E. Wittman, T. Thorsen, Water treatment, in Nanofiltration. (eds. A.I. Schafer, A.G. Fane, T.D. Waite), Elsevier: Oxford, UK (2004), 263–286.

39 ROPRO, Fluid Systems, San Diego, CA.

40 RO Membrane Design, Hydranautics, Oceanside, CA.

41 R. Singh, Removal of volatile organic compounds by ultrafiltration, Desalination, 104 (1996), 203–214.

42 D.J. Shaw, Introduction to Colloidal and Surface Chemistry, Butterworths: London, UK (1986).

43 S.K.S. Razavi, J.L. Harris, F. Sherkat, Fouling and cleaning membranes in the ultrafiltration of the aqueous extract of soy flour, J. Memb. Sci., 1996, 114(1), 93–104.

44 R. Singh, Characteristics of a chlorine resistant reverse osmosis membrane, Desalination 95 (1994), 27–37.

45 A. Collentro, M. Costello, G. Zoccolante, Hot water sanitisation of reverse osmosis systems for pharmaceutical water production, Ultrapure Water, October (2000), 1–6.

46 B. Wise, Hot-water sanitisation membranes tackle microbes in RO permeate water, Chemical Processing, 15 August (2002).

47 W.V. Collentro, Hot water sanitisation of RO and double-pass RO/EDI systems, Executive Forum, Ultrapure Water Journal's Expo 2003, Philadelphia, April 2003.

48 M. Manttari, et al., Effect of temperature and membrane pre-treatment by pressure on the filtration properties of nanofiltration membranes, Desalination 145 (2002), 81–86.

49 L. Mandelkern, An Introduction to Macromolecules, Springer-Verlag: New York, NY (1972).

50 R. Singh, A rubber heat engine for ground water irrigation in India, Ag. Ecosys. Environ. 25 (1989), 271–278.

51 C. Starr, Sustaining the human environment, Daedalus 125 (3) (1996), 235–253.

Chapter 3

Hybrid membrane systems – applications and case studies

'It's tough to make predictions, especially about the future'.

Yogi Berra

Ultrafiltration (UF) and microfiltration (MF) membrane processes were traditionally used as stand-alone separation processes with nominal pre-treatment, for example ultrafiltration (UF) processing of industrial wastewater or dairy products. In the case of desalination and water purification by reverse osmosis (RO), however, it was always a hybrid process integrated with other unit operations as discussed in Chapter 2. The term hybrid or integrated membrane processes refers to the integration of one or more membrane processes with or without conventional unit operations to increase performance depending on the type of feed and product quality required.[1–3] The main goal of these systems is to increase purity, for example in the case of water and gas purification, and productivity and yield of the product, for example in the case of fermentation processing. Similarly, pervaporation (PV)-based hybrid processes can be more economical than traditional separation processes for the removal of volatile organic compounds from industrial, municipal water, or groundwater contaminated with carcinogenic halogenated hydrocarbons instead of catalytic combustion.[4,5] Hybrid membrane systems also reduce operating costs and environmental pollution, and make the overall process more efficient. For a perspective on membrane system design and operation, the reader should review Chapter 4 and the relevant portions in Chapter 2.

Hybrid membrane systems for several applications are illustrated in this chapter followed by a discussion of a few case studies specific to water treatment. All examples highlight the centrality of membrane technology; implicit in their performance is the physicochemical properties of each membrane and process in the integrated membrane design. Mechanisms that control separation by various membrane processes were discussed in Chapter 1. Briefly, depending on the physical and chemical characteristics of the membrane and the solute, separation is achieved by one or more mechanisms: (a) physical selectivity based on charge repulsion, size exclusion, or steric hindrance, and/or (b) chemical selectivity, i.e. solvation energy, hydrophobic interaction, or hydrogen bonding. Hence, an integrated membrane system becomes an effective and efficient tool utilising membranes and membrane processes appropriate for separating solutes or purifying a particular feed stream.

3.1 Novel Applications

The hybrid systems described below include membrane technologies for processing a wide spectrum of feed streams in various applications. A small sampling of current and potential applications is discussed. The list of examples selected is intentionally limited; the intention is to illustrate the extent of process alternatives possible with various hybrid membrane processes.

3.1.1 Industrial wastewater treatment

Integrated membrane systems include MF or UF as pretreatment for NF or RO processing. A laundry wastewater treatment shown in Figure 3.1 is a MF–NF hybrid system.[6] Industrial laundry wastewater has a high concentration of detergents and surfactants, and is high in BOD and COD. In order to recycle water in the wash cycle or to meet wastewater pollutant requirements, the contaminants must be removed. The integrated membrane process schematic shown in the figure is used to remove oil, grease, detergents, surfactants, and other impurities such as heavy metals (e.g. 99.9% removal of lead, chromium and others) from wastewater. MF removes suspended solids including lint, oil, grease, and dirt. The permeate flows to a three-stage NF system for removing total dissolved solids (TDS), detergents, and surfactants. The face velocity in the MF membrane channels is kept high to reduce concentration polarisation by a high recycle flow rate. The NF permeate is reused in the hot water wash cycle.

Another application of an integrated membrane system is treating meat and poultry plant effluent for reuse as shown in Figure 3.2. The effluent TDS typically is 2.0–2.5%. The treated effluent should be low in sulphate concentration with a salt concentration in the 1.0–1.5% range. RO is not suitable because some calcium hardness needs to be present. Both MF and UF are effective in removing bacteria, cysts, and other microbiological

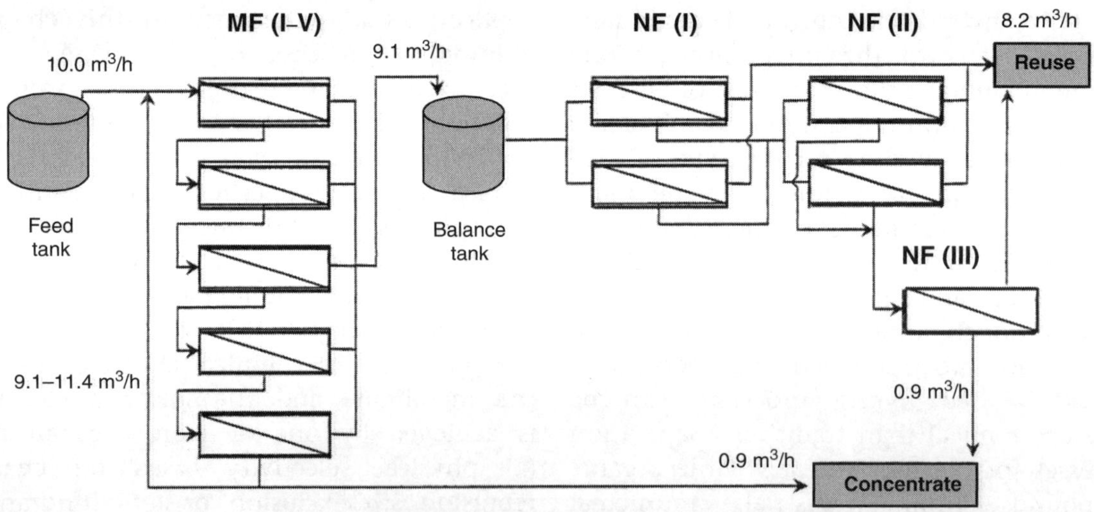

Figure 3.1 A dual membrane system for laundry wastewater treatment. Source: Tanninen et al.

compounds whereas UF also removes some viruses.[6]

Hybrid membrane systems using RO or NF for pre-concentration followed by mechanical vapour compression (MVR) is more efficient than the conventional physico-chemical and/or biological treatment systems for treating dilute industrial wastewater streams.[7] In the hybrid process, the pre-treated very dilute feed stream (to prevent membrane fouling) is concentrated by RO/NF. The concentrate is fed to a MVR evaporator to further remove the remaining water whereas the RO/NF permeate (which is of high quality) is recycled back to the process. The MVR concentrate is transferred to the wastewater treatment system. An example of this application is the treatment of wastewater from de-inking processes in the pulp-and-paper industry as shown in Figure 3.3.

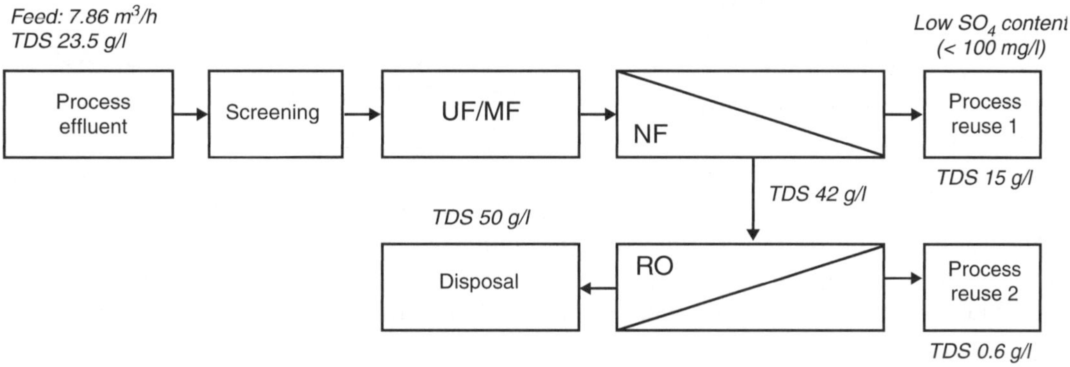

Figure 3.2 Flowsheet of a multi-membrane system for treating meat and poultry wastewater. Source: Tanninen et al.

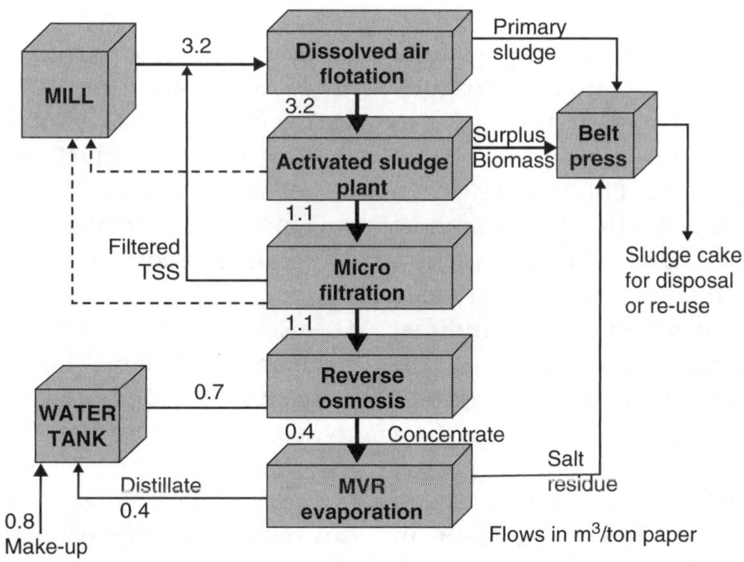

Figure 3.3 A hybrid membrane system for treating wastewater at a pulp-and-paper plant. Source: Nyström et al.

3.1.2 Arsenic and nitrate contaminated water treatment

Arsenic is a naturally occurring element that can be present in relatively high concentrations in surface water and groundwater. Inorganic arsenic is a toxic and carcinogenic compound that causes skin and various forms of internal cancer. The maximum allowable limit standard set by the World Health Organisation (WHO) is 10 ppb (μg/l). It is estimated that 33–77 million Bangladeshis are at risk due to arsenic contamination of their drinking water supplies where concentrations range from 10 to 14,000 ppb. High concentrations of arsenic have also been found in other countries, viz. India, Taiwan, northern China, Thailand, Vietnam, Hungary, Argentina, Chile, Mexico and south-west USA.[8]

Arsenic removal methods include adsorption, IX, chemical precipitation, coagulation, and membrane separation.[9] Since arsenite is uncharged, it is not removed by IX, and only slightly removed by coagulation. It is, therefore, necessary to oxidise the dissolved arsenic to the arsenate state before removal. RO membranes, on the other hand, reject both the arsenate and the arsenite. A hybrid membrane process that has been successfully demonstrated in the US combines coagulation with MF.[10] The coagulation/MF process consists of the addition of an iron-based coagulant such as ferric chloride (10 mg/l) at pH of 6.8 followed by MF. The membrane hybrid process eliminates the flocculation step. A high MF flux of 90 gallon/ft^2/day (153 l/m^2/h) was achieved with minimal fouling and long cleaning intervals.

The problem of high nitrate content in many groundwaters was mentioned in Chapter 2. Nitrate concentrations greater than 50 mg/l in water are carcinogenic, and is becoming a serious health problem in heavily fertilised and irrigated farmlands. A NF/ion exchange (IX) water treatment system is very effective in removing virtually all nitrates from drinking water.[6] In the NF/IX hybrid system, pre-filtered water flows through a NF membrane system, which removes all the multivalent ions, and some of the monovalent nitrate ions. The NF permeate is then polished in an anion IX column where the nitrates are exchanged with chloride ions. NF pretreatment increases the efficiency of the IX process since it removes dissolved organic carbon and divalent ions such as sulphate ions.

3.1.3 A multipurpose NF–RO–MSF seawater desalination process

Membrane (MF or UF) pretreatment of RO desalination plants is now a viable option for removing suspended solids, fine particles, colloids, and organic compounds. NF pretreatment of seawater in RO and multi-stage flash (MSF) plants is also being used to soften RO feed water instead of IX softening as shown in Figure 3.4 to prevent or decrease scaling, thereby, increasing product water recovery.[6] NF has several advantages over water softening by IX:

- No chemical consumption as in the case of regeneration of IX resins.
- No chemical wastewater.
- Consistent product water quality and reliable operation.
- Operation independent of feed water hardness and flow rate.

The NF–RO rejected brine stream is typically discharged to the sea. However, the brine can be recovered and used by the chemical industry, e.g. Mg salts in the NF reject stream and chloride salts in the RO reject stream as shown in Figure 3.5.[11] The recovery and reuse of salts can reduce the cost of potable water production from seawater by more than 50%. Typical rejection of dissolved ions by NF membranes

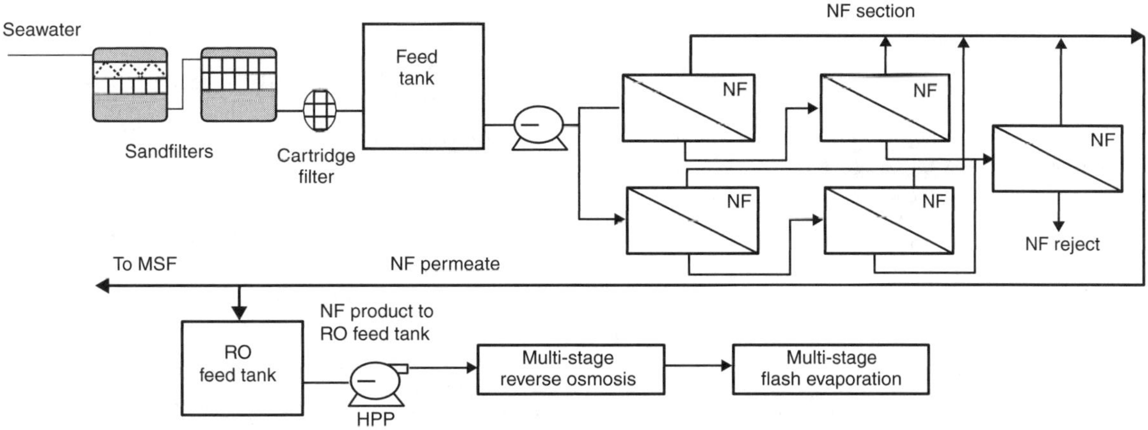

Figure 3.4 Seawater desalination using a NF–RO–MSF staged process. Source: Tanninen et al.

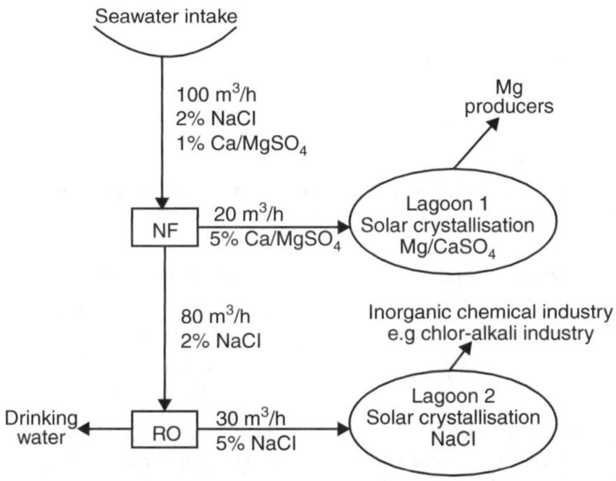

Figure 3.5 A multipurpose integrated membrane process applicable for seawater desalination. Source: Kyburz and Meindersma

is as follows: sulphate, 99.9%; magnesium, 8%; calcium, 91%; bicarbonates, 56%.

3.1.4 Purification of produced water

During offshore oil recovery, injection of steam into the oil wells leads to water production referred to as produced water. Once brought to the surface, petroleum is separated from water before the petroleum component is transported to refineries. Produced water is the single largest volume of wastewater generated by the oil and gas industry.[12] Approximately 11.7 billion barrels of produced water were generated in the US in 1997. Depending on the well site, the produced water may be high in salts (TDS), and also contain hydrocarbons, high levels of suspended solids, iron, silica, and boron. Toxic petroleum hydrocarbons found in oilfield produced water is a major environmental problem.

The hybrid scheme shown partially in Figure 3.6 used an integrated system of UF–NF–RO–IX to treat produced water.[11] The multi-treatment system had a dual purpose: recycle and disposal. Specifically, this was achieved in several steps shown in the figure:

- UF for oil/water separation from the pretreated feed
- NF for removing hardness and TOC
- RO for reducing the TDS from 10,000 ppm to 1,000 ppm, and
- IX for reducing the boron concentration from 5–10 ppm in RO permeate to <1 ppm.

The purified water is suitable for land irrigation but not for drinking. Boron is an important micronutrient for living organisms.

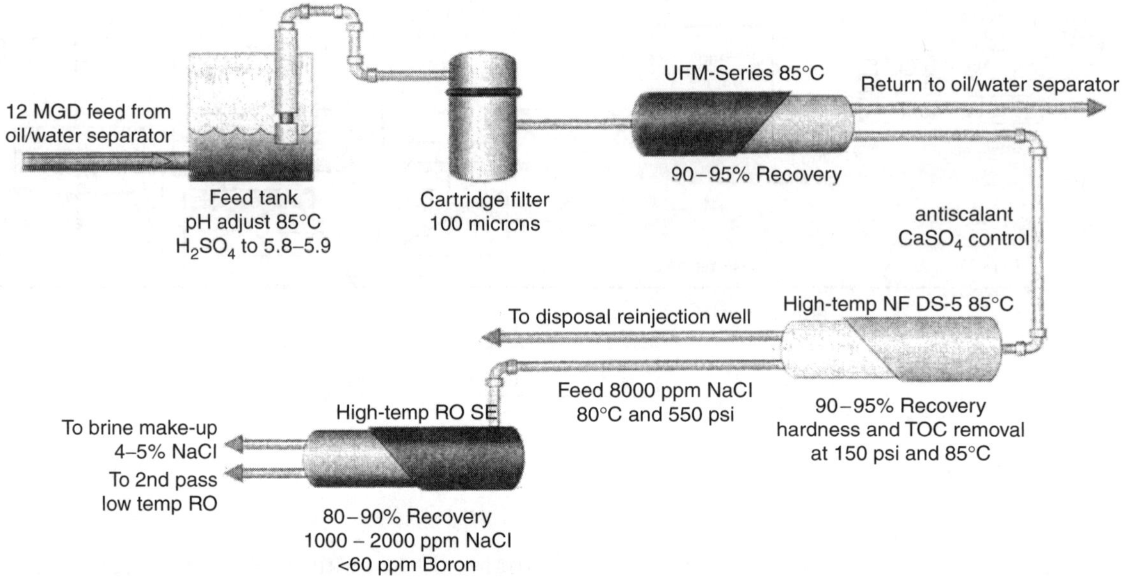

Figure 3.6 Process flow diagram for treating oilfield produced water. Source: Kyburz and Meindersma

However, the maximum concentration for drinking water recommended by the WHO is 0.3 ppm.[13]

3.1.5 Landfill leachate water treatment

Leachate from a landfill can be a major source of contamination of groundwater and surface water in the vicinity of the landfill by hazardous substances. The leachate contaminants are a combination of organic and inorganic compounds shown in Table 3.1. The organic compounds are mostly fatty acids, fulvic acids, and humic substances. The composition of the leachate changes with time; is acidic and has comparatively high proportion of organic compounds during the first 3–5 years.

The leachate is usually treated by a combination of several processes: (a) biological treatment; (b) chemical oxidation; (c) adsorption; (d) precipitation; (e) evaporation; (f) drying; (g) stripping; and (h) membrane processes such as UF and RO. An overview of the various treatment methods is given in Table 3.2.[14] For example, granular activated carbon adsorption in combination with biological pre-treatment is the leading process for removing COD and AOX. The process flow sheet with a membrane bioreactor (MBR) is shown in Figure 3.7a. Biological treatment is required for removing ammonia, COD, and BOD. In the MBR system, MF or UF membranes are used to retain and concentrate the biomass. In addition to removing particulate matter such as pathogens, this

Table 3.1 Impact category of landfill leachate compounds

Compound/Parameter	Environmental category
Cd, Ni, AOX	Human toxicity
Cu, Zn, Pb, Hg	Ecological toxicity (surface/groundwater)
Salts, Chlorine, Sulphide, Ammonia	Ecological toxicity (surface/groundwater)
Nitrogen	Eutrophication
COD	Eutrophication

Source: Melin et al.

Table 3.2 Process combination for landfill leachate treatment

Option #	Treatment units						
1	BIO	UF	OX				
2	BIO	UF	ADS				
3	BIO	UF	RO		EVA	DRY	
4	BIO	UF	NF		EVA	DRY	
5	BIO	UF	NF	OX			
6	BIO	UF	NF	ADS			
7	FIL	RO	RO		EVA	DRY	
8	FIL	RO	NF	RO		DRY	N_2
9	FIL	RO	RO			EXT	

FIL: Filtration, e.g. sand filtration
BIO: Biological treatment and nitrification/denitrification
UF: Ultrafiltration for biomass retention
ADS: Adsorption, e.g. activated carbon adsorption
OX: Oxidation, e.g. ozonation
RO: Reverse osmosis
NF: Nanofiltration
IX: Ion exchange
EVA: Evaporation
DRY: Drying
EXT: External concentrate removal
N_2: Nitrogen recovery, e.g. ammonia stripping
Source: Melin et al.

configuration enhances the performance of the biological system. First, the membrane unit increases the sludge residence time; and second, the membrane unit promotes the stability of the autotrophic microbial population that carries out the nitrification – the microorganisms display a slow rate of growth, and unlike in the conventional systems, their concentration cannot be reduced to an unacceptable level. Additional advantages of the MBR process include lower ammonia level in the UF permeate (down to 5–15 ppm from 1,200 to 1,500 ppm in the raw leachate), and nitrogen concentrations up to 50–150 ppm. The COD in the permeate is high (500–1,500 ppm) since UF membranes do not reject low MW compounds.

RO is the most widely used membrane process for leachate treatment because it rejects almost all the contaminants. Membrane processing made an early headway (1984) since the conventional processes could not meet the new stringent regulations.[14]

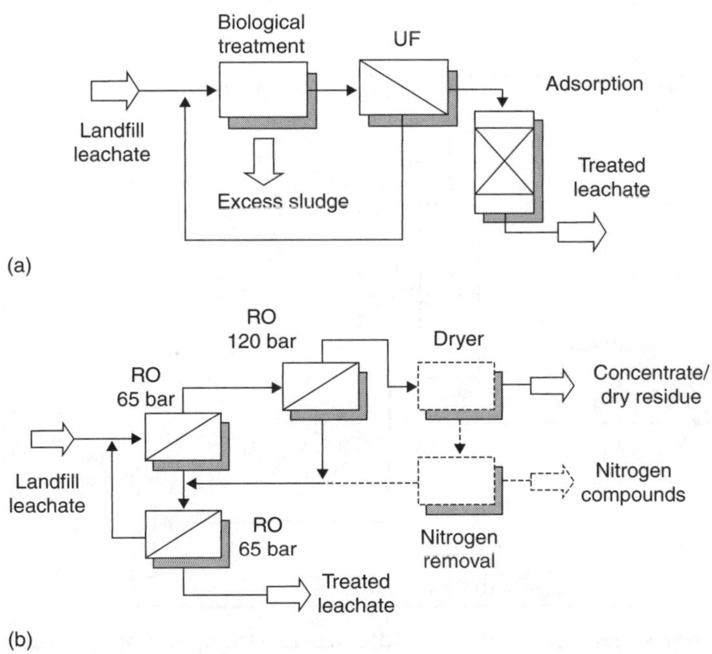

Figure 3.7(a) A membrane bioreactor-based system for treating landfill leachates. (b) A multi-staged RO and reject RO integrated membrane system for treating landfill leachates. Source: Melin et al.

Pre-treatment is essential to prevent membrane fouling that can be severe with these waters. The RO concentrate is further processed by another RO system operating at 120 bar g to reduce the concentrate volume before it is disposed off. These RO plants operate at 75–90% recovery. The flow sheet for the RO hybrid system is shown in Figure 3.7b.

3.1.6 Treatment of acidic streams

Acidic solutions are commonly used in the winning and processing of metals resulting in the waste streams containing aids and metals.[15] Membrane technologies such as RO and ED are used commercially for metal and acid recovery from waste streams, for example RO is used for recovering Cr, Cu, Cd, Zn, and Ni from plating rinse waters, and ED is used for acid recovery from acidic pickling, anodising, and etching solutions. NF technology, on the other hand, is unique; it recovers the metals in the retentate while passing the acids in the permeate. This is because NF rejects acids poorly, while retaining divalent ions and some monovalent ions. Because of low acid rejection, the osmotic pressure is low so that the acid plus water can be considered as the solvent.

A patented hybrid process that combines NF with IX for separating heavy metals from acids is shown in Figure 3.8. The process can treat both strong and weak acids with high or low metal concentrations. The waste feed is a spent 20 wt.% sulphuric acid solution containing 9.6 g/l aluminium. The NF permeate contains purified acid when operating at ~50% recovery. The NF concentrate contains 99.9% of the aluminium in the original feed. The main advantages of the hybrid process include much lower osmotic pressure for a stand alone NF process, and higher purity achieved than with IX stand alone process.

3.1.7 Dairy processing

The earliest application of membrane processes in the food industry and one of the

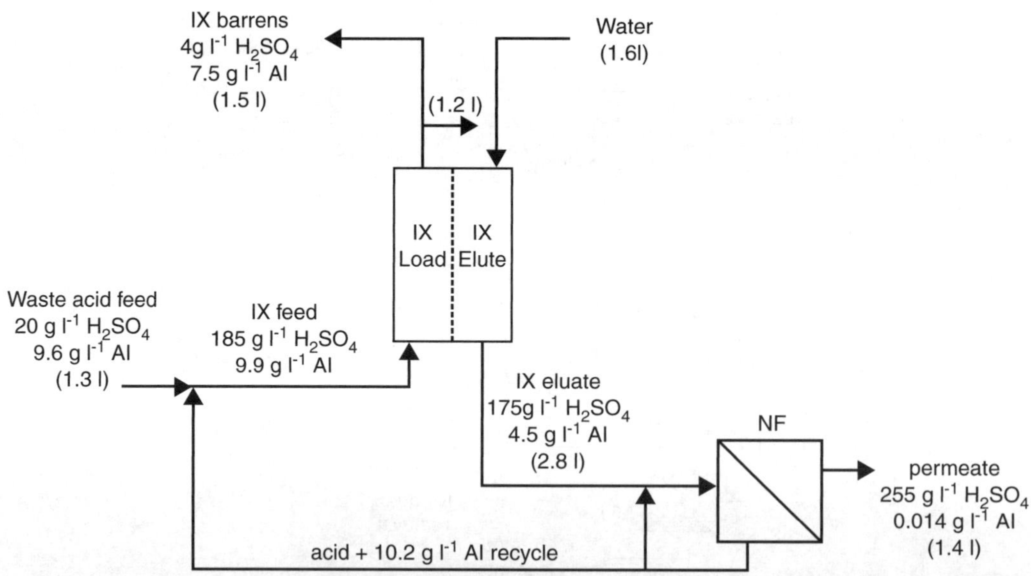

Figure 3.8 Process flow sheet for treating acidic waste streams. The NF–IX hybrid process treats acidic waste containing 20 wt.% sulphuric acid and ~1% aluminium, separating the heavy metal from the acid. Source: Soldenhoff et al.

largest, is in the dairy industry where it is used for (a) the pre-concentration of milk for cheese manufacture and (b) the fractionation of cheese whey. Milk is a complex liquid consisting of 12.5 wt.% total solids, 3.3% fat, 4.9% lactose and 0.7% ash. RO of cheese whey was commercialised in 1971, and the first UF/RO plant for fractionating cottage cheese was commercialised in 1972. By 1996, approximately 80,000 m^2 RO and 240,000 m^2 UF membrane area was installed for dairy processing,[16] of which whey processing constituted 75%. In 1984, nanofiltration (NF) was introduced for simultaneous concentration and desalination.

A hybrid membrane system used for processing milk may entail MF, UF, NF/RO and ED. Cheese whey, a by-product during cheese production, was traditionally concentrated by evaporation and spray drying. Membrane filtration was successful because it not only reduced energy and processing costs, it also improved product properties; minimal protein denaturation as a consequence of lower process temperatures. Further, since UF both concentrated proteins, and removed lactose and salt simultaneously, specialty products such as whey protein concentrates (WPC) and protein isolates could be produced.[16,17] Mass balance for an integrated UF/RO membrane process for cheese whey processing is shown in Figure 3.9.

In membrane processing of cheese whey, the whey is pretreated with MF prior to UF instead of the less efficient centrifugation for removing small quantities of fat (soluble globules, 0.2–1.0 μm) and casein (fine particulates, 5–100 μm). Complete removal of these fines by MF nearly doubles the UF flux due to reduced concentration polarisation and fouling. Demineralisation of whey by removing monovalent salts in whey powders is necessary because they cause negative sensory quality whereas the divalent ions in the powder have a positive health benefit.[16] NF is well suited for demineralisation since it retains divalent ions but not the monovalent ions. NF displaces other processes – evaporation (EV), RO and ED – as shown in Figure 3.10.

A typical commercial process for the production of whey protein concentrates (WPC) powder is shown in Figure 3.11. In the process shown, cheese whey (5 wt.% solids) is

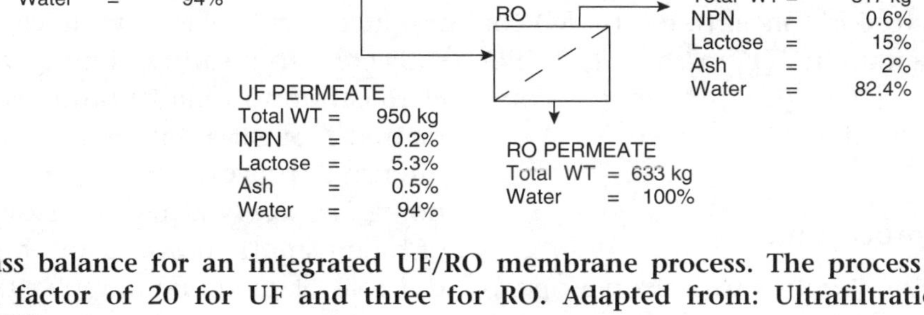

Figure 3.9 Mass balance for an integrated UF/RO membrane process. The process is based on a concentration factor of 20 for UF and three for RO. Adapted from: Ultrafiltration Handbook, Technomic, 1988

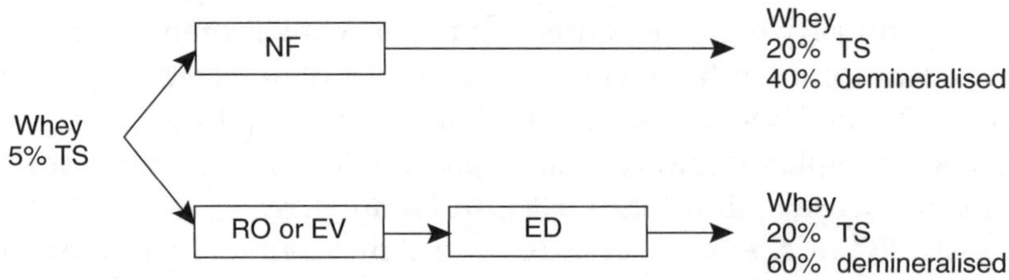

Figure 3.10 Various options for whey processing

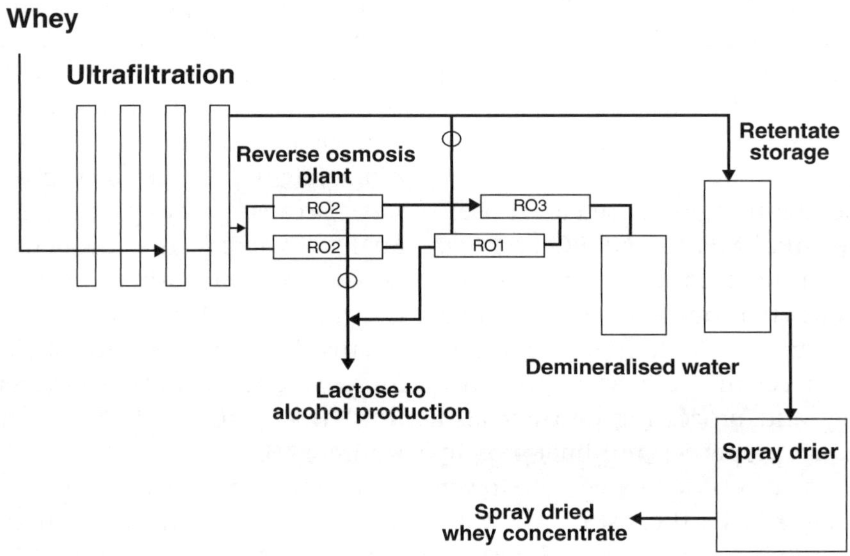

Figure 3.11 Production of whey protein concentrate (WPC) by membrane processing. Source: Koseglu and Guzman

concentrated by UF to 15–20 wt.% solids. The UF retentate is evaporated and/or spray dried producing WPC powder. UF simultaneously reduces the liquid feed volume by 90–95% prior to evaporation and drying, thereby, substantially reducing the energy load of the drying processes. The UF permeate (35 m^3/h) containing lactose is concentrated by RO for alcohol fermentation (17 m^3/h). The RO permeate flows (20 m^3/h) through a second RO producing demineralised water.[18]

3.1.8 Sugar processing

Sugar production involves two distinct operations; (a) processing sugar cane or sugar beets into raw sugar, and (b) processing the raw sugar into refined sugar. Cane and beet sugar extracts contain sucrose, and undesirable amounts of polysaccharides, lignins, proteins, starches, gums, waxes, and other colloidal impurities that contribute colour and/or taste to the crystalline product, and reduce product yield. The raw juice, therefore, is subjected to heating, liming addition, and clarification to remove proteins and colloidal matter.[19] Sugar refining is a highly energy-intensive process; hence, membrane technology is highly attractive. However, because of high osmotic pressure and high viscosities, the role of membrane application is limited to dilute streams for clarification and purification at the juice extraction stage. Possible operations that could be replaced by UF or MF are shown in the dotted areas in Figure 3.12.[17] UF and MF could be used for removing these colloidal and macromolecular impurities with little or no addition of lime, carbon dioxide, or sulphite before the clarified juice is evaporated and crystallised. Further, removing macromolecules and reducing lime levels reduces fouling and scaling of the evaporators.

The cane sugar processing consists of the following steps: sugar cane is crushed, the juice is heated and filtered, sent to a series of crystallisation steps to create crystals of raw sugar, followed by centrifugation to remove any remaining juice or syrup. The last step produces a small stream of remaining syrup called cane mill molasses containing up to 55% (wt.) sucrose and substantial amounts of invert sugar (glucose–fructose mixture) impurities. Sucrose is the fraction that becomes crystallised sugar, and has considerable commercial value but cannot be extracted economically because of the impurities. Several US patents[20] describe a process to recover sucrose from molasses and other syrups incorporating NF and MF. In this process, the molasses is pretreated by MF and/or UF to remove colloids, polysaccharides, and other high molecular weight impurities to prevent fouling of NF membranes. NF membranes retain more than 75% sucrose and small amounts of invert sugar while the permeate contains most of the invert sugar and less than 10% sucrose. The retentate flows back to the first crystalliser/centrifuge stage where it crystallises into raw sugar, and is then processed in a refinery to make refined sugar. The desugarised molasses is useful as animal feed. Although beet sugar contains no invert sugars, molasses can still be treated by the NF process.

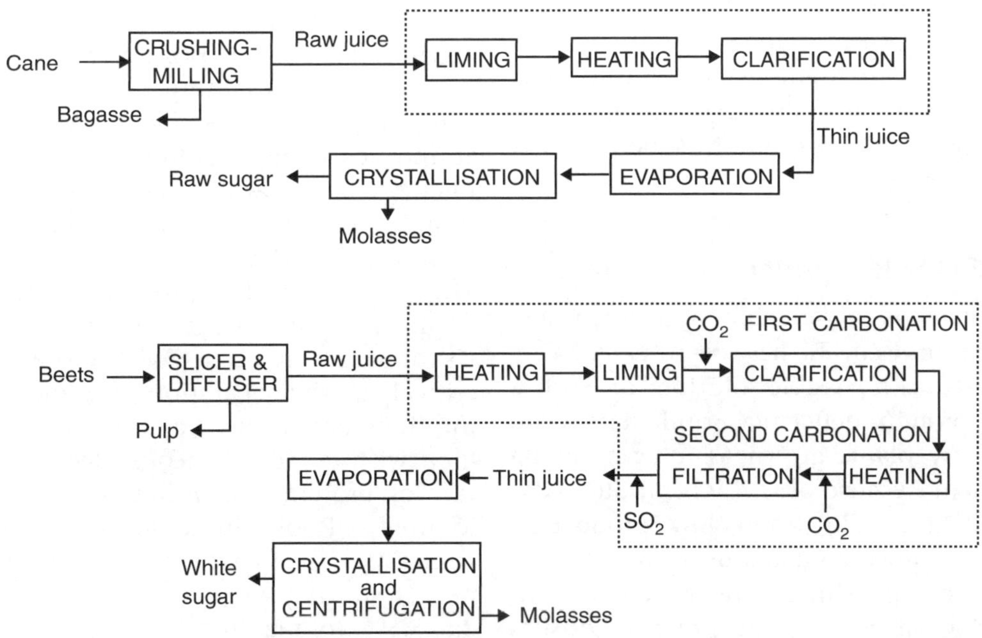

Figure 3.12 Sugar refining by membrane processing. Possible operations that could be replaced by MF or UF are shown in dotted areas. Source: Cheryan, Copyright® 1998 from Ultrafiltration and Microfiltration Handbook by M. Cheryan. Reproduced by permission of Routledge/Taylor & Francis Group, LLC

3.1.9 Vegetable proteins

Conventional methods for producing soy protein concentrates and isolates consist of extraction, heat treatment, and centrifugation to separate the protein and fat from undesirable compounds such as oligosaccharides, phytic acid, and trypsin inhibitors. Typically, the traditional methods are time consuming and the products sometimes have inferior functional properties possibly due to the denaturation of proteins during heating. Since the undesirable compounds are smaller in molecular size than proteins and fat components, proteins can be easily recovered and isolated by using UF membranes. Further, since UF separation does not involve heat and chemicals (used in extraction), the functional properties of the soy protein are superior. Typically, the UF membranes used are in the high MWCO range of 50,000–200,00 Da. Although the composition of the final soy product seems to be independent of the MWCO of the membrane, the flux is much higher. Furthermore, membrane processing results in higher yields because the whey proteins are also recovered. These proteins possibly contribute to superior functional properties.

3.1.10 Edible leaf protein concentrates

Leaf protein is a valuable and renewable source of protein. It has been extensively investigated as a possible solution for world's hunger, protein deficiency, and starvation. Of the leafy plants grown as forages, alfalfa is dominant because of its high productivity, high nutritive value, and low production costs.[21] Leaf juice contains up to 5% protein (20% dry basis), which is recovered by heat coagulation at 80°C. Leaf protein consists of a chloroplastic or green protein fraction (65–75%) and a mixture of cytoplasmic and chlorophyll-free protein or white protein (25–35%). The chloroplastic fraction is a mixture of proteins with MW > 65,000 Da, while the cytoplasmic protein is a mixture of two proteins exhibiting a MW between 10,000 and 65,000 Da.[22] Whole leaf protein concentrate (LPC) is a dark green product, which is very low in fibre, and high in protein and xanthophyll.[21,23] Xanthophyll gives the LPC special value as a chicken feed ingredient, since it is this pigment which gives egg yolks and broiler skins their golden yellow colour.

Whole LPC has been unacceptable as a food because of its colour and flavour characteristics.[21] The green protein can be separated from the white protein by selective heat coagulation.[23] Fractionation by heat, however, denatures the protein. Preparing the white proteins in soluble form is most desirable, since solubility and other functional properties of undenatured protein are important for many food applications. Recovery and fractionation of alfalfa protein by membrane UF is, therefore, a preferred method.[22–24] The UF process removes 80–90% of the water containing non-protein soluble components such as low molecular weight amino acids, sugars, minerals, vitamins, and phenolic bitter-tasting pigments. Ultrafiltered protein concentrates also contains more amino acids than heat precipitated ones. The spray-dried white protein is water soluble, cream-white in colour and tasteless. During UF of whole juice, a 7–8% protein level is critical for the separation efficiency. Exceeding these levels causes a rapid decrease in the flux with a simultaneous decrease in the rate of protein concentration due to severe fouling. UF of chloroplastic-free juice, by contrast, was efficient even up to 19% dry-matter starting with a clear juice containing 2.5–3.5% dry matter.[22]

An alternate approach, therefore, has been to separate the green protein from the white protein by heat coagulation at 55°C, centrifuge the product to separate the coagulated green protein from the brown liquor containing white protein, and recover the white protein by UF using a 10,000 Da MWCO membrane. One hundred gallons of fresh juice squeezed from 1.1 tons of fresh crop, produces 4.5 kg of food grade spray-dried LPC and 22 kg of feed grade LPC.[22] However, even a mild heat treatment causes some denaturation of the protein.

An integrated membrane process is based on a MF–UF–DF model as shown in Figure 3.13.; a back-washable dead-end MF with a pore size of $<10.0\,\mu m$ to separate the green fraction at $pH=8.0$ followed by UF to recover the white fraction, and purification of the white fraction using diafiltration (DF). MF with a pre-coat traps the chloroplastic fraction producing a chlorophyll-free brown liquor. The pre-coat material can be diatomaceous earth or corn starch. The latter is preferred since the filter cake is useful as cattle feed. The efficiency of the process can be increased by adding a body-feed of the coating material at very low concentrations (5%). Because of its unique design, the MF cartridge is not subject to internal pore blockage like the traditional dead-end MF cartridge. The backwash cycle is very fast (<1 min) and efficient. The filter elements are durable with a 7–10 year life, and use a variety of materials for making serrated disk elements including nylon, polypropylene and stainless steel.

Green plants and especially alfalfa also hold promise for growing therapeutic proteins genetically.[25] Genetic engineering makes it possible to custom design a green plant to produce a desired product in the seed, leaves, fruit, root, or sap, and grow these modified plants on farms. The modified plant alfalfa crop is harvested before the plant's blooming season and seed production stage followed by extraction and recovery of the protein.

3.1.11 Vegetable and corn oil processing

In conventional chemical processing of vegetable oils, crude oil is recovered from the mechanically expressed vegetable seed by solvent extraction using hexane, as shown in Figure 3.14.[16] The micella (a mixture of extracted oil and solvent) from the extractor contains 60–70% solvent, which is recovered by distillation and reused. Purification of crude oil consists of removing components that are harmful and are not wanted in the final product, the pure oil. These components are phosphatides (gums and lecithin), free fatty acids, pigments, and volatile odouriferous materials. These materials are removed successively in steps termed degumming, refining, bleaching, and deodourising.

The conventional technology is energy intensive, uses large amounts of water and chemicals, and generates heavily contaminated effluents. Membrane processing tackles these drawbacks, and also removes phospholipids and free fatty acids. Further, hexane removal by distillation after degumming increases the distillation efficiency. Energy savings and wastewater (high BOD acidic waste) elimination are estimated to be 10–20 trillion BTU/year and 20–25 million gallons/year, respectively. Additional energy savings are possible when RO is used to remove hexane instead of distillation;[18,26] the solvent-rich stream that exits the RO unit is recycled to the extractor and the oil-rich stream is processed by distillation to recover the remaining solvent. Energy savings of 2 trillion BTU/year are estimated for the hybrid-membrane-distillation operation in the US domestic edible oil industry.

Phospholipids (4.5 wt.% of crude oil) are among the most abundant surfactants found in nature. At concentrations higher than the critical micellar concentration, surfactants form micelles, which are then easily rejected by UF membranes, as discussed in Chapter 1.

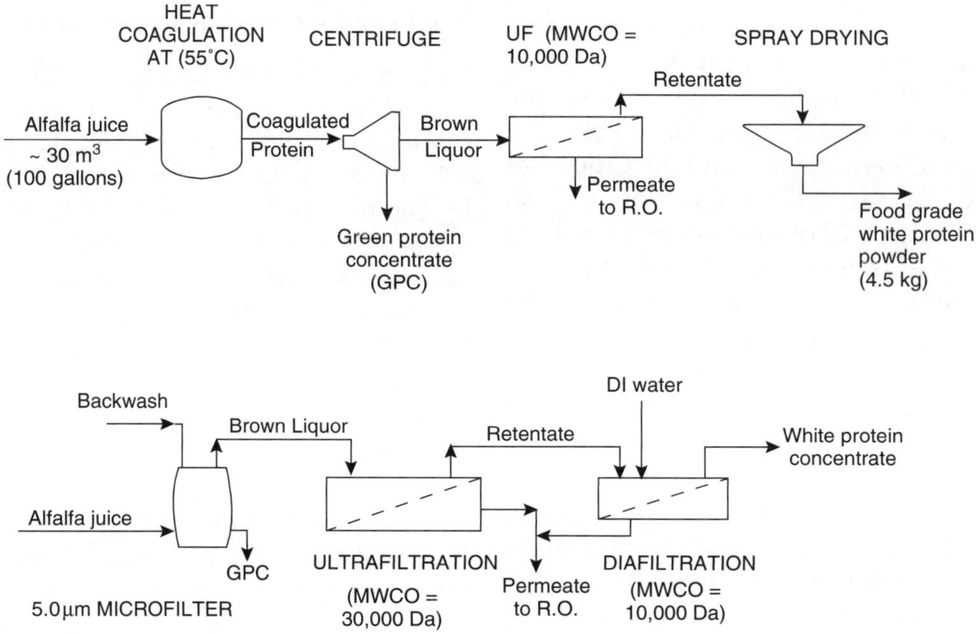

Figure 3.13 Hybrid membrane process schemes for recovering leaf proteins

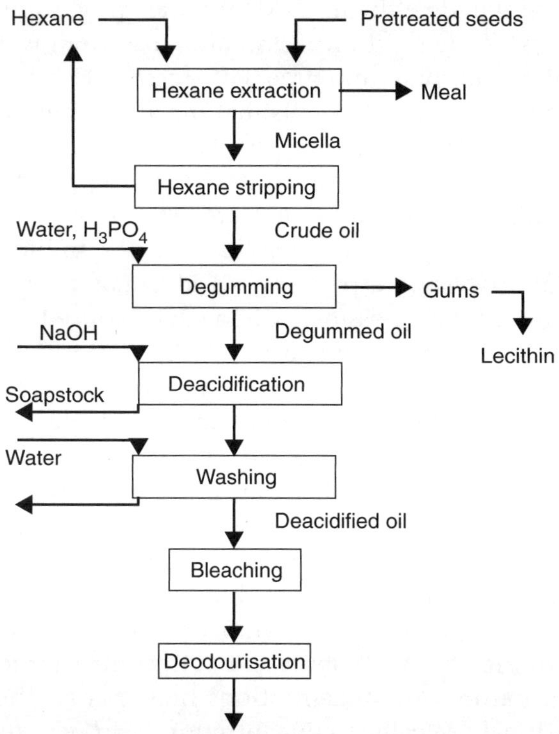

Figure 3.14 Schematic diagram of conventional chemical processing of vegetable oils. Source: Bargemann et al.

Phospholipids have hydrophilic polar heads and hydrophobic non-polar tails, and form reverse micelles in organic solvents such as hexane. Moreover, the micelles bind low molecular weight impurities such as free fatty acids, sugars, amino acids, and colour compounds.[27] Because of the formation of micelles, triglycerides are also fractionated from phospholipids. The latter is then easily purified to produce a valuable product, lecithin, oil-free. Since the micelles are estimated to have pseudo-molecular weights of 200,000, it should be possible, therefore, to recover lecithin oil-free using UF membranes with MWCO of ≥20,000 Da.

An integrated membrane process may involve the following unit operations:[17]

- Degumming – Hexane-resistant UF membranes (PA, PS, PVDF, PI, PAN or inorganic) with a MWCO of 20,000 Da and molecular sizes of 20–200 nm.
- Deacidification – Hydrophilic UF membranes (PAN) with a MWCO of 30,000 Da

in the first stage to remove free fatty acids (permeate) followed by a hydrophobic membrane in the second stage to process the retentate containing triglycerides from the first stage.

- Removal of metals – Ceramic MF membranes with pore sizes of 0.01–0.2 μm for recovering the nickel catalyst, and reducing the nickel level in the permeate to 1–10 mg/l.
- Dewaxing – Ceramic MF filters with pore diameters of 0.05–1 μm operated in conjunction with periodic timed backwashing.

A processing scheme shown in Figure 3.15[28] includes two or more dead-end MF stages using increasingly fine filters to prevent fouling of the downstream UF membranes, followed by a feed-and-bleed (F&B) filter. The retentate from the F&B filter is recycled back to the extractor whereas the permeate flows to several UF stages to separate the phospholipids from the oil–solvent mixture. The permeate from the first UF stage contains less than 200 ppm phospholipids, and is evaporated to remove the solvent and produce pure vegetable oil. The phosphoplipid content may be reduced to less than 5 ppm by adding a third UF stage. The second UF stage retentate contains concentrated phospholipids (25 wt.%), and is further processed to make lecithin, a food additive.

Wet milling of corn accounts for more than 75% of the corn processed for food, feed, and industrial chemicals in the United States. The corn wet milling process is quite energy and capital intensive. Corn is also the main feedstock of ethanol ($>1.5 \times 10^9$ gallons per year). Since ethanol is used as an oxygen additive in petrol for clean burning, an energy-efficient technology for processing corn makes the overall process more efficient. A conventional corn refining process is shown in Figure 3.16. The unit operations that could be replaced with membrane processes, MF, UF, NF, RO are marked as I, II, III and IV, respectively.

3.1.12 Fruit juice and wine processing

Conventional process of producing final juice product is shown in Figure 3.17. In the manufacture of clear juice, the pressed fruit juice is pasteurised, and treated with pectinase (an enzyme) to hydrolyse the pectin and

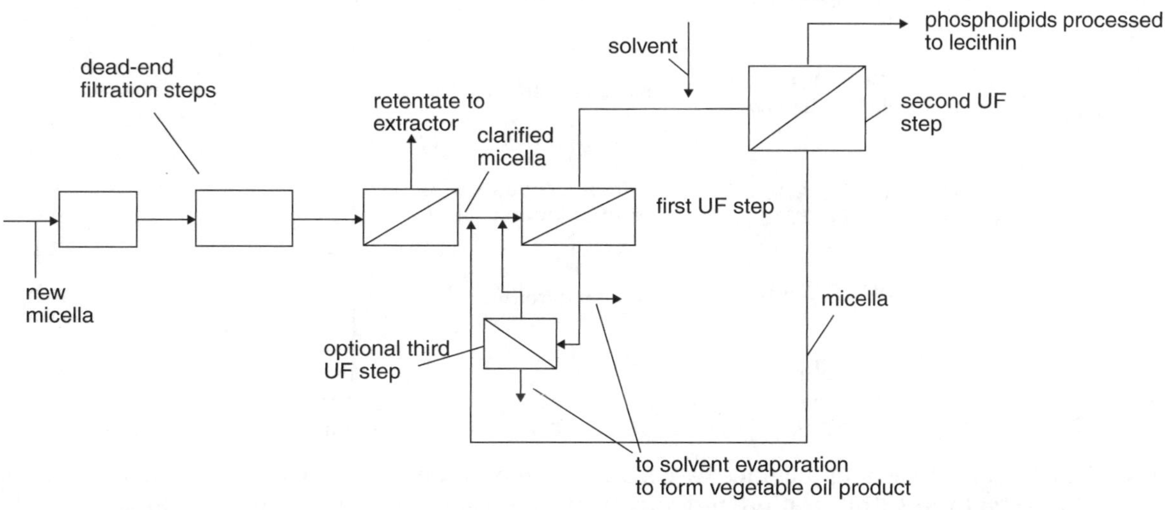

Figure 3.15 Processing of vegetable oil micella in a hybrid membrane system

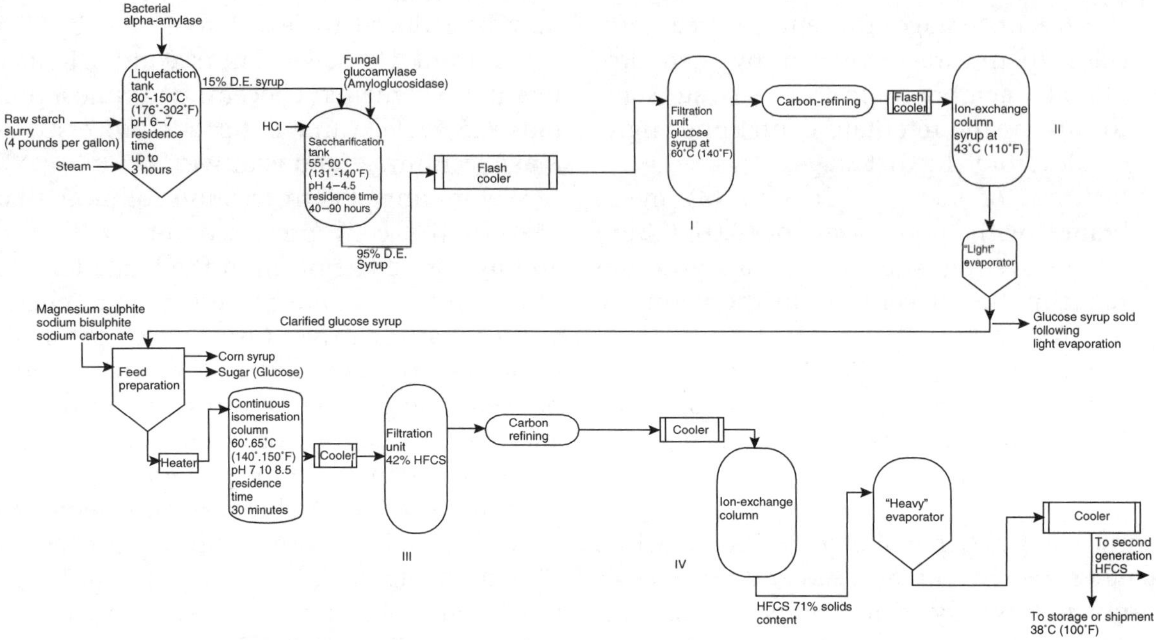

Figure 3.16 Corn refining process steps. I, II, III and IV operations could be replaced with membrane processes. Source: Biotechnology in Industry, Butterworth, 1983

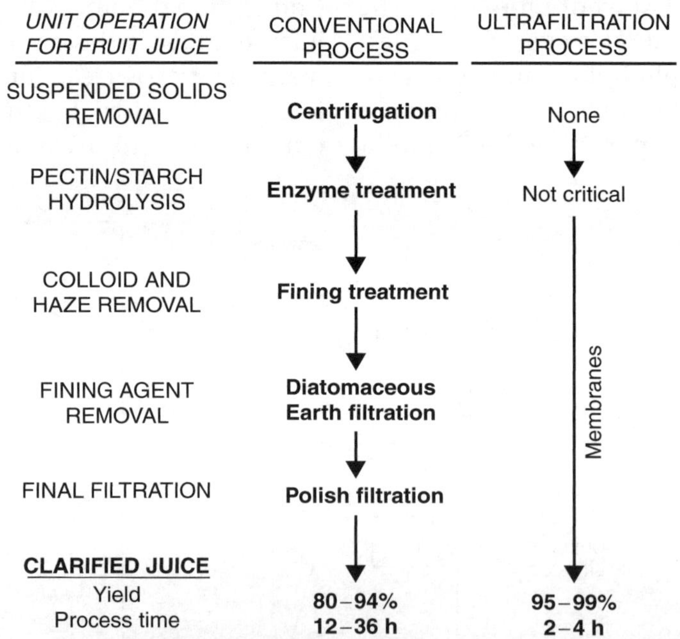

Figure 3.17 Manufacture of apple juice by conventional and membrane processes. Source: Cheryan, Copyright® 1998 from Ultrafiltration and Microfiltration Handbook by M. Cheryan. Reproduced by permission of Routledge/Taylor & Francis Group, LLC

reduce cloudiness. Next, a fining agent such as gelatin is added during the preliminary clarification step followed by holding and decanting. The juice is filtered using diatomaceous earth as a filter aid to remove suspended solids, colloidal particles, proteins, and condensed polyphenols. The final clear juice is pasteurised again, if necessary. The process involves several batch operations that are both time consuming and labour intensive[17]. Membrane processes are used to replace the holding, filtration, and decantation steps as shown in the above figure. Membrane processing produces higher quality juice due to reliable and consistent removal of haze-forming components including suspended solids, colloidal particles, proteins and polyphenols with a turbidity of 0.1–0.3 NTU as compared to 2–5 NTU for conventional processing.

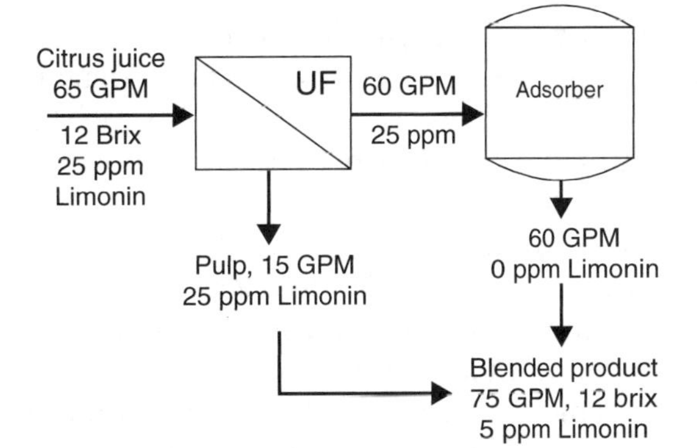

Figure 3.18 Processing of citrus juices by a hybrid membrane process. Source: Romicon

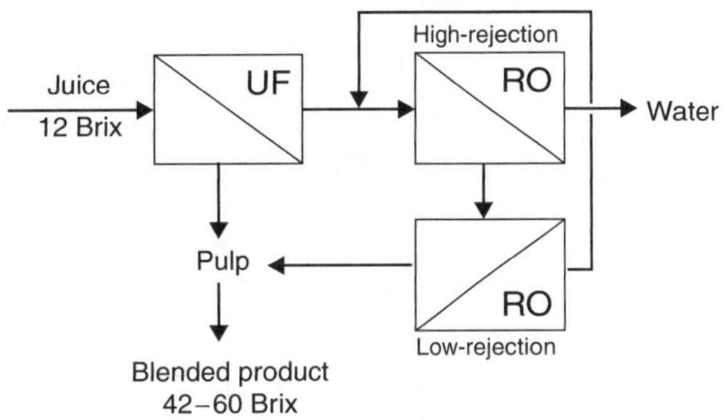

Figure 3.19 Concentration of fruit juices by an integrated membrane process. Source: Cheryan and Alvarez, Memb. Sci. Tech., Academic Press (1995).

Processing of citrus juices by integrated membrane processes are shown in Figures 3.18 and 3.19. In the case of debittering (Figure 3.18), fresh or reconstituted citrus juice that has been de-oiled and pasteurised is ultrafiltered to remove the pulp.[17] The clarified permeate containing sugars, bitter compounds (limonin, naringin, hesperdin, polyphenols) and off-flavour compounds flows through an adsorbent resin column which contains an adsorbent macroreticular resin with no functional groups. The resins are designed to remove these compounds. The debittered juice is recombined with the UF retentate containing the pulp to produce the blended and refined citrus juice product. The hybrid process is cheaper than the traditional process using centrifuge-cum-hydrophilic adsorbent system.

Another application of membrane processes is in the production of citrus juice concentrates as shown in Figure 3.19. UF is used to separate the pulp (retentate) from the serum (permeate). The serum fraction contains sugars and flavour compounds. The pulp fraction contains some soluble solids, all insoluble solids, pectins, enzymes, orange oils, and the microrgansims. The UF permeate (90–95% of the feed voloume) is concentrated by a two-stage RO unit. The concentrated serum is blended with the UF concentrate to produce a highly concentrated juice (42–60° Brix). Prior to mixing with the RO concentrate, the UF pulp fraction (5–10% of the feed volume) is pasteurised for improving stability by destroying spoilage microorganisms.

Membrane processes have also been used for producing juice concentrates without loss of quality that occurs during evaporation.[17,18] Specifically, evaporation results in the loss of various volatile alcohols, esters, and aldehydes resulting in a significant deterioration in quality, aroma and flavour of the concentrated product. A schematic flow diagram of a hybrid membrane process for producing superior juice concentrate is shown in Figure 3.20.

Wine processing by a hybrid membrane process is shown in Figure 3.21. The first UF step removes microorganisms, colloids, and high molecular weight materials. The MF step removes yeast used for fermentation. The final UF step is used for sterilising prior to bottling. Membrane pore size plays a crucial part in the retention of colour and aroma compounds. UF membranes with MWCO of between 10,000 and 500,000 Da are best-suited.[17]

3.1.13 Recovery of flavour compounds by osmotic distillation

Osmotic distillation (OD) or sometimes called 'isothermal membrane distillation' is a membrane process in which a liquid phase (usually an aqueous solution) containing one

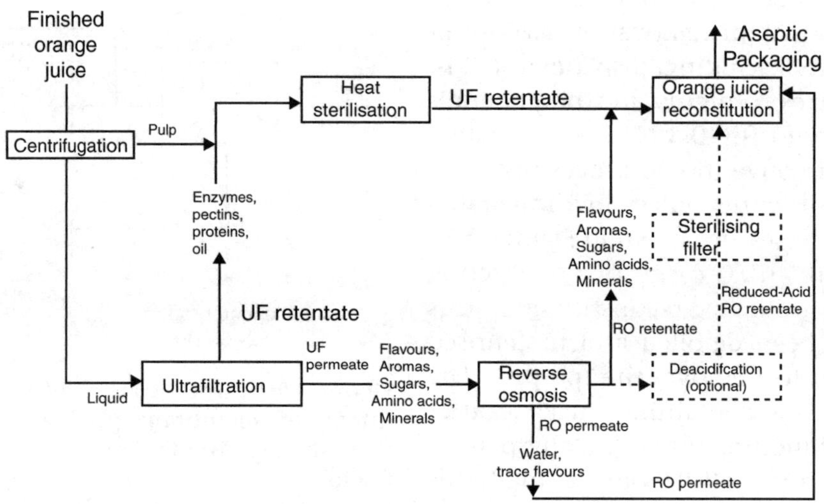

Figure 3.20 Schematic flow diagram for producing reconstituted orange juice[18]. Source: Koseglu and Guzman

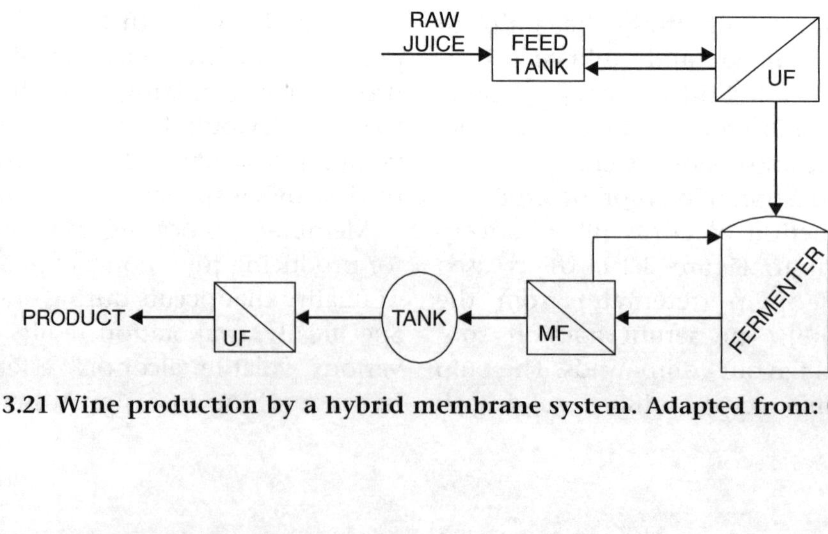

Figure 3.21 Wine production by a hybrid membrane system. Adapted from: Cheryan

or more volatile components flows across one surface of a microporous membrane whose pores are not wetted by the liquid, while the opposing surface is in contact with a second non-wetting liquid phase (usually an aqueous solution) in which the volatile components are soluble or miscible.[29] The device is similar to the membrane contactor (MC) discussed in Chapter 1 containing hollow-fibre membranes that are hydrophobic (non-wetting). Unlike the MC used for degassifiation, the feed to be concentrated flows in the shell side of the membrane element while the brine strip solution flows in the lumen side.

The device or process is best suited for concentrating dilute solutions containing non-volatile solutes (relatively low concentration) of moderate to high molecular weight such as sugars, polysaccharides, carboxylic acid salts, and proteins with limited stability to high temperature and shear stress. Process flow sheet for a juice concentration unit is shown in Figure 3.22. The process incorporates partial batch recycling on the feed side to minimise large feed viscosity changes in the MC, and continuous countercurrent recycling with evaporative re-concentration of the brine strip stream.[29] The MC modules can be arranged in a series, parallel, or series–parallel array.

The OD process is more costly than evaporation (EV), UF, or RO for concentration of aqueous mixtures. However, unlike these other relatively inexpensive processes, OD can selectively remove water relative to other low-molecular weight solutes yielding products of superior quality especially when volatile solutes such as flavours and fragrances are involved. Usually with most liquid foodstuffs, maximum acceptable concentration achieved by (EV/UF/RO) is 30%. Hence, a hybrid process involving pre-concentration of the feed by RO followed by further

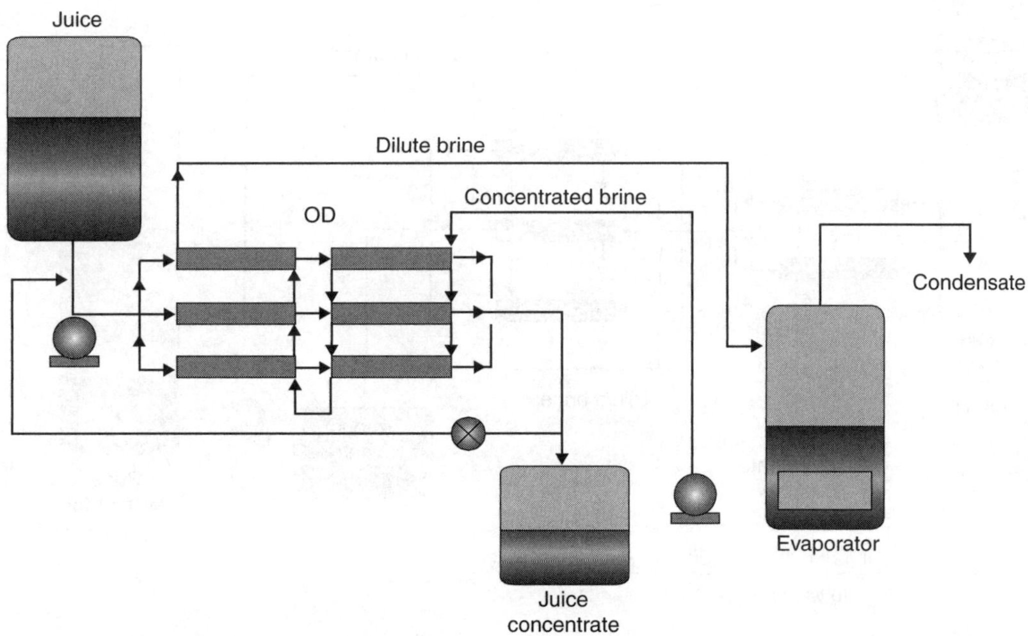

Figure 3.22 Hybrid osmotic distillation schematic for juice concentration. Source: Hogan et al., Reproduced with permission from Chemical Engineering Progress, July 1998. Copyright 1998, AIChE

concentration of the RO retentate by OD is a logical extension for recovering a higher concentration product at lower processing cost. A typical hybrid process is shown in Figure 3.23. The process flow sheet includes an MF or UF stage to pre-treat the feed for removing suspended solids and colloids thereby minimise fouling of RO or OD membranes. Another potential application includes dealcoholisation of ferments – removal of ethanol from fermented beverages such as beer and wine. In this hybrid process food-grade (190 proof) ethanol can be recovered from the strip solution using a rectifying column. Similarly, OD can be used for the selective removal of organic volatiles from aqueous solution in the pharmaceutical industry, for example removing or recovering a part of the organic solvent from the product-rich supernatant, used in the extraction of intracellular products from fermentation broths.

3.1.14 Bioprocessing

Bioprocessing or biotechnology is used in the production of pharmaceuticals, foods, flavours, fuels, and chemicals with the aid of a biocatalyst such as an enzyme, microorganisms, plant cell, or animal cell in a bioreactor. It also involves genetic engineering for the manipulation of plants, animals, and microorganisms such as yeasts, bacteria, and fungi. Downstream processing is required to remove impurities, bulk-volume reduction and simultaneous concentration of the desired product from the bioreactor.[31] Various steps deployed during downstream processing for recovering the final product in protein production are illustrated in Figure 3.24.

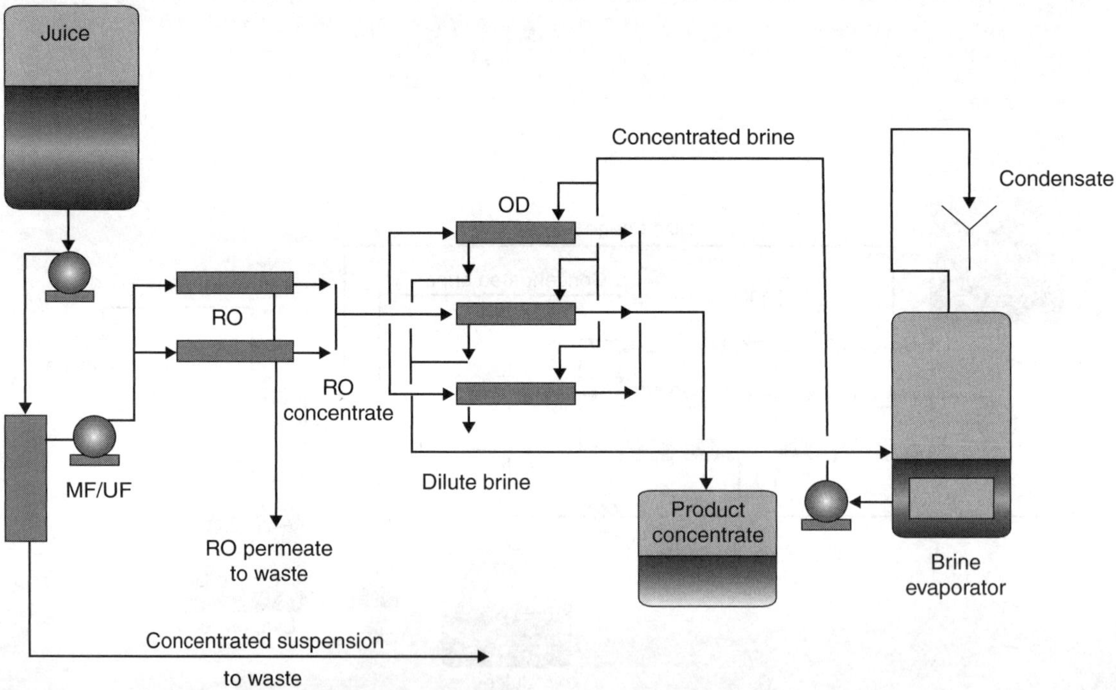

Figure 3.23 Hybrid osmotic distillation membrane schematic for concentration of biological products. Source: Hogan et al., Reproduced with permission from Chemical Engineering Progress, July 1998. Copyright 1998, AIChE

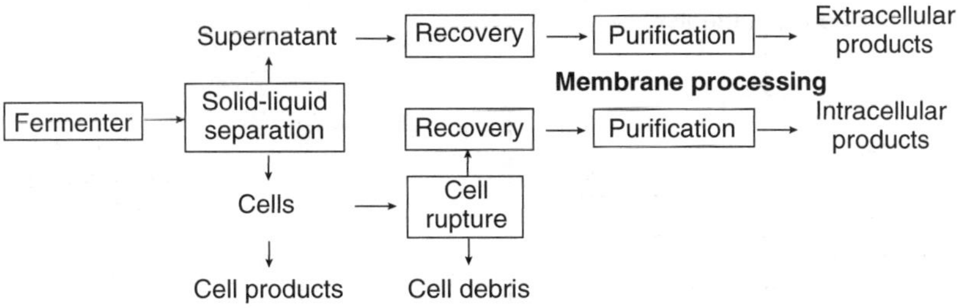

Figure 3.24 Major process steps in downstream processing. Membrane processes include MF, UF, RO and ED for concentration, separation and purification

Protein recovery is sensitive to operating conditions because their function depends on the integrity of the delicate 3D tertiary structure.

Most pharmaceutical substances are manufactured in batch processes by (a) chemical synthesis, (b) fermentation, (c) isolation and recovery from natural sources, and (d) a combination of the above. Fermentation broths are usually very dilute and contain many complex compounds as given in Table 3.3. Because of their dilute and impure nature of the broths, and sensitivity to operating conditions, membrane processes may be well suited for downstream processing by MF, UF, and NF, i.e. separation, isolation, purification, and recovery of the product,[17,31] as illustrated in Figure 3.25. Membrane processes such as RO and gas separation (GS) are shown for use upstream of the bioreactor. NF technology is well-established in pharmaceutical and biotechnology applications including the recovery of antibiotics from fermentation processes. A schematic flow diagram of NF separation for recovery of 6-APA is shown in Figure 3.26.[11] 6-Aminopenicillanic acid (6-APA, MW 216 Da) is an intermediate in the manufacture of synthetic penicillin, and can be manufactured by an enzymatic process. A medium-sized plant produces 15–20 tons of mother liquor per day. NF (solvent stable membrane is used) separates the 6-APA (in the retentate) from the other liquor at 60–80 kg/day, and is recycled to the extraction unit, thereby, minimising product losses (recovery is 90–95%).

Table 3.3 Fermentation broth components concentration

Component	Concentration, %w/w
Bacteria or yeast (ethanol, lactic acid broth)	1–5
Fungi (citric acid, penicillin production)	1–3
Animal cells (mammalian tissue culture)	0.1–5
Acetic acid in fermentation broth	0.2–5
Lactic acid in fermentation broth	8–10
Citric acid in fermentation broth	5–10
Extracellular enzymes	0.5–1
Vitamins	0.005–0.1
Antibiotics	1–5
Ethanol	7–10

Source: Cheryan, Copyright® 1998 from Ultrafiltration and Microfiltration Handbook by M. Cheryan. Reproduced by permission of Routledge/Taylor & Francis Group, LLC

3.1.15 Membrane recycle bioreactors

A process flow diagram for producing ethanol and lactic acid from cheese whey by fermentation with the aid of a membrane recycle bioreactor (MRB) is shown in Figure 3.27.[17] The UF permeate contains lactose, non-protein nitrogen, and dissolved salts, and is then concentrated by RO to the desired sugar concentration for fermentation (2.3 times

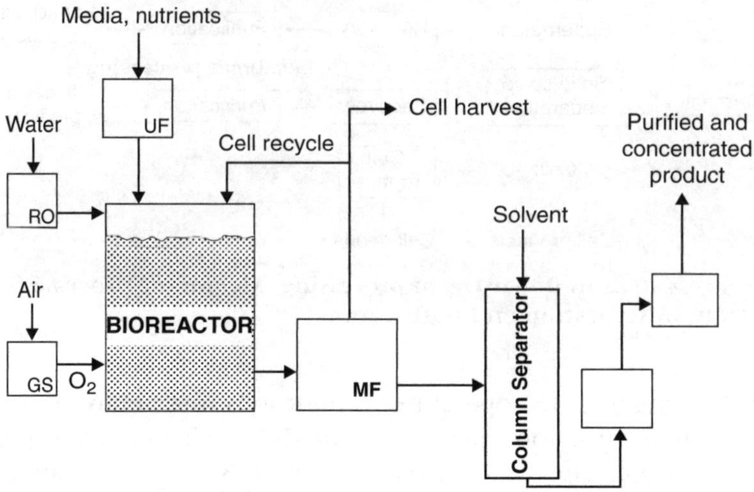

Figure 3.25 Bioprocessing by hybrid membrane systems showing upstream and downstream membrane processing steps. Adapted from: Cheryan

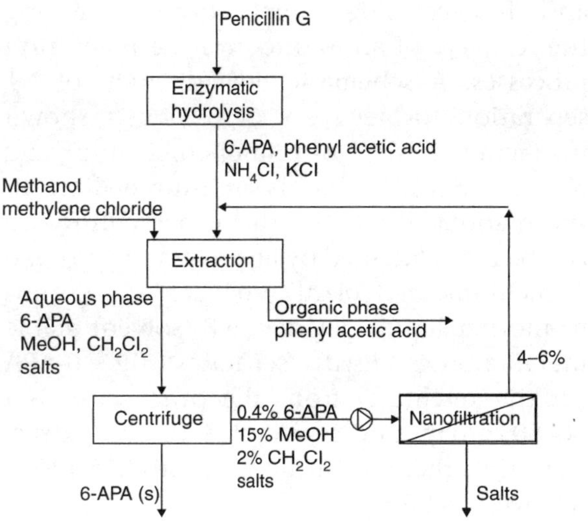

Figure 3.26 Pharmaceutical drugs recovery with a hybrid membrane system. Source: Kyburz and Meindersma

for lactic acid, and 4–5 times for ethanol). The bioreactor or fermentor contains a suspension of the microorganism (biocatalyst). In the case of ethanol production by yeast, a cell concentration of 90 g/l is optimal whereas lactic acid production requires a cell concentration of 60 g/l. MRBs offer the advantage of conducting fermentations in a continuous mode instead of a batch mode resulting in lower capital costs, better process control, and improved productivity.[17,31]

The MRB shown in Figure 3.27 consists of a reaction vessel and a membrane (UF or MF) module. The biocatalyst is suspended in the main reaction vessel, and the feed is pumped into the reaction vessel at the same rate as the permeate (product) flow rate. The reaction mixture is pumped continuously through the membrane module where the products of the reaction and other low molecular weight compounds are removed in the permeate, while the biocatalyst and other rejected compounds are recycled back to the reaction vessel for further reaction. MRBs generate higher product concentration because of higher cell concentrations that allow high dilution rates with complete substrate utilisation. UF membranes with a MWCO of 10,000–100,000 Da are used to retain the enzymes, and MF membranes when microbial cells are used.

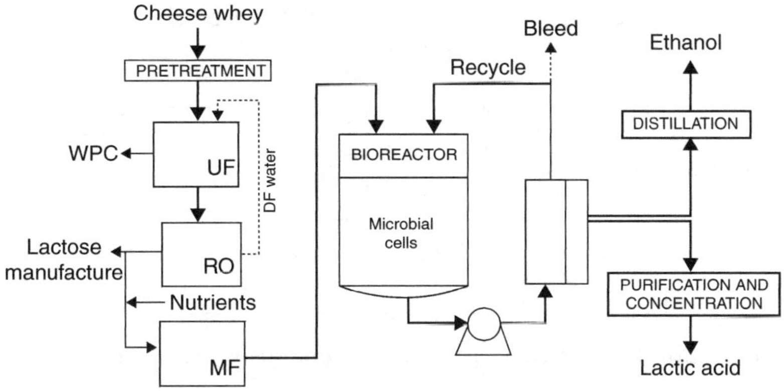

Figure 3.27 Processing of cheese whey using multiple membrane systems and a membrane recycle bioreactor. Source: Cheryan, Copyright® 1998 from Ultrafiltration and Microfiltration Handbook by M. Cheryan. Reproduced by permission of Routledge/Taylor & Francis Group, LLC

3.1.16 Separation of organic liquids by pervaporation

Separation of organic/organic mixtures as in the case of separation of methanol from methyl tert-butyl ether (MTBE) was one of the first industrial applications of pervaporation (PV).[3] MTBE is produced by reacting methanol with isobutene over an acidic ion-exchange catalyst. In commercial practice, the isobutene conversion of 87–94% per pass is set with a molar ratio of methanol to isobutene slightly greater than 1.0. When an excess amount of methanol is fed into the reactor, an increase in the isobutene conversion is expected, and 5–6% additional conversion is possible. Adding extra methanol, however, results in the formation of azeotropes in the reaction product. Two variations of a PV-based hybrid process to break the methanol azeotrope are shown in Figures 3.28 and 3.29. In the first version (Figure 3.28), a PV membrane unit is added immediately downstream of the reactor. Methanol: isobutene ratio in the reactor is increased from 1.0 to 1.2, thus increasing the isobutene conversion about 5% per pass. The PV unit removes methanol from the reactor effluent, reducing the methanol concentration from 5% to 2%, which is fed to the distillation (debutaniser) column without causing the formation of azeotropes.[32]

In the second scheme shown in Figure 3.29, a side-draw from the distillation column is fed to the PV unit, to remove methanol. By taking a side-draw, removing methanol and returning the residual, the column operation becomes more efficient. The composition of the PV permeate is 65–90 wt.% methanol, the rest being mostly C_4s, when the feed methanol concentration is 1–10 wt.%. The use of MTBE as a fuel additive is being phased out because of environmental considerations.

3.1.17 Dehydration of ethanol

Dehydration of ethanol has been the most successful and important application of membrane pervaporation.[2,3] Another purpose of PV is to remove the ethanol product from fermentation broth thus preventing product inhibition. The effect of product removal by PV is shown in Figure 3.30. The result of ABE (acetone, butanol, and ethanol) production

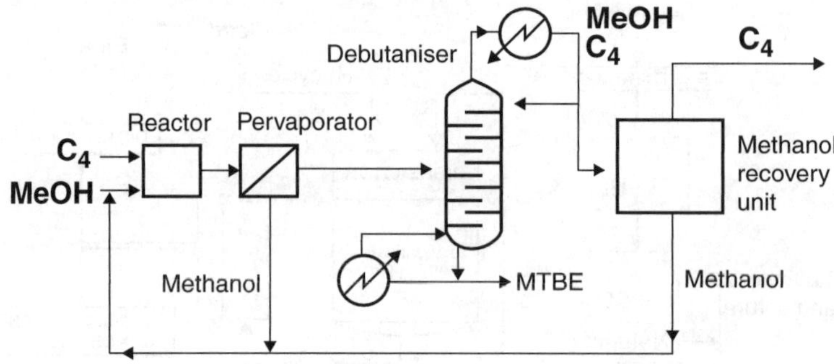

Figure 3.28 A pervaporation-based hybrid process for separation of organic liquid mixtures. Source: Matsuura

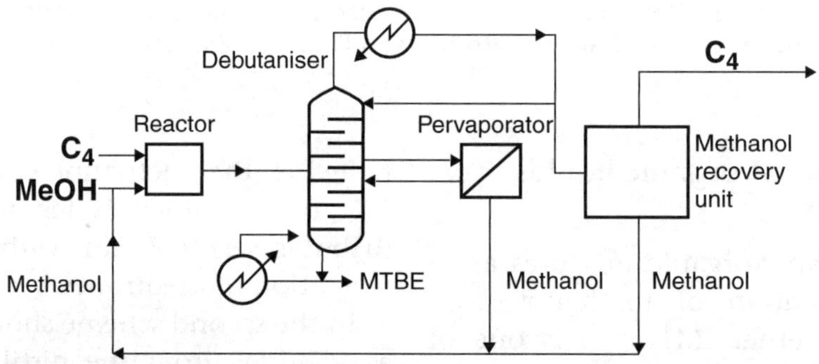

Figure 3.29 An alternate scheme for separation of organic liquid mixtures by a hybrid membrane process. Source: Matsuura

from xylose substrate using C-acetobutylicum in a conventional fermenter is shown in Figure 3.30a. The data indicates that butanol production reached a maximum value after 180 h and dropped after 260 h. Figure 3.30b shows that in the case of a bioreactor-PV integrated process, butanol production reached a maximum after 300 h, and the high value was maintained due to the continuous removal of the product by the PV membrane unit.[32]

Ethyl alcohol concentration in the fermentation broth varies between 5 and 10%. Ethyl alcohol is concentrated and dehydrated by a distillation process. However, ethyl alcohol forms an azeotropic solution with water at ethyl alcohol concentration of 95.6 wt.%, and distillation becomes ineffective at removing the trace amount of water. Because of the formation of azeotropes, azeotropic distillation with an additive, normally either cyclohexane or benzene, needs to be applied. The energy consumption of this dehydration process is very high. The PV process, therefore, has been considered as an alternative separation process for breaking the azeotrope. However, the one-step PV membrane concentration of ethyl alcohol is not considered economically feasible. According to the vapour phase–liquid phase ethyl alcohol concentration diagram for a commercial polyvinyl alcohol/polyacrylonitrile composite PV membrane, distillation is more favourable than PV for the concentration of ethyl alcohol up to 85 wt.% while PV is more effective for concentrating ethyl alcohol from 85 to 99 wt.% or higher. A hybrid distillation–PV system is shown in Figure 3.31.

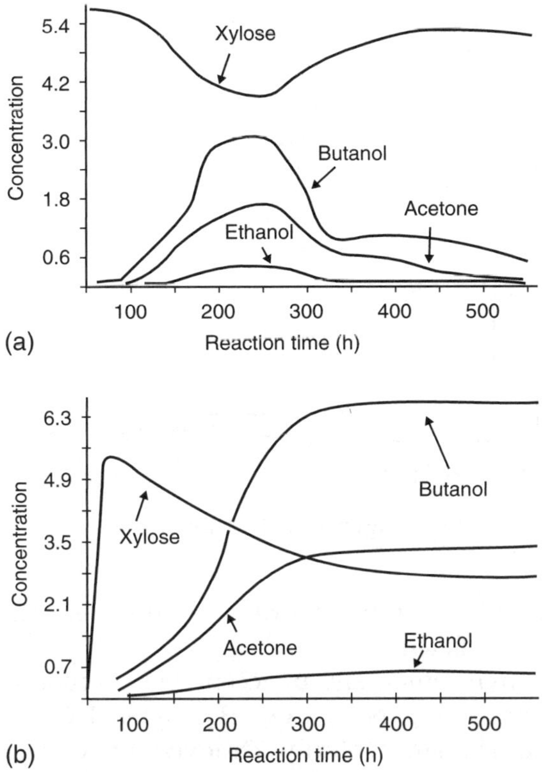

Figure 3.30(a) Performance curves for dehydration of ethanol by conventional processing. (b) Performance curves for dehydration of ethanol by a distillation/pervaporation process. Source: Matsuura

The steam consumption of the above process is 2.7 kg per kg of dehydrated ethyl alcohol as compared to 4.0–6.0 kg per kg of dehydrated alcohol.[32]

3.1.18 Fuel cell hydrogen gas purification

An integrated membrane-pressure swing adsorption (PSA) hybrid system for the purification of hydrogen gas, for on-site fuel cells from synthesis gas is shown in Figure 3.32. In the process shown, synthesis gas is produced by either steam gas reforming or by partial oxidation (POX) of hydrocarbons.[33] Steam reforming of natural gas produces a product stream with a H_2:CO ratio of approximately 3:1, whereas POX of hydrocarbons gives a product stream with a H_2:CO ratio of 2:1.[19] It is desirable to adjust the H_2:CO ratio of the synthesis gas produced in the first stage to 1:1, and simultaneously recover pure hydrogen.

The synthesis gas produced in the steam reformer or the POX reactor is scrubbed to remove CO_2, which is either recycled as feed to the reactor or removed from the system. The CO_2-free stream enters the gas

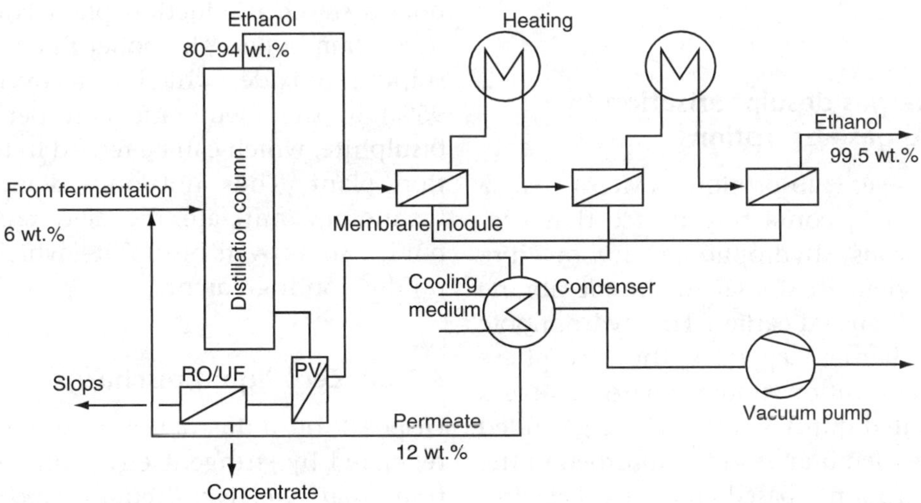

Figure 3.31 Flow schematic of a distillation/pervaporation process for dehydration of ethanol. Source: Matsuura

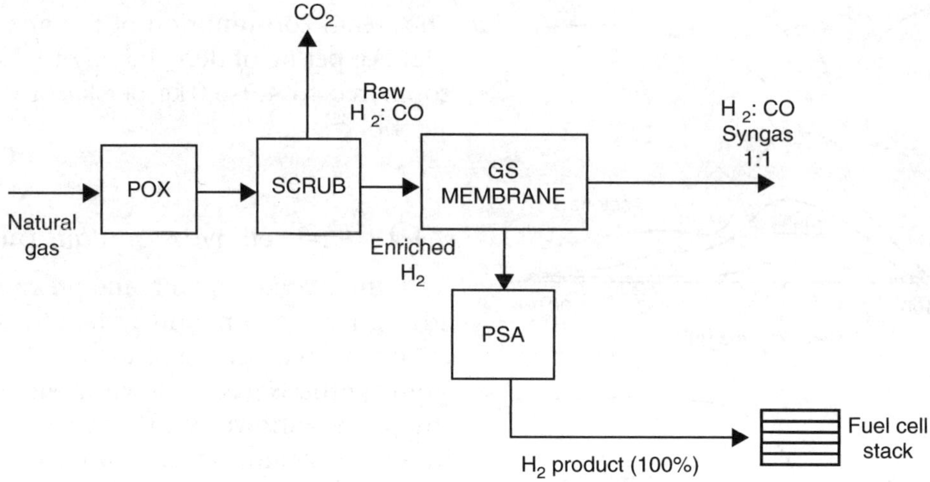

Figure 3.32 A hybrid membrane process for hydrogen purification

separation membrane module where hydrogen permeates preferentially, leaving carbon monoxide enriched stream on the high pressure (feed/reject) side. Hydrogen-rich permeate (>99% purity) flows to the PSA unit where CO is adsorbed, and a high-purity hydrogen (~100%) is produced for a fuel cell power plant.[33] The tail gas from the PSA unit contains CO and traces of hydrogen, which is combined with the membrane retentate to adjust the H_2:CO ratio to 1:1.

3.1.19 Flue gas desulphurisation by membrane gas absorption

Membrane gas absorption (MGA) is a gas–liquid (G–L) contacting device that uses a microporous hydrophobic hollow-fibre membrane element similar to the membrane contactors discussed earlier. The hydrophobic membrane barrier separates the gas phase from the absorption liquid phase. The gas to be separated diffuses through the gas-filled pores of the membrane, and is absorbed in the liquid. Absorption is based on physical absorption or by a chemical reaction. Both phases should not mix in order for the operation to be efficient.

Environmentally unacceptable compounds such as SO_2, NO_x, CO_2, H_2S, NH_3, HCl, O_3, and Hg are typically removed by conventional G–L contactors such as packed towers. MGA is an alternative to conventional G–L contactors.[34] A process flow sheet of a MGA pilot plant with a capacity of $100\,m^3/h$ is shown Figure 3.33. Biogas containing hydrogen sulphide from an anaerobic digester at a potato starch production plant is combusted in a steam boiler. The boiler flue gas contains sulphur dioxide, which is recovered (up to 95%) in the membrane absorber unit as a bisulphite, which can be reused in the production plant. Thus not only does the MGA technology mitigate the acid rain problem but it also recovers bisulphite, which is used as production ingredient.

3.1.20 Zero liquid discharge

Disposal of industrial wastewaters is being regulated by stringent environmental regulations that effectively require zero liquid discharge. Zero liquid discharge means industrial

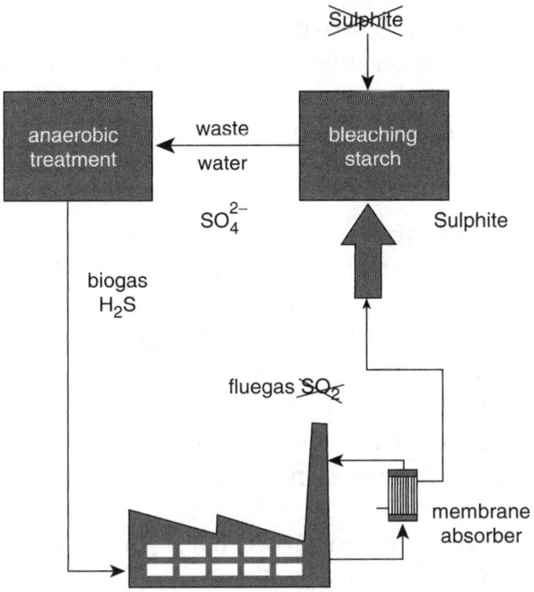

Figure 3.33 Flow schematic of flue gas desulphurisation by membrane gas absorption. Source: Klassen and Jansen

wastewater is recovered as a clean stream for re-use, and the concentrate stream can be disposed off in an environmentally benign manner, or further reduced to solid. Typical waste streams that produce large volumes of wastewater include cooling tower blowdown, gas scrubber blowdown, IX regeneration effluent and rinses, plant wash down, and rain water runoff, and process wastes.[35] Organics are first removed from wastewater by several technologies depending on the type of organics in wastewater before water flows to an RO unit. The wastewater treatment for removing organics may include anaerobic or aerobic digestion, carbon adsorption, UF, oxidation (UV, ozone, sodium hypochlorite), dissolved air flotation, oil/water separation, and/or stripping either with steam or air. This is followed by RO, evaporation, and crystallisation. The RO concentrate is treated with ED, which is capable of achieving a maximum concentrate of 80,000 ppm. Evaporators achieve a concentration of 300,000 ppm or 30%. The evaporator clean water can be blended with the RO or ED permeate for re-use as cooling tower make-up or feed to boiler feed water IX demineraliser.

3.2 Water desalination

In 2004, more than 11,000 desalination plants were in operation producing more than 20 M m^3/day (~5300 million gallons per day) with 60% of the capacity in West Asia, 11% in North America, 7% in Europe and North Africa, and 4% in Central an South America.[36] Plant sizes vary between 20–100 m^3/day and 500,000 m^3/day (~132 MGD). Brackish water desalination was the first successful application of RO with the first large-scale plant built in the late 1960s using cellulose acetate membranes. The first seawater RO (SWRO) was built in 1973 with the advent of high permeability membranes (and reduced operating pressure). By 1993, SWRO total plant capacity had reached 15 MGD (~57,000 m^3/day).

The market share for RO at the beginning of the 1980s was about 23% compared to about two-thirds for multistage flash distillation (MSF). Today, RO controls more than 85% of the seawater desalination market with MSF's share dwindled to 3%.[37,38] There are several reasons for this turnaround: (a) RO requires about 50% of energy required by thermal plants, and 25% for RO systems using energy recovery turbines, (b) low product water costs – 0.5–0.8 \$/m^3 (~2–3 \$/1,000 US gallon) for RO vs. 1.0–1.4 \$/m^3 for thermal processes, and (c) improvements in RO membranes and systems design resulting in more reliable operation with >95% on-line factor, and lower membrane costs (SWRO membrane replacement costs are now 5% of the total plant cost). Capital and operating costs for pretreatment are up to 50%

of the overall cost of a RO system.[38] Additional cost savings are possible with improved RO feed water treatment, for example, conventional pretreatment is being displaced by flocculation and UF.

Some of the recent developments in desalination processing include:

- Hybrid membrane systems combining RO and MSF for seawater desalination in dual-purpose plants for the cogeneration of water and electricity where the RO plant can provide water when the electricity needs are low. Further, by blending higher TDS RO product water (300–400 ppm) with MSF distillate (25 ppm), water production can be more economical for potable water production.[37,38]
- Integrated membrane system systems deploying MF, UF, NF, or some combination of these for RO feed water pretreatment instead of conventional pretreatment. For example, it has been shown that hollow-fibre UF membrane systems are more efficient and cost effective for RO brackish water and seawater plants.[38,39] MF is very effective in removing fines and colloidal particles, oil is effectively removed by UF, and NF is very effective in removing hardness and low molecular weight organics.
- Electrodialysis/Electrodialysis Reversal (ED/EDR) represents 3% of all the desalination capacity in the world.[38] It is, however, used mainly for desalinating brackish water. It can achieve 95% water recovery with minimal chemical feed. However, it can only remove ionised matter, and substances such as colloids and silica are not removed. A triple membrane system using UF–EDR–RO has been very effective in producing purified water for power plants.[40] In this integrated system, UF is used for removing suspended solids and macromolecules; EDR removes bulk of dissolved ions; and RO removes the remaining dissolved ions, silica, and the smallest organics.

Some of the salient differences between brackish water RO (BWRO) and SWRO desalination processes that determine plant design, and influence capital and operating costs are summarised below. Several hybrid RO desalination plant case studies are subsequently discussed to illustrate systems design and operation.

1. Feed pressure: 50–80 bar g for SWRO vs. 20–30 bar g for BWRO.
2. Feed water TDS: Seawater = 33,000–42,000 mg/l; brackish water = 1,000–10,000.
3. Feed water quality: Seawater – fairly consistent; brackish water – highly variable.
4. Feed water treatment: Simple for SWRO because of low water recovery, high ionic strength and low concentration of bicarbonate ions; complex unless for BWRO.
5. Product water recovery: 40–50% in the case of SWRO vs. 70–80% in the case of BWRO. SWRO permeate flux: 8 GFD vs. 10–12 GFD for BWRO.
6. Product water TDS: 300 ppm at 99% rejection (33,000 ppm seawater) vs. 100 ppm at 96% rejection (2,500 ppm brackish water). Salt rejection >99% for SWRO is required to meet WHO standards of <500 ppm TDS.
7. Product water specifications: TDS <400 mg/l; chloride level <100 mg/l; and WHO standard boron level of <0.5 mg/l for seawater desalination product water.[41] Other than TDS, there is no issue for BWRO.

8. Membrane array design: Usually one-pass, single-stage for SWRO because of low recovery vs. single-pass, two-stage for BWRO.[42]
9. Ancillary equipment: Energy recovery for SWRO for systems larger than 40 m^3/h product water capacity reduces energy consumption by more than 40%. Energy recovery is not economical unless the reject pressure is >20 bar g and product water recovery is <80%.
10. Product water cost: 0.5–0.8 \$/m^3 (2.00–3.00 \$/kgal) for SWRO vs. 0.2–0.35 \$/m^3 (0.75–1.30 \$/kgal) in the case of BWRO.
11. Energy consumption: 16–18 kWh/kgal for SWRO with energy recovery vs. 3–5 kWh/kgal for BWRO.
12. Critical parts material: Duplex stainless steel for SWRO plants.

3.2.1 Seawater desalination

MSF desalination had been the mainstay for producing potable water from seawater in western Asia and the arid regions of the Mediterranean since the early 1960s. During the last 20 years, however, SWRO membrane desalination has made significant inroads gradually becoming the dominant technology.[42] For example, SWRO plants with total capacity exceeding 100,000 m^3/day (>26 MGD [million gallons/day]) provide more than 60% of water needs in the Mediterranean nation of Malta (population ~0.5 million).[43] SWRO plants exceeding 200,000 m^3/day (~53 MGD) are now commonplace. In the United States, SWRO desalination is being increasingly used in California and Florida.[40,41]

Membrane desalting costs have dropped nearly 50% in the last 15 years due to higher performance membranes, improved process controls, and lower energy consumption with the deployment of energy recovery turbines. Process flow scheme of a typical hybrid SWRO plant with 35,100 mg/l feed water TDS operating at 60 bar g and 50% recovery is illustrated in Figure 3.34. The complexity of the seawater desalination process increases with the product water requirements: TDS <400 mg/l; chloride level <100 mg/l; and boron level of <0.5 mg/l TDS.[41]

A new potential technology for desalination is membrane distillation (MD), which is a thermally-driven membrane process that uses hydrophobic microporous membranes similar to the ones used in osmotic distillation discussed earlier in this chapter. MD combines the advantages of conventional thermal processes with those of membrane processes. However, fouling affects the performance quite adversely.[44] One possible solution is to use an integrated membrane system using MF/UF pretreatment followed by RO and MD. Since, RO seawater desalination product water dissolved solids content is typically 300–400 mg/l (as compared to about 10 mg/l with thermal distillation), post-treatment of RO product water by MD is an attractive option. A product water recovery of 87.6% was achieved.[45]

Case Study I. The process flow diagram of a SWRO plant is shown in Figure 3.35. The plant design and operating conditions are as follows:

- Feed water flow rate: 94 m^3/h at 69 bar g
- Permeate flow rate: 42 m^3/h
- Product water recovery: 45%
- Feed water TDS: 38,000 mg/l
- Product water TDS: 325 mg/l at >99% salt rejection
- Energy consumption: 5.8–6.0 kWh/m^3

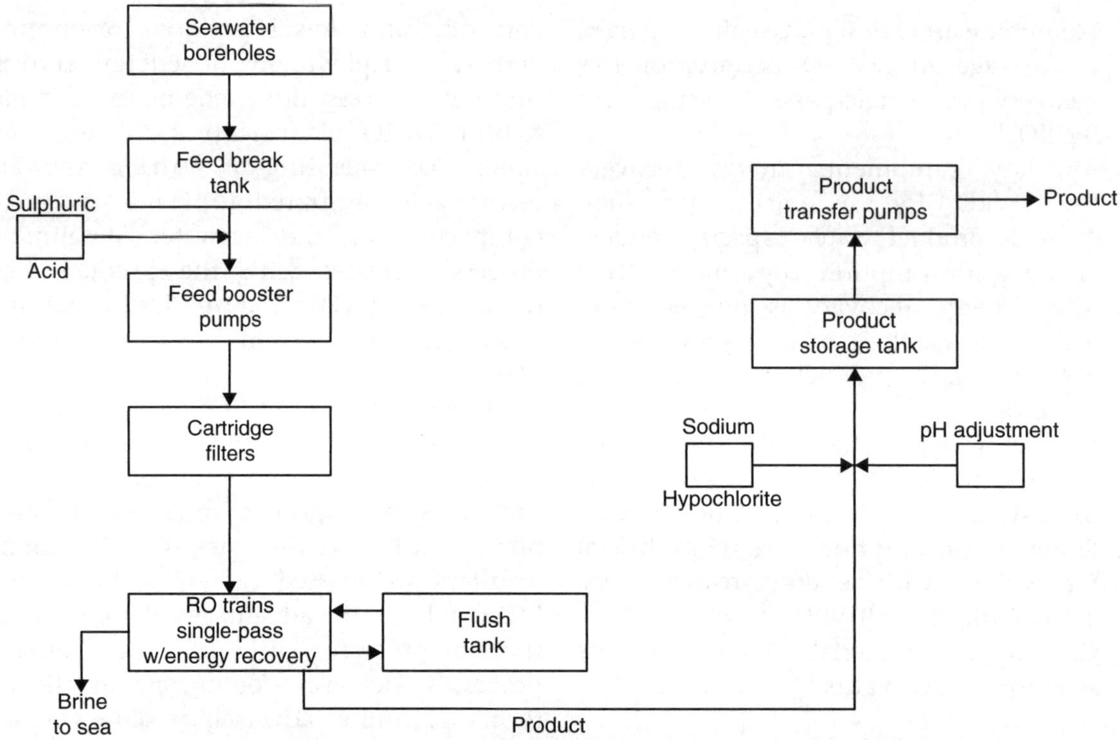

Figure 3.34 A seawater RO plant process flow schematic. The membrane plant uses boreholes for raw seawater intake. Sulphuric acid and 5.0 μm cartridge filter are used for pretreatment. Post-treatment includes decarbonation, lime addition for pH adjustment and sodium hypochlorite injection for chlorination. The plant includes an automatic system for flushing the membranes. Source: Rahman

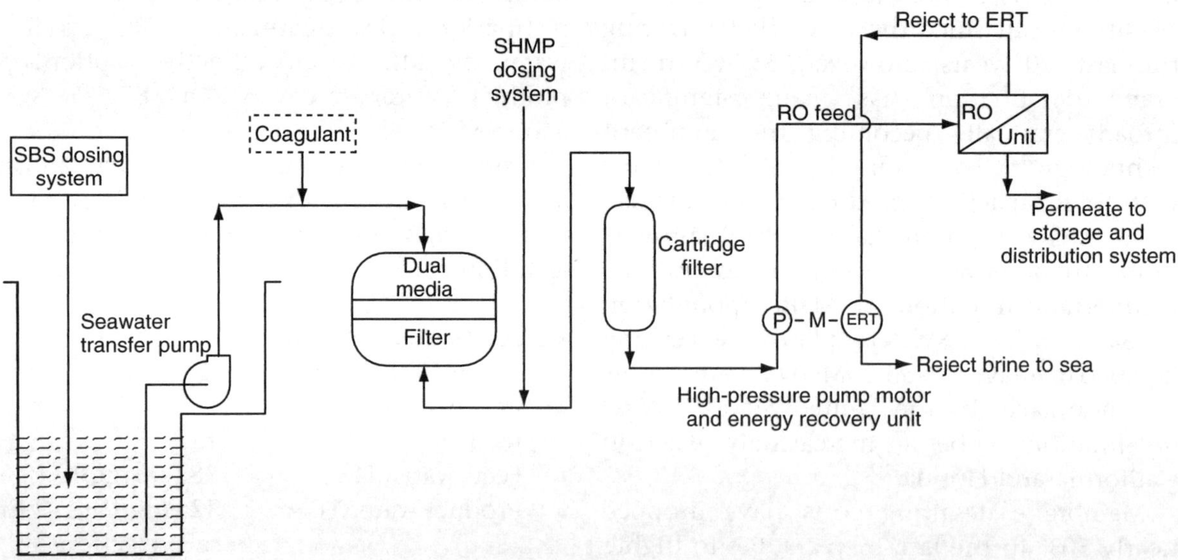

Figure 3.35 Engineering process flow diagram of a 42 m^3/h seawater RO membrane plant integrated with a hydraulic energy recovery turbine. Source: Desalination 82 (1991), 31

The feed water source is a seawater infiltration well located 100 m from the shoreline. The well is dosed with sodium bisulphite (SBS) for disinfection (30–40 ppm). Seawater is transferred from the well to three media filters operating in parallel (one filter on standby) by two centrifugal pumps. Anti-scalant sodium hexametaphosphate (SHMP) is injected upstream of the 5.0 μm cartridge microfilter (MF) in the media filters effluent to prevent precipitation of magnesium sulphate. The SHMP dosage is based on a 5–10 ppm level in the RO feed water. Coagulant dosage upstream of the sand filters is 1–2 ppm, if required. The silt density index (SDI), an indicator of colloidal fouling, was consistently in the range of 0.5–1.0, indicating the feed seawater is very clean. This is because the seawater feed intake is a well.

The RO feed water is transferred by the high-pressure pumps (one pump is in operation, and the other is on standby) to the RO membrane unit. The high-pressure, multistage, submersible centrifugal pump is connected to an energy recovery turbine (ERT). The pump-motor-turbine-set enables the recovery of energy from the high-pressure brine reject stream. The high-pressure brine drives the turbine (reverse-running pump) mounted on a common shaft with the electric motor and the RO high-pressure pump. The high-pressure feed seawater is piped to the RO vessel rack. The membrane unit is a two-stage array (12:6); stage one consists of 12 pressure vessels in parallel, and stage two contains six pressure vessels also in parallel. The rejected brine is piped to the ERT from which it discharges to the sea.

Each pressure vessel contains six polyamide TFC seawater membrane (Film-Tec 30) spiral-wound elements, 8-in. diameter × 40 in. long, producing 1000 m^3/day (~0.25 MGD). The feed seawater TDS is about 38,000 mg/l. Since 45% is recovered as product water with a TDS of 325 mg/l, the TDS of the rejected brine is 69,000 mg/l, and the osmotic pressure of the rejected brine stream is nearly 54 bar. This means the effective driving force at the end of the RO unit is only 16 bar, since the feed pressure is 70 bar g.

Case Study II. The process and instrumentation diagram (P&ID) of a SWRO system is shown in Figure 3.36. The one-pass RO unit is a single-stage array (6:0) consisting of six parallel pressure vessels with each pressure vessel containing six spiral-wound TFC membrane elements connected in series.[42] The permeate flows to the product water storage tank. High-pressure reject flows to the hydraulic ERT before it flows back to the sea. The ERT transfers the recovered energy to the high pressure RO pump. The system design conditions are as follows:

- Feed water flow rate: 60 m^3/h at 75 bar g
- Permeate flow rate: 21 m^3/h
- Product water recovery: 35%
- Reject flow and pressure: 39 m^3/h at 72 bar g
- Product water TDS: 300 mg/l at >99% salt rejection.

The feed water pretreatment for this SWRO plant in the Mediterranean is minimal; multimedia filtration for removing particulate matter larger than 20.0 μm, acid injection for reducing pH to less than 6.0 to prevent calcium carbonate scaling, sodium metabisulphite injection for dechlorination, and 5.0 μm cartridge filtration for protecting the RO modules from fouling. Post-treatment includes a product water drawback tank to protect the membranes after shutdown, and addition of lime solution to the product water for raising the pH to 7.5–8.0 to prevent corrosion of downstream piping and equipment since the pH of the permeate is typically in the range of 5–6.

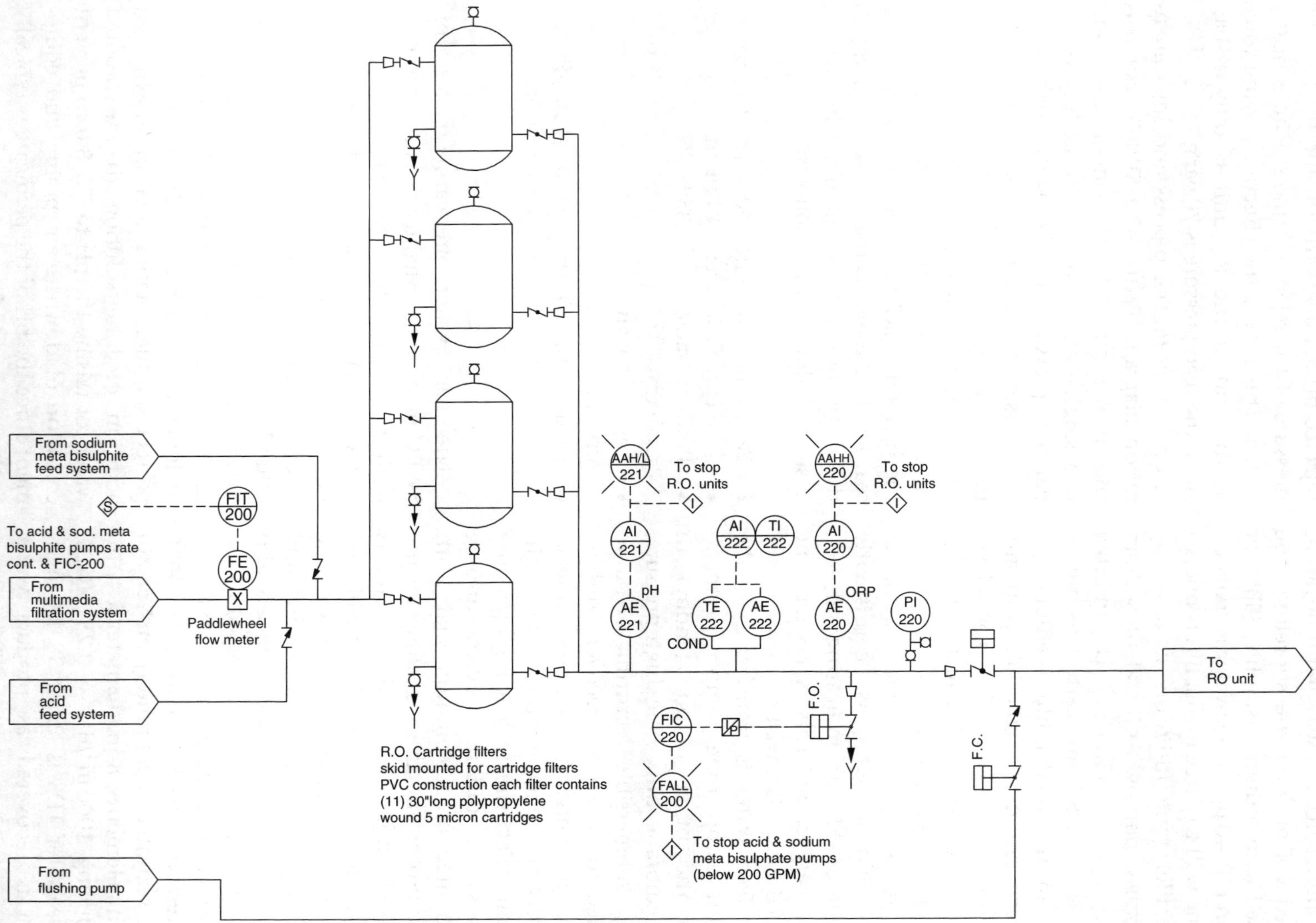

Figure 3.36a A P&ID of a dead-end cartridge microfiltration pretreatment system for a seawater RO plant. Source: Rahman

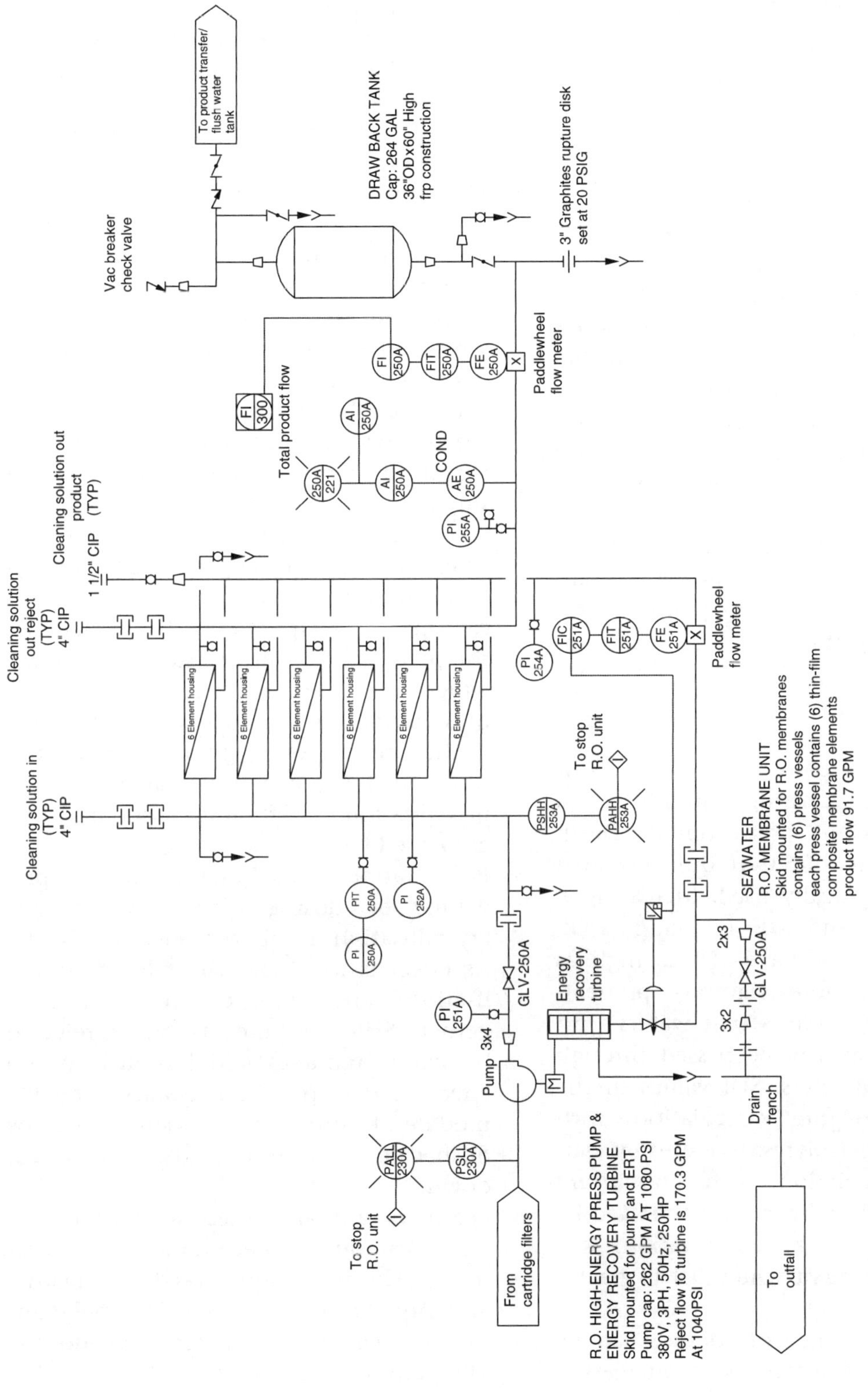

Figure 3.36b A P&ID of a 94 m^3/h seawater RO one-pass single-stage membrane system. Source: Rahman

3.2.2 Brackish water desalination

Nearly all brackish water desalination plants are either RO or ED/EDR. More than 1,200 desalination plants operate in the United States, which has 16% of world's desalination capacity.[38] With the membrane technology for brackish water plants fully established, feed water pretreatment and brine disposal are the major current challenges. In Saudi Arabia and Florida, for example, brine disposal is a major problem when desalinating inland aquifers. Several options are available for brine disposal including deep well injection, pond evaporation, near zero liquid discharge, solar energy ponds, shallow aquifer storage, or return to a saline water body by pipeline.[38]

Feed water pretreatment was discussed in Chapter 2. Since most brackish water desalination plants require extensive pretreatment, the main features that affect pretreatment design are summarised below.[46,47] Case studies are presented later.

1. Groundwaters usually have very low SDI values, and cartridge filtration is a generally sufficient pretreatment. Dead-end MF elements of 5.0–10.0 μm nominal pore size are typically used. Increasingly, cross-flow MF and UF is being used. Surface waters may have SDI as high as 200 requiring more rigorous pretreatment. A rough guide is that waters with SDI less than six can be passed through deep-bed media filters. SDI values in the 6–50 range require coagulation with cationic polyelectrolytes, followed by clarification. Coagulation with aluminium sulphate (alum) or ferric salts, followed by sedimentation and media filtration, is a suitable pretreatment when SDI is greater than 50.
2. The fouling tendency of colloids can be reduced in some instances by increasing their zeta potential, thus reducing their tendency to adhere to surfaces as discussed in Chapter 2. Methods used to achieve this include softening the feed water by IX, NF, and/or dosing the feed with a polymeric dispersant.
3. The most common precipitates encountered in brackish water desalination are calcium and magnesium carbonate, sulphates of calcium, barium, and strontium, and silica. Well water may have special problems associated with components that are soluble under reducing conditions but form insoluble substances when oxidised by contact with air. Dissolved ferrous salts and hydrogen sulphide fit in this category.
4. Since brackish waters usually contain high levels of calcium, sulphuric acid dosing is commonly used to prevent calcium carbonate precipitation. This increases the potential for calcium sulphate scaling, which may become the limiting factor in determining product water recovery. If the calcium or magnesium concentrations are high, cation exchange or lime-precipitation softening should be employed before acid dosing.
5. Precipitation of sulphates can be prevented by dosing with polyelectrolyte crystallisation inhibitors such as SHMP. Its effect is to hinder the rate of crystallisation, rather that to prevent it altogether. SHMP is almost totally rejected by RO membranes and has little or no effect on the permeate quality. The RO module feed/reject channels must be flushed before shutting the RO system down.
6. Polymers are also used to inhibit calcium carbonate precipitation, for example, Flocon®, Aqua Feed®, Belgard®, and Aquakreen®. Use of these polymers can reduce the need for acid dosing. The advantages over SHMP are that

they are more stable against thermal and hydrolytic degradation, making dilution procedures less critical. These advantages are particularly useful in small installations, but the use of such polymers has become established in large plants as well.

7. Silica precipitation is a serious problem. The formation of insoluble silica compounds is not easy to predict accurately, as the solubility is strongly dependent on temperature and pH as discussed in Chapter 5. Silica compounds, once formed are difficult to remove – particularly from CA membranes, which are not tolerant to high pH cleaning solutions. Silica, soluble and colloidal, can be removed by the lime-precipitation process, but because of the relative cost and complexity of the process, this is generally avoided if possible.
8. The concentration of ferrous ions that can be tolerated by most membranes depends on the concentration of dissolved oxygen. If dissolved oxygen is less than 0.1 mg/l, the brine may contain up to 4 mg/l ferrous ions. If the brine is oxygenated, the iron concentration should be below 0.05 mg/l. Contact with air should be avoided or the water should be aerated followed by filtration or clarification.
9. Well water may contain dissolved hydrogen sulphide. At $pH \geq 8$, a significant portion of the hydrogen sulphide is present as sulphide ions, which may be oxidised to elemental sulphur by either oxygen or chlorine. This can be prevented by acidifying the water. Under acid conditions, the hydrogen sulphide passes through the membrane into the product water, which then requires degasification. The release of hydrogen sulphide gas can cause corrosion and environmental problems.

Case studies. Feed water composition of several brackish waters in the south-west United States are given in Table 3.4. This region of the US is drought prone, and with the rapid population growth, the situation is likely to get worse.[38,48] Membrane pretreatment options for RO and ED designed plants operating at 70% and 80% water recovery are given in Table 3.5.[49] RO and ED plant process flow sheets based on various pretreatment systems for these feed waters are shown in Figures 3.37–3.40. Sample RO performance process flow sheets based on the membrane manufacturer's projections at 70% product water recovery for Welton-Mohawk and Rio Grande sites are shown in Figure 3.41, illustrative of RO unit design calculations.

The Welton-Mohawk RO unit is a two-stage array (29:15) comprised of 44 pressure vessels. Each pressure vessel contains six polyamide TFC water spiral-wound elements, 20 cm diameter × 100 cm long, producing 3820 m^3/day (700 gpm) permeate. The membrane (Hydranautics CPA2) average flux is 17.9 l/m^2/h (GFD). The initial operating pressure is 17 bar g. The reject brine flow rate is 1636 m^3/day (~300 gpm). The product water TDS is 90 mg/l, and the pH is 4.9 units. The feed water pH is lowered to 6.0 with sulphuric acid to prevent calcium carbonate scaling. An anti-scalant (5–10 ppm) is also required to prevent calcium sulphate scaling.

The Rio Grande RO unit is a two-stage array (26:14) comprised of 40 pressure vessels. Each pressure vessel contains six polyamide TFC water spiral-wound elements, 20 cm diameter × 100 cm long, producing 3820 m^3/day (~1 MGD) permeate. The membrane (Hydranautics ESPA4) average flux is 17.9 l/m^2/h. The initial operating pressure is 14 bar g. The reject brine flow rate is 1636 m^3/day (~300 gpm). The product water TDS is 127 mg/l, and the pH is 5.5 units.

Table 3.4 Composition of representative brackish waters in the US and seawater

Item/Ion*	WM[1]	CL[2]	TUL[3]	FM[4]	RG[5]	OC[6]	SW[7]
Cations							
Calcium	204	129	420	299	163	140	400
Magnesium	91.7	89	163	36	51	10	1,252
Potassium	8	-	2.3	-	-	35	10,561
Sodium	804	521	114	216	292	300	
Strontium	3.1	-	-	-	-	-	140
Anions							0
Bicarbonate	433	161	270	334	275	275	-
Boron	-	2.8	0.14	-	0.23	0.8	2,650
Chloride	1,090	262	170	87	492	350	18,980
Fluoride	-	-	-	-	0.08	0.8	1.5
Iron	0.35	1.2	-	0.02	0.03	0.3	1.4
Manganese	0.48	-	0	-	0.03	0.1	4
Nitrate	-	-	10	25	1.5	1.0	
Phosphate	0.13	-	-	-	-	1.0	-
Sulphate	902	1,260	1,370	960	336	350	-
Other						-	-
Silica (SiO_2)	32	49	22	-	35	10	-
Ph	7.95	7.7	7.2	7.4	8.1	8.0	-
TDS	3,628	2,478	2,410	1,880	1,611	1,400	33,990
Alkalinity**	355	133	221	274	226	226	
Hardness	850	692	1,750	898	616	391	-

* Ion as mg/l; ** Ion as $CaCO_3$
[1] Welton-Mohawk, Arizona; [2] Coalinga, California; [3] Tularosa, New Mexico; [4] Fort Morgan, Colorado; [5] Rio Grande, Texas; [6] Orange County, California; [7] Seawater
Source: Cabibbo et al.

3.3 High purity water production

High purity waters include ultra pure water (UPW) and pure water. The definition of UPW differs from pure water or deionised water. Generally speaking, pure water refers to water that meets the specific needs of a given process or product, for example, in pharmaceutical applications removal of organisms or pyrogenic substances is of utmost importance, whereas in textile washing, removal of hardness, iron and manganese is essential. UPW refers to water that is free of 'all' impurities. Power plants are the single largest user of high purity water. Water treatment for the power generation industry for both utility and non-utility producers is estimated at more than 30% of all industrial water treatment sales, or about $6 billion. Other industrial users of high purity water include microelectronics, pharmaceuticals, beverages, and research laboratories. The microelectronics high purity water requirements are the most stringent of all the water production requirements in the general

Table 3.5 Pretreatment options for brackish water desalting plants*

Feedwater/ Product Water Recovery	Reverse osmosis (RO)	Electrdodialysis (ED)
Tularosa, NM		
70%	Ion exchange, Acid, SHMP	Ion exchange
80%	Lime, soda ash, SHMP	Lime, Soda ash, Acid
Rio Grande, TX		
70%	Acid	Acid
80%	Acid, SHMP	Acid
Fort Morgan, CO		
70%	Ion exchange, acid, SHMP	Minimum
80%	Ion exchange, acid, SHMP	Ion exchange, acid
Coalinga, CA		
70%	Acid	Minimum
80%	Acid, SHMP	Minimum
Welton-Mohawk, AZ		
70%	Lime, acid	Lime, acid
80%	Lime, acid	Lime, acid
Orange County, CA		
70%	Acid	Acid
80%	Acid	Acid

*Based on feed water chemical analysis given in Table 3.4
Source: Cabibbo et al.

industry. High purity water (resistivity 1–10 MΩ-cm) is also used in metals finishing industry to rinse the metal surface clean of any dirt or chemicals prior to the plating operation.

The water source of most high purity water plants is city water or municipal water, which has undergone pretreatment. The city raw water source may by groundwater or surface water. Municipal treated wastewater TDS is typically 200–400 mg/l. Feed water pretreatment for high purity water plants, therefore, is minimal unless brackish water is used as raw feed water.

3.3.1 Boiler water treatment

High purity water is required as boiler feed water for steam generation during electricity production using steam turbines to protect the boiler tubes and turbines from corrosion, scale formation and structural defects. In conventional power plants – oil, coal, natural gas, nuclear – steam is generated at a pressure of 40–200 bar g and fed to a turbine, which converts the heat energy to mechanical energy that in turn is converted to electrical energy. The power generation efficiency increases with steam temperature and pressure. These conditions, in turn, require water of very high purity to protect the equipment from corrosion and scaling. Steam generation in chemical process plants such as oil refineries, pulp and paper production, and food processing, on the other hand, does not require UPW since steam is used for heating where precipitation of silica and corrosion of turbine blades is not an issue. High purity water usage in non-conventional power plants such as fuel cell power plants is markedly different as it is a reactant for synthesis gas.[50] Hence, resistivity of 1.0 MΩ-cm is adequate instead of 10.0 MΩ-cm or higher.

In the last fifteen years, RO has replaced demineralisation by two-bed (cation–anion) IX (DBIX) prior to mixed-bed IX polishing (MBIX). Higher operating costs for IX systems combined with lower operating costs and increased confidence in operating membrane systems helped RO systems to be adopted. RO–IX treatment is more cost effective than the DBIX–MBIX especially when the feed water TDS is greater than 350 mg/l.[51] Hybrid systems based on feed water and product water purity may include RO, other membrane methods, such as MF, UF, and/or NF, MBIX, EDI, ozone, and UV. A typical process flow schematic diagram for producing high purity water for steam generators is shown in Figure 3.42. Condensate treatment

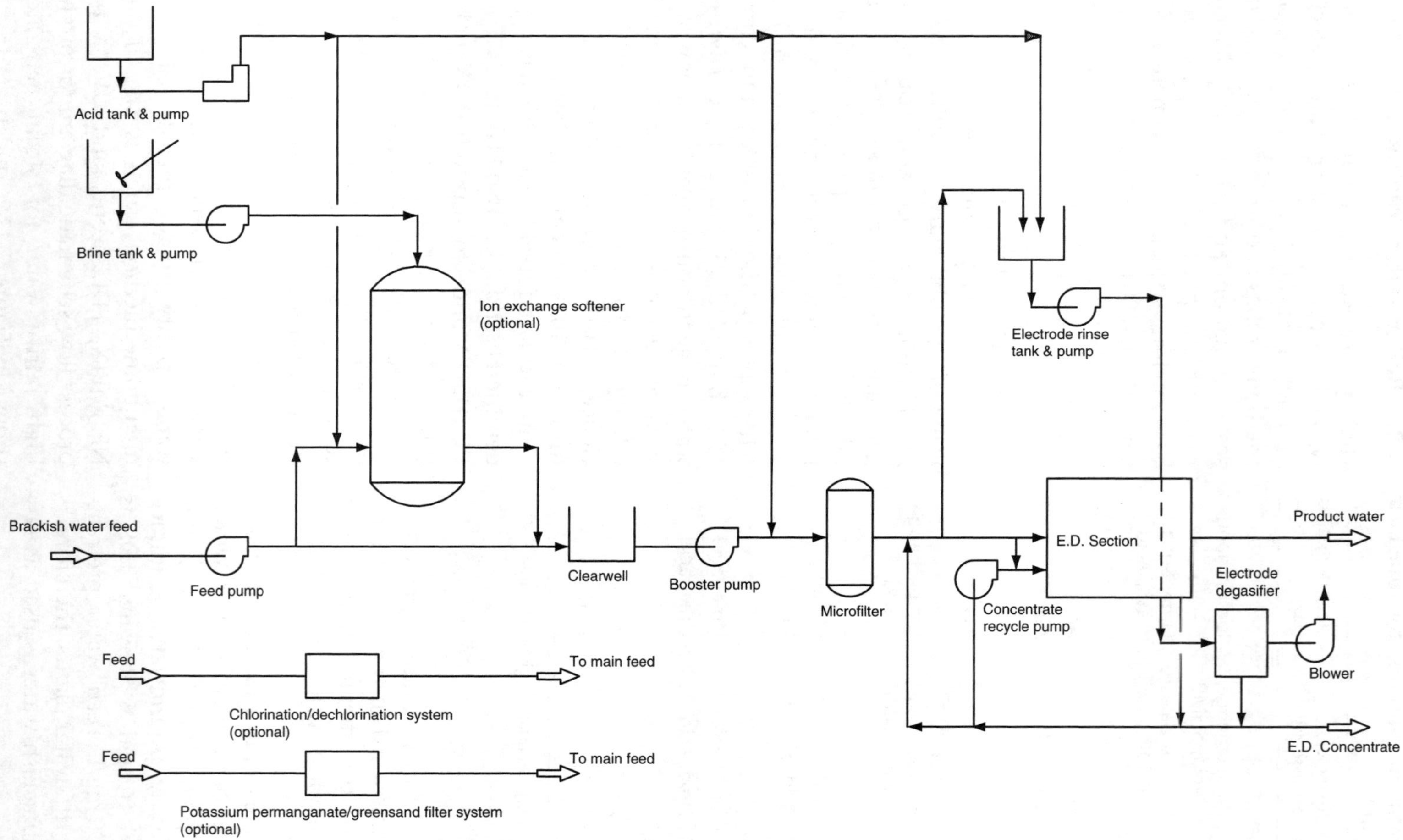

Figure 3.37 Acid pretreatment system plus an ED membrane brackish water desalination plant process flow sheet. Source: Cabibbo et al.

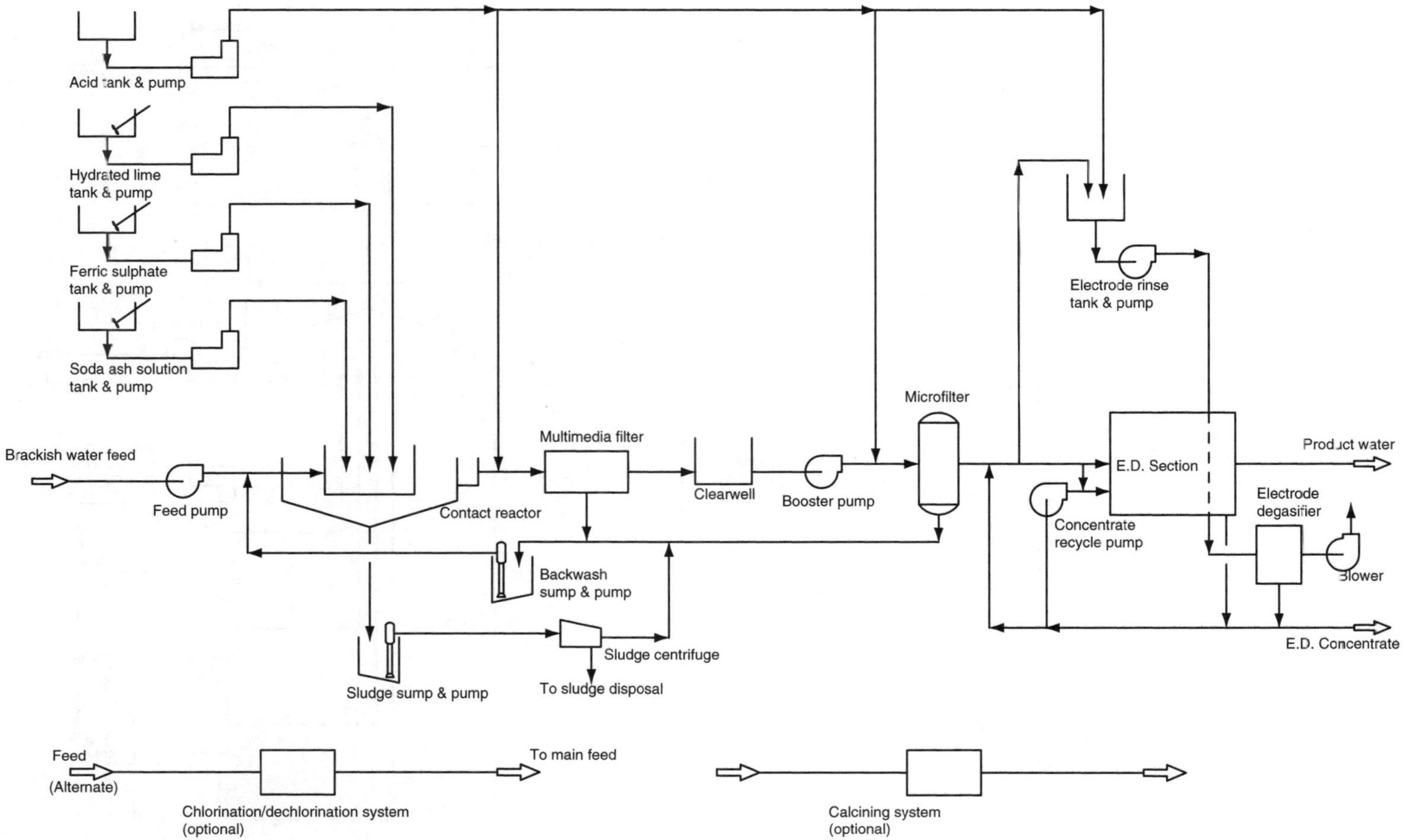

Figure 3.38 Lime pretreatment system plus an ED membrane brackish water desalination plant process flow sheet. Source: Cabibbo et al.

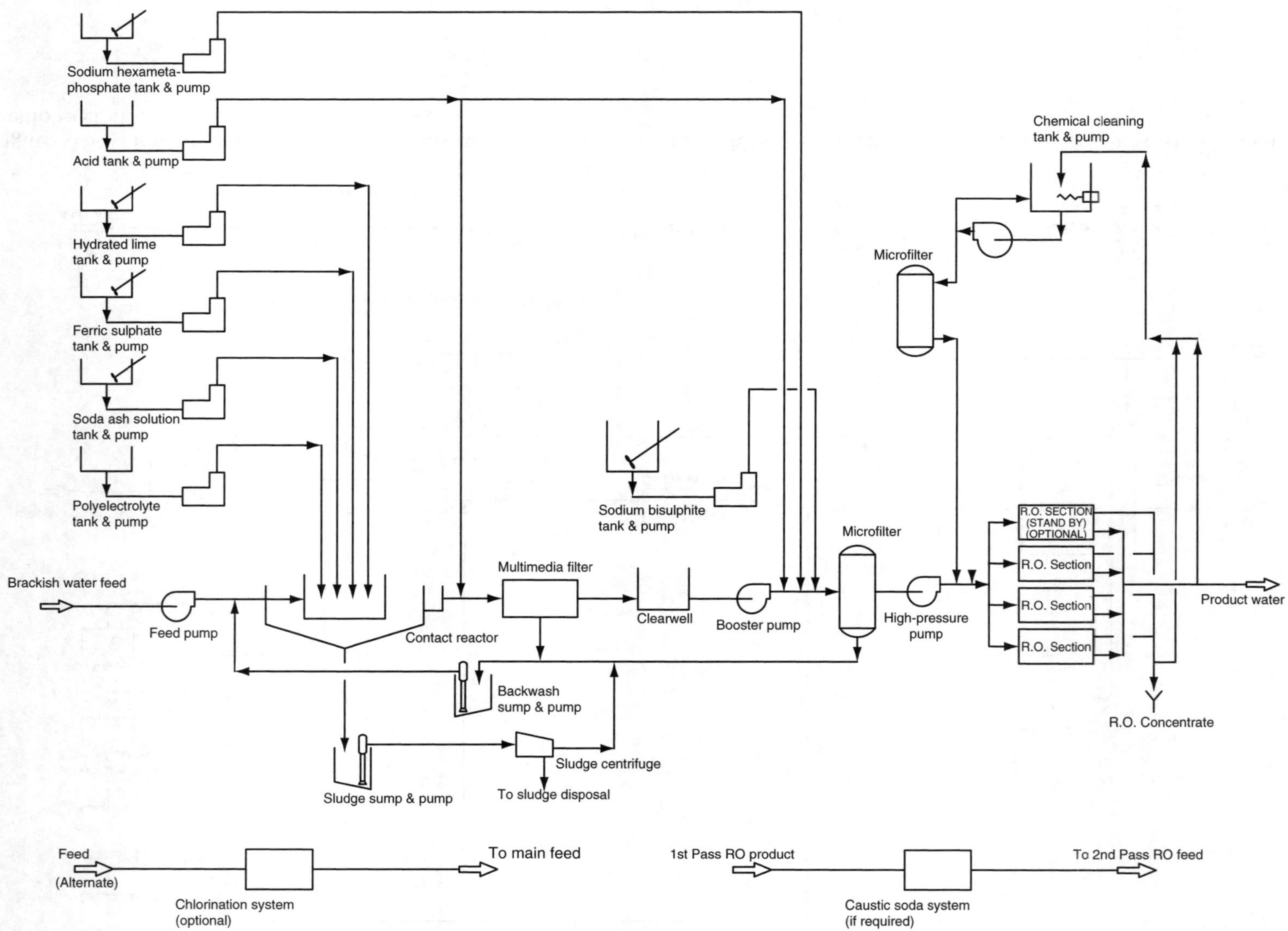

Figure 3.39 Lime pretreatment system plus a RO membrane brackish water desalination plant process flow sheet. Source: Cabibbo et al.

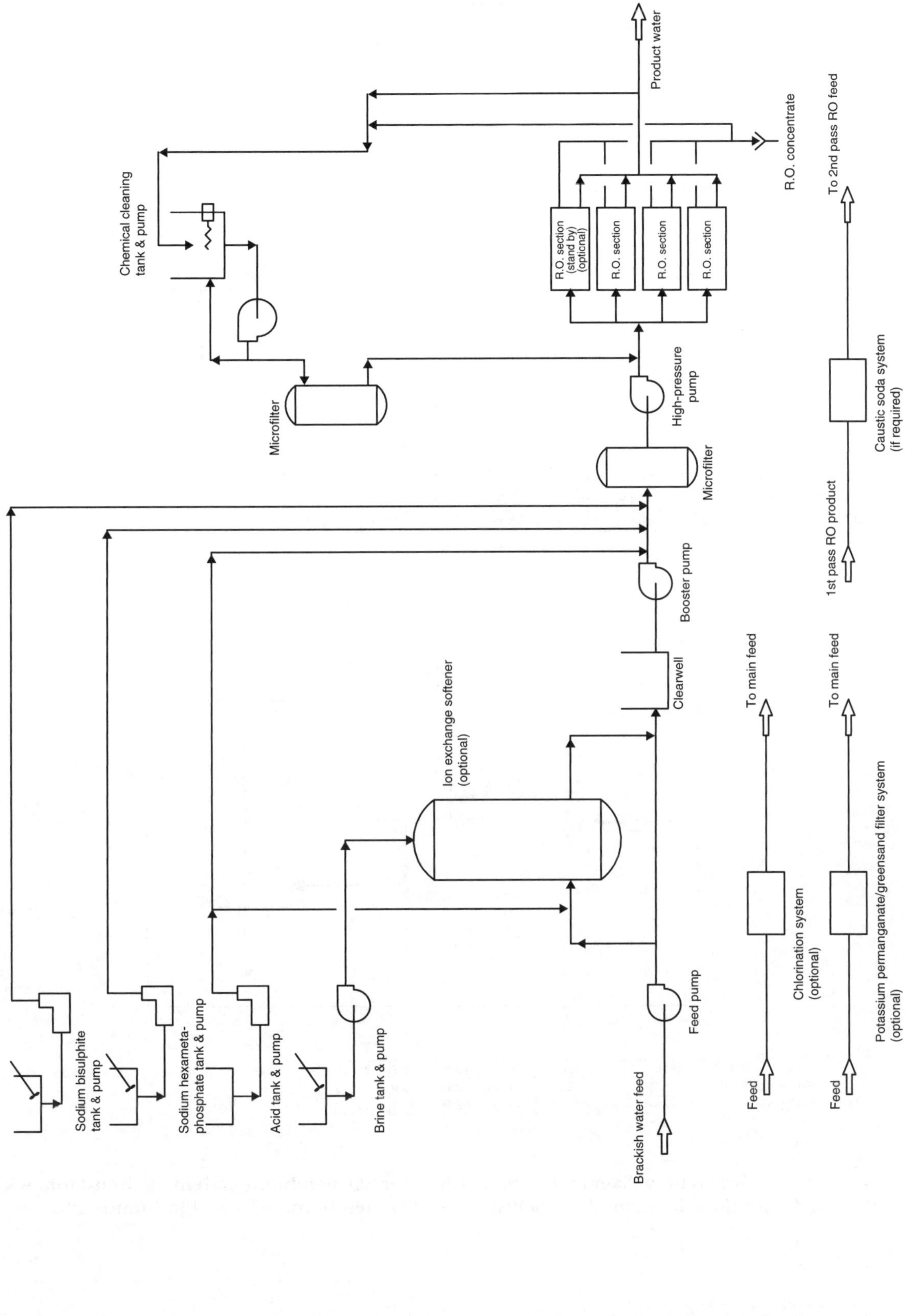

Figure 3.40 Acid pretreatment system plus a RO membrane brackish water desalination plant process flow sheet. Source: Cabibbo et al.

Welton Mohawk

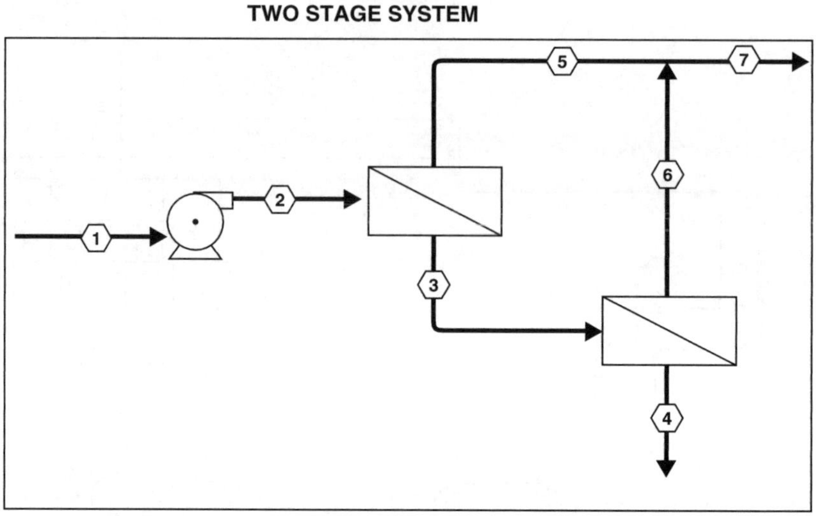

		1	2	3	4	5	6	7
Flow	(GPM)	1000.0	1000.0	428.3	300.0	571.7	128.3	700.0
Pressure	(psi)	0.0	154.2	140.4	123.8	0.0	0.0	0.0
TDS (ppm)		3728.4	3694.8	8553.8	12105.7	55.2	245.9	90.1

(a)

Rio Grande

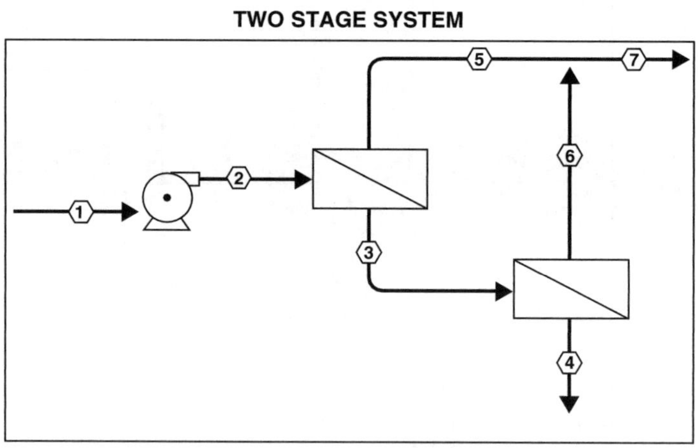

		1	2	3	4	5	6	7
Flow	(GPM)	1000.0	1000.0	386.7	300.0	613.3	86.7	700.0
Pressure	(psi)	0.0	73.3	58.2	41.6	0.0	0.0	0.0
TDS (ppm)		1656.5	1622.4	4097.3	5111.9	62.0	586.1	126.9

(b)

Figure 3.41(a) Process design flow diagram of a brackish water RO membrane system. Welton-Mohawk site. (b) Process design flow diagram of a brackish water RO membrane system. Rio Grande site

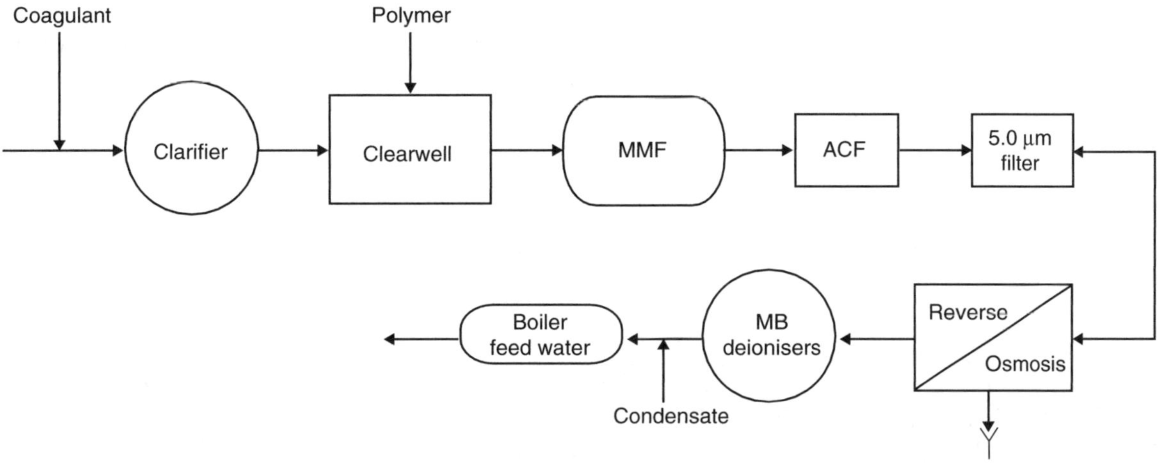

Figure 3.42 A typical boiler water treatment hybrid membrane system process flow schematic

or condensate polishing is often the major portion of the water treatment plant; the spent steam from the turbine is usually condensed. Most high-pressure boilers (70 bar g) deploy a condensate polishing IX system to minimise problems associated with corrosion products such as iron and copper.

Case study. It takes about two cubic metres of high purity water to produce one megawatt-hour of electricity. With increasing focus on water conservation, and tighter pollution control regulations, power plants have to rely on reclaimed water in many cases. Reclaimed water generally means treated effluent from municipal wastewater treatment plants. Process flow schematic of one such water treatment plant producing high purity water from reclaimed water at a 550 MW combined cycle power plant in New Jersey (USA) is shown in Figure 3.43.[52]

Waste treatment effluent is filtered and sent to the cooling towers. A small portion of the pre-filtered water flows to multimedia filters at 110 m^3/h. Sodium bisulphite is added to filtered water and pumped through the RO system where more than 98% of dissolved solids are removed, operating at 70% product water recovery. The RO permeate flows through a bank of membrane degasifiers, which remove dissolved carbon dioxide to reduce the ionic load on anion resins of MB deionisers downstream. The second RO unit is used to further reduce the TDS of water before it is polished by MB ion exchange. The hybrid membrane system supplies 70 m^3/h high purity water to the steam generator with resistivity greater than 10.0 MΩ-cm, and silica less than 10 ppb. Reusing reclaimed water reduces the power plant's dependence on potable water, and saves the local community nearly 2200 m^3/day (~0.6 MGD) of city water.

3.3.2 Microelectronics rinse water

UPW is used in the microelectronics industry for applications such as rinsing and washing wafer surfaces between chemical etching steps. A large semiconductor fabrication facility typically uses 160 m^3/h (1 MGD) high purity water. The amount of UPW required to produce one 150-mm wafer is about 5 m^3. A method of manufacturing a VLSI (very large integrated circuit) is shown in Figure 3.44.[32]

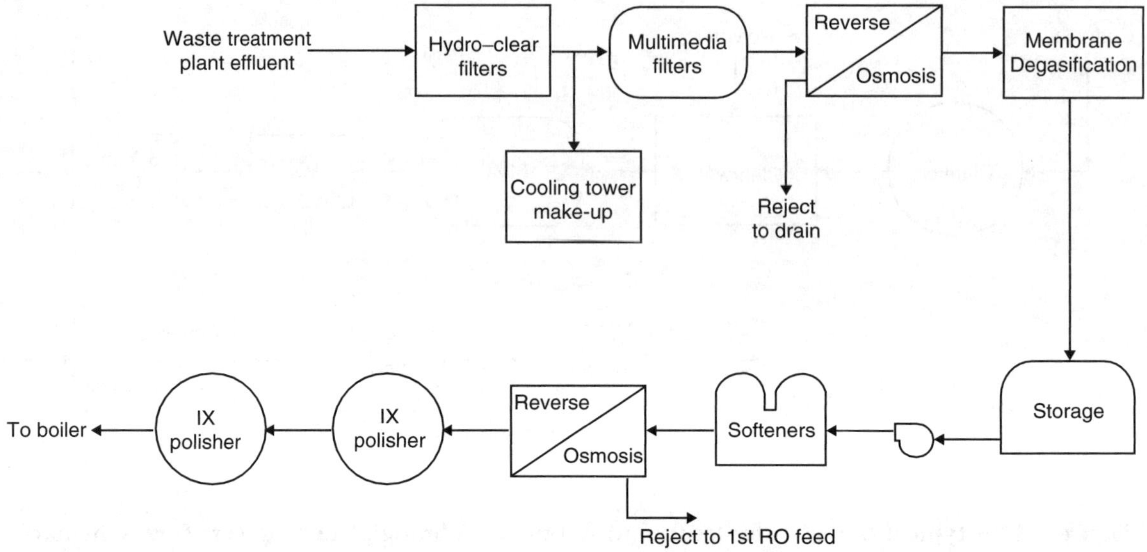

Figure 3.43 Process flow schematic for power plant boiler water production from municipal wastewater effluent plant. Source: USFilter

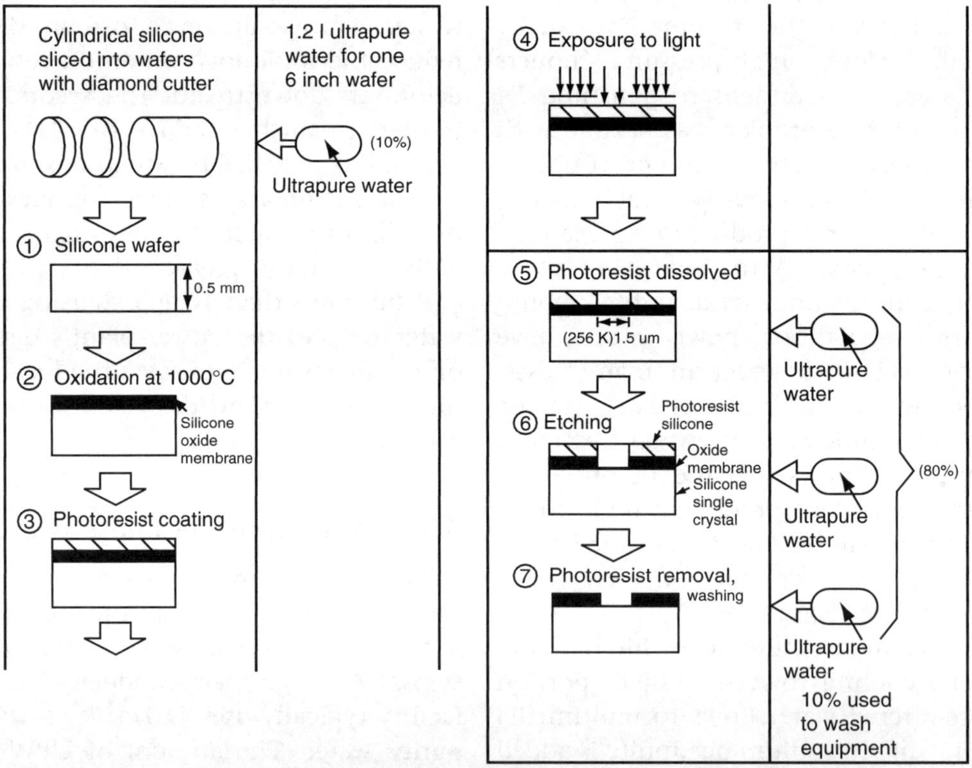

Figure 3.44 Ultrapure water applications during the manufacture of a VLSI wafer. Source: Matsuura, Copyright® 1994 from Synthetic Membranes and Membrane Separation Processes by T. Matsuura. Reproduced by permission of Routledge/Taylor & Francis Group, LLC

There are four steps where UPW is used as shown. Why is water of extremely high quality required in the washing process of the semi-conductor industry? A simple explanation is that without very high purity water, short-circuiting is a real risk resulting in lower yields of the wafers. In 1984, the level of water purity was based on microcircuits with 2.0 μm line spacings. By 1995, the spacing had decreased to 0.3 μm, and currently, they are less than 0.13 μm. The Semiconductor Industry Association projection for future sizes is 0.07 μm by the year 2010.[53] Current UPW requirements for 300-mm wafer producing fabs are as follows:[54] resistivity ≥ 18.2 MΩ-cm TOC < 0.5 ppb, silica < 0.1 ppb, boron 10–50 ppt and particle count (0.05 μm) < 200/l.

Case study. The hybrid membrane plant described in this case study provides 15 m^3/h UPW to a fab manufacturing facility in the US. The high purity water plant consists of a pretreatment section, a two-pass RO unit, and a post-treatment section similar to the one shown in Figure 3.45. The UPW requirements are given in Table 3.6. Raw (city) feed water is surface water with TDS = 492 mg/l and the following major constituents: bicarbonate = 183 mg/l; chloride = 61 mg/l; sulphate = 83 mg/l; silica = 14 mg/l; potassium = 3 mg/l; sodium = 147 mg/l.

City water is pumped to the carbon filter at 13 m^3/h and 5 bar g. Activated carbon media removes organic compounds and free chlorine from water. The dechlorinated filtered water flows to the IX softener at 13 m^3/h. The strong acid cation IX resins softener reduces the total hardness to less than 2 mg/l by removing calcium, magnesium, barium and strontium ions thereby preventing RO membrane

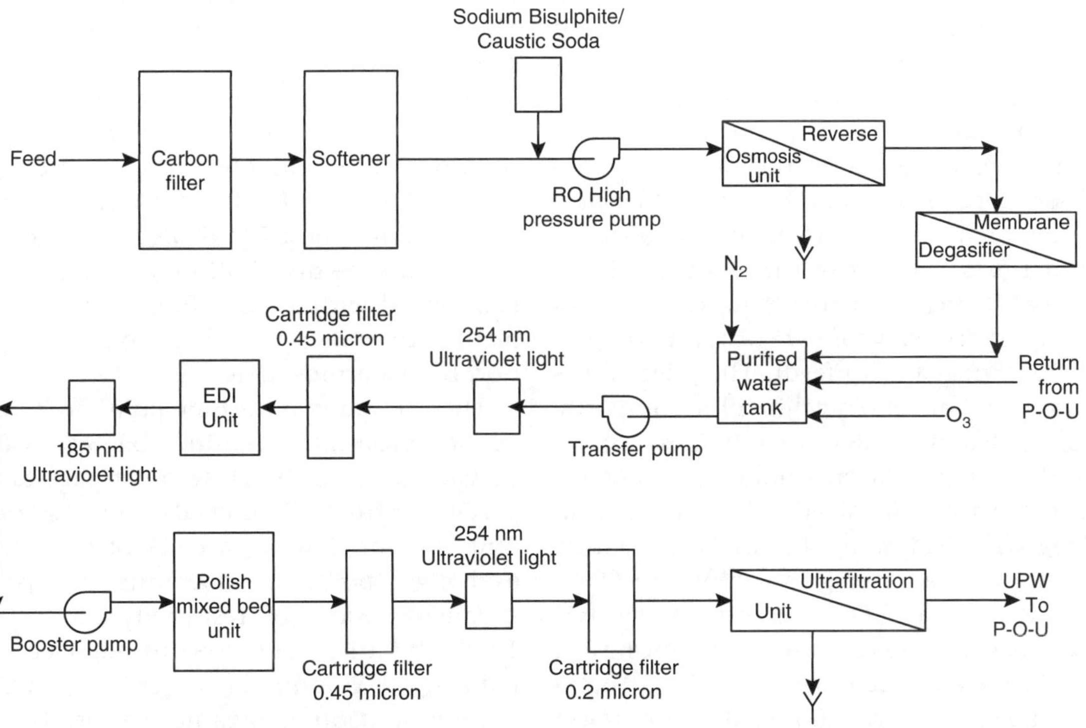

Figure 3.45 Process flow schematic of a typical ultrapure water system used in semiconductor manufacturing. (5.0 μm cartridge filter upstream of RO pump not shown).

Table 3.6 UPW plant water specifications

Parameter	Value
Resistivity	≥18.2 MΩ-cm
Particles	
>0.2 μm	<10 count/ l
>0.5 μm	<1 count/ l
Bacteria	<1 cfu/ 100 ml
Total organic carbon (TOC)	<2 μ /l
Dissolved oxygen	≤2 mg/l
Calcium	<2 μg/l
Cations (each)	<0.2 μg/l
Chloride	<0.1 μg/l
Anions (each)	<0.2 μg/l
Silica (dissolved)	<3 μg/l

scaling. Softening by NF instead is planned for future expansion. The softener IX resins are regenerated with brine (10–12% NaCl solution).

Softened water flows to the RO pre-filter at 13 m^3/h. The RO pre-filter contains 5.0 μm (nominal pore size) polypropylene dead-end cartridges for removing resin fines, particles and complexed colloids thereby protecting the RO membranes from particulate fouling. The pre-treated water is transferred by the high-pressure RO pump to the first-pass RO membrane unit at 16 m^3/h and 35 bar g. The first-pass RO unit is a three-stage (2:1:1) array producing 12 m^3/h permeate at a product water recovery of ~75% (12 m^3/h product/16 m^3/h make-up feed). The reject flows to drain at 4 m^3/h. The thin film composite (TFC) polyamide RO membranes (spiral wound, 20 cm × 100 cm nominal element) reduce the total dissolved solids (TDS) from 492 mg/l in feed water to less than 3 mg/l in permeate at an average rejection of ~99%.

The second-pass RO unit is a two-stage (1:1) array that produces 9 m^3/h permeate at a product water recovery of ~75% (9 m^3/h product/12 m^3/h make-up feed). The reject flows back to the RO pump inlet at 9 m^3/h. The second-pass TFC membranes reduce the conductivity from 6.0 μS/cm in feed water to less than 0.2 μS/cm in permeate corresponding to a rejection of ~97%.

The double-pass RO system produces 9 m^3/h permeate at an overall product water recovery of ~69% (9 m^3/h product/ 13 m^3/h make-up feed). Caustic soda solution is injected into the second-pass feed water line (first-pass permeate) to raise the pH from 5.4 to 7.5–8.3. At alkaline pH, free CO_2 gas in water is converted to bicarbonate ions, which are easily rejected by the membranes.

RO product water flows through a 185 nm rated ultraviolet (UV) unit at 9 m^3/h and is degasified in the membrane contactors (MC). The UV light reduces the total organic carbon (TOC) in water. The three MC units operate in series, and remove dissolved gases and TOC from water using a vacuum pump, with nitrogen flowing in the permeate side as sweep gas. Degassed water flows to the MB deionisers operating in series at 9 m^3/h. The DI product water has a resisitivity of 16.0 MΩ-cm.

The DI water flows through a 1.0-μm (absolute pore size) post-filter. The polysulphone dead-end MF cartridges trap resin fines before the DI flows to the DI water storage tank. The DI tank also receives return water from the distribution loop. The DI tank is provided with a HEPA filter and a nitrogen blanket to protect the water from contamination by the atmosphere.

The DI water is further polished to meet the specifications given in Table 3.6. Water is transferred from the DI water storage tank to a TOC destruct UV unit at 17 m^3/h. The UV unit effluent flows to a bank of five MB ion exchange polishers operating in parallel producing water of resisitivity 18.2 MΩ-cm. The high purity water flows through a bacteria reducing UV unit in order to deactivate bacteria and other organic matter. The irradiated water flows through four UF elements operating in parallel. The UF membranes are rated for a MWCO of 10,000 Da. The UPW is monitored continuously for silica, dissolved oxygen, particle count and TOC in addition to resistivity before it flows to the point-of-use (POU) at 15 m^3/h. A portion of the POU supply water recycles to the DI water tank to ensure there is no stagnant water in the distribution loop.

3.3.3 USP grade water

The United States Pharmacopoeia (USP) has three general specifications for water quality that are applicable to medical and pharmaceutical uses, namely 'USP Water for Injection' (WFI), 'USP Purified Water', and 'Drinking Water'.[55] WFI is the most demanding and expensive to produce, and is generally used when necessary (e.g. for the final purification steps of parenteral products) or is cost-effective. USP water specifications for pharmaceutical manufacturing are: resistivity = 1.0–5.0 MΩ-cm; TOC level <500 ppb, and bacteria count <10 cfu/100 ml. Water quality requirements recommended for various pharmaceutical applications and operations are given in Table 3.7.[56] WFI water is not recommended for laboratories since WFI water can contain metals leached from stainless steel piping that will contribute to conductivity even though it is generally within the USP requirements.[56]

Traditionally, the pharmaceutical plants used evaporation stills to obtain water free of organisms and pyrogens. Currently when large amounts of USP purified water and small amounts of WFI is required, RO/EDI system is used for the production of the USP purified water, and a small distiller for the production of the WFI, as shown in Figure 3.46.[57] The European Agency for the Evaluation of Medicinal Products (EMEA) has designated a new grade of water called 'Highly Purified Water' to be used 'for the preparation of products where water of high biological quality is needed except where WFI is required'.[56] This type of water is generally produced using two-pass RO followed by deionisation and UF.

Table 3.7 USP water quality by application

Application	Water quality
Media preparation	USP purified
Seed and production bioreactor batching	USP purified
Cell or mycelia separation	USP purified
Cell or mycelia washing	USP purified
Buffer preparation	USP purified
Diafiltration	USP purified
Column chromatography (pre-columns)	USP purified
Column chromatography (final column)	WFI
Viral clearance	USP purified
Final filtration	WFI
Lyophilisation	WFI

Case Study. The case study describes a high purity water plant similar to the one shown in Figure 3.46. The water treatment plant supplies 9 m^3/h of USP purified water to a drug manufacturing facility in the US. Purified water is used in the preparation of topical and oral medications. The hybrid membrane plant consists of a pretreatment section, a two-pass RO unit with inter-pass pressure booster pump, and a post-treatment section similar to the one shown in the above figure. Raw (city) feed water has a pH = 7.6 and TDS = 140 mg/l with the following major constituents: bicarbonate = 19 mg/l; chloride = 19 mg/l; sulphate = 47 mg/l; silica = 7 mg/l; calcium = 9 mg/l; magnesium = 4 mg/l; sodium = 30 mg/l).

City water flows to multimedia filters operating in parallel at 18 m^3/h at 3–4 bar g. Multimedia filtration improves clarity and reduces turbidity by removing particles, from coarse sediment down to 10.0 μm.

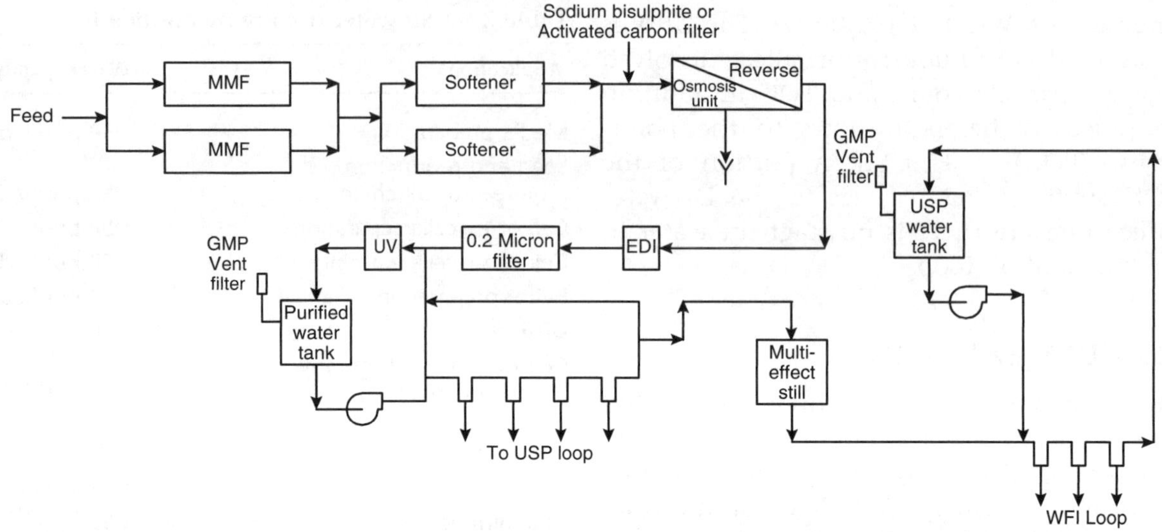

Figure 3.46 Process flow schematic of a typical high purity water system for producing USP water and WFI in pharmaceutical manufacturing. Source: Gsell

Low turbidity water is required to protect the equipment downstream from particulate fouling. The filtered water flows to the IX softeners at $18\,m^3/h$. Softened water is required to prevent RO membrane scaling. The application of anti-scalants to prevent scaling is not recommended for use in pharmaceutical water plants. The duplex water softeners operate in series with the primary unit removing most of the hardness ions and the polishing unit removing remaining hardness ions. The softeners produce water with hardness less than 1 mg/l. The softener IX resins are regenerated with brine (10–12% NaCl solution).

The softened water flows to a holding tank at $18\,m^3/h$. The tank also receives up to $9\,m^3/h$ EDI system product water when the system is in re-circualtion mode as discussed below. The holding tank provides storage for pretreated water and for recycled water thereby preventing various upstream and downstream equipment from cycling on/off frequently. Thus the tank allows the RO/EDI system to operate in a continuous mode during periods when the water is not required by the product water storage tank. The holding tank is provided with vent filters – 0.2 μm hydrophobic – that prevent water from getting contaminated by particles from the atmosphere.

Water is transferred from the holding tank to the carbon filter by either booster pump (one pump is on standby) at $15\,m^3/h$ and 3 bar g. Each pump motor is provided with a variable frequency drive to maintain constant pressure. Activated carbon media bed removes organic compounds and free chlorine from water. Since, the polyamide RO membranes are damaged by oxidants such as chlorine, dechlorination of RO feed water is mandatory. The carbon filter effluent flows through a steam heat exchanger before it flows to the RO skid. The heat exchanger maintains the RO feed water temperature at 25°C during the normal service run. It is also used to heat water to 80°C during the hot water sanitisation cycle. A rule of thumb is that the RO product water (permeate) flow rate increases 3% per °C rise in water temperature due to lower viscosity. A constant water temperature is required to compare RO flux data; otherwise normalisation of flux data to a constant temperature is required.

RO feed water flows through two 5.0 µm pore size (nominal) cartridge filters operating in parallel at 15 m^3/h. The cartridge microfilters remove carbon fines and particles thereby protecting the RO membranes from particulate fouling. The RO pre-filters effluent flows through a UV light. The 254 nm UV lamps destroy organic matter and render biological organisms inactive, thus protecting the RO membranes from biological fouling, which is often irreversible.

The pretreated water is pumped by the first-pass RO high-pressure pump to the RO membrane array at 20 m^3/h and 8 bar g (TDH). The RO system is designed to run continuously and supply water to the CDI unit. The first-pass RO unit is a three-stage (2:1:1) array designed to produce 11 m^3/h of purified water at an overall recovery of 75% (11 m^3/h product/15 m^3/h make-up feed). The reject flows to drain at 3 m^3/h, and the remaining reject recycles to the RO pump inlet at 5 m^3/h. The first-pass RO permeate flows to the second-pass RO high-pressure pump at 11 m^3/h. The second-pass RO pump transfers water to the second-pass RO membrane array at 11 m^3/h (11 m^3/h first-pass permeate + 8 m^3/h second-pass reject recycle) and 8 bar g. The second-pass RO unit is a two-stage (2:1) array designed to produce 10 m^3/h of purified water at an overall recovery of ~90% (10 m^3/h product/11 m^3/h feed). The reject recycles to the second-pass RO pump inlet at 8 m^3/h. The TFC membranes (spiral wound, 20 cm × 100 cm nominal element) reduce the TDS content from 144 mg/l in softened feed water to 3 mg/l in final RO permeate at an average rejection of ~98%.

RO system product water flows to the EDI skid at 10 m^3/h via an automatic valve. The EDI skid houses two TOC reduction 185 nm rated UV lights operating in parallel, and four EDI modules (24 cell pairs/module) operating in parallel. The EDI system performance specifications are: product water recovery = 95% (9 m^3/h product/10 m^3/h feed); rejection >99% with the product water resistivity equal to 5.0 MΩ-cm (conductivity = 0.2 µS/cm). The product water flows to the purified water storage tank at 9 m^3/h for use.

The RO and EDI systems are sanitised with hot water as required after the membranes have been cleaned. The sanitisation cycle consists of three basic steps: heat-up, hold, and cool-down as discussed in Chapter 2. Water is typically heated to 80°C using the steam heat exchanger, and re-circulated through the RO/EDI system loop during the heat-up step. Once the water temperature reaches 80°C, it is re-circulated in the RO/EDI loop typically for one hour during the hold step. After one hour, the steam shuts off, and the cool down phase starts when chilled water is re-circulated through the RO/EDI loop and sent to drain. The system continues to flush water to drain until the water temperature has dropped to below 30°C.

3.3.4 Concluding remarks

Current high purity water production methods such as those discussed in the case studies above are based on certain well-established rules such as RO product recovery of 70–80%, high fluid velocities in the UPW loop piping, hot water sanitisation of USP water systems, hot water sanitisation of carbon filters, and use of stainless steel piping in USP water systems. Modifications to these design guidelines have been suggested by some,[56] and are visited below. The merits of these observations may be worth debating.

- Operate the RO units systems at 50% recovery instead of 70–80%. This reduces the frequency of membrane cleanings substantially, increases reliability, and results in consistent product water quality. In addition, the reject water is better suited for reuse applications such as for cooling tower make-up. It can also be used as feed for vapour compression stills and clean steam generators provided the silica content is less than 15 mg/l.

- There is no reason to operate the RO/EDI system in a USP water plant continuously. This wastes water and there is no benefit in microbiological control. Similarly, hot water sanitisable RO and EDI systems are no better at controlling bacteria.
- Frequent backwashing of carbon media filters is more effective than hot water or steam sanitisation since the microbial load reduction by sanitisation is temporary due to the rapid recolonisation of the sanitised biofilm in the carbon bed.
- Minimum fluid velocities in the high purity water recirculating loops are 2–3 m/s. Apparently, there is no single velocity based on which a loop system can be designed or operated to control the formation of biofilms.
- WFI operation is just as effective at 60°C as it is at 80°C. If this is true, perhaps hot water sanitisation of the RO/EDI systems should also be done at 60°C instead of 80°C.
- As in the case of semiconductor plants, PVDF piping should be used in the high purity loops in USP plants. It is 30–35% cheaper than 316L SS, and reduces steps such as cleaning and passivation of SS components.

3.4 Water reclamation and recycling

Reclaimed water is an important and reliable resource for industrial and agricultural applications. In semi-arid areas, for example southern California, reclaimed water is an indispensable water source for the petrochemical industry. In rapidly industrialising countries such as India, water reclamation and recycling seems to be the only viable solution to the perpetual and worsening water crisis in urban areas. Water reclamation refers to the recovery of reusable high-grade water from municipal or industrial effluents by the removal of pathogens, dissolved organics and salts for applications, in for example, power plants and semiconductor fabs (discussed earlier). The beneficial uses of reclaimed water besides agricultural, urban, and industrial applications include restoration of groundwater quality following industrial contamination and augmentation of traditional water sources such as groundwater or surface water.[58]

In reuse applications, the feed water is usually biologically treated municipal wastewater. Generally speaking it is required that municipal wastewater is either secondary or tertiary treated water before it can be reclaimed for health reasons. Composition of commonly used wastewater as feed water and reclaimed water are given in Table 3.8. Several examples of major water reclamation projects are given in Table 3.9.

The treatment processes used for producing high grade industrial and potable reuse applications consist of a pretreatment stage followed by advanced wastewater treatment (AWT) processes such as granular activated carbon, RO, IX ED, or a combination of these processes. The most common type of AWT is the addition of tertiary media filter. Conventional pretreatment for AWT is being increasingly displaced with cross-flow MF or UF for providing superior quality feed water for RO systems as shown in Figure 3.47.[59] MF is used for clarification and disinfection by removing suspended solids, colloidal particles, cysts, and bacteria.

Table 3.8 Typical feed water and product water quality targets for high grade industrial and potable water

Parameter	Feed water[a]	Product water[b]
Total suspended solids (TSS)	2–20 mg/l	< 1 mg/l
Total coliforms	10^6–10^6 cfu/100 ml	<1 cfu/100 ml
Total organic carbon (TOC)[c]	8–12 mg/l	0.1–1 mg/l
Total nitrogen[d]	1–40 mg/l	1–10 mg/l
Total dissolved solids (TDS)[e]	350–1200 mg/l	50–200 mg/l

[a] Typical range of values for secondary effluent (post activated sludge settling tank) used in municipal water reclamation projects
[b] Indicative requirements for high grade and potable reuse projects. Actual requirements determined by local conditions
[c] Final TOC requirements and nature of TOC are very site specific and determined by needs of industry or regulatory agencies
[d] Feed water nitrogen values vary depending on the level of nutrient removal. Final nitrogen concentration depends on industrial water requirements or environmental protection regulations
[e] Final TDS levels depend on ultimate use of reclaimed water. Often minerals are added to the water to stabilise and mitigate against the aggressive nature of the reclaimed water (typical of RO permeate)
Source: Fane et al.

As compared to multimedia filtration that removes particles down to 10 μm, MF removes particles down to 0.1 μm particle size. Further, the MF permeate typically has an SDI of less than three and turbidity consistently less than 0.1 NTU thereby significantly reducing the potential for fouling of RO or NF membranes. MF membranes commonly used in water reclamation applications are backwashable hollow-fibre membranes – PVDF and polypropylene (PP) – with a pore size of 0.1–0.2 μm.

3.4.1 Industrial water usage

In the industrial sector, water is often recycled for generating high purity water for power plants, oil refineries, and semiconductor manufacturing plants. According to one survey, membrane filtration for recycling water has been beneficial in all cases with overall savings ranging between \$102,000 for a 6 m^3/h water treatment plant and \$640,000 for a 50 m^3/h plant.[60] For example, at a leading semiconductor manufacturing plant in UK, financial savings are nearly 20% since 83% of the treatment cost is associated with bringing new raw water for the DI water treatment plant. Spent water from the manufacturing plant is passed through an activated carbon filter for removing hydrogen peroxide, which is present in some of the acid baths. The main difference between the reclaimed water and the raw water quality is the very low cation and bacterial content in the reclaimed water. Next, weak anion resin is used to remove free ions such as sulphate, phosphate, and fluoride. The IX effluent flows through 10.0 μm filters and 254 nm UV sterilisers before entering the RO system. The RO product water mixes with the pretreated potable water, and the combined water has a conductivity of 3.0 μS/cm.[61] Several hybrid membrane water recycling plants in operation at various industrial sites are discussed.

Case study I. Secondary treatment effluent from a sewage plant is processed by an integrated membrane system to supply reclaimed water for high purity water production at a

Table 3.9 Examples of water reclamation projects

Application	Location	Purpose/Benefit
I. Irrigation – Agriculture	Santa Rosa, California	Reclaimed water is used to irrigate food and non-food crops.
II. Irrigation – Urban	Irvine Ranch, California; Rouse Hill, NSW, Australia	Uses include landscape irrigation and toilet flushing.
III. Industrial	West Basin, Los Angeles	The West Basin recycling project was built to serve the 'big 3' oil refineries in Los Angeles with reclaimed water in lieu of groundwater. Reclaimed water is used in cooling towers and as make-up for high pressure boilers.
	Luggage Point, Brisbane	The Luggage Point reclamation project supplies potable water at a clean fuel refinery. The cost to supply potable water was significantly greater than the treatment and conveyance costs for the reclamation project.
	Wafer Fabrication Plants, Singapore	UPW produced from reclaimed water with consistent levels of TOC was substituted for surface water which had wide variations in the level of organic carbon that caused problems for the UPW system.
IV. Groundwater management	WF 210, Orange County, California	Reclaimed water is injected into a potable aquifer to prevent ingress of seawater into the groundwater basin. The use of reclaimed water will also be used to recharge the groundwater basin.
V. Augmenting surface waters	Republic of Singapore	Use reclaimed water to augment the island's drinking water reservoirs thereby reducing the dependence on importation of fresh water.

Source: Fane et al.

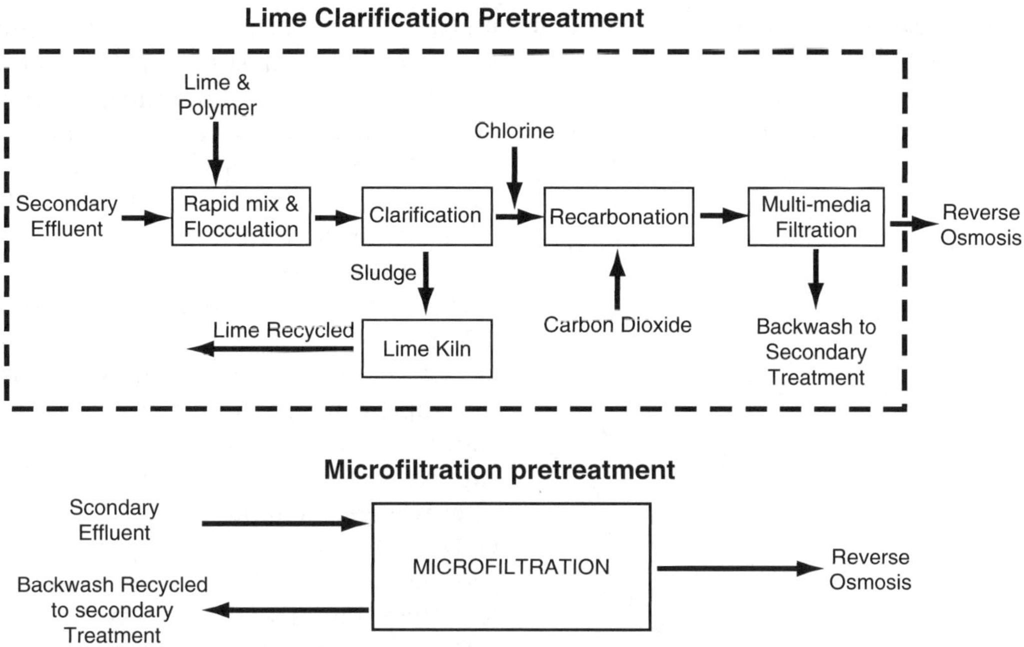

Figure 3.47 Simplified RO pretreatment with MF instead of lime clarification. Source: Mierzejewski and Guendert

wafer manufacturing facility in Singapore.[62] The effluent from an activated sludge treatment facility is dosed with sodium hypochlorite, which combines with ammonia present in the feed to form chloroamines to control membrane fouling. The pretreated water flows to a cross-flow MF unit that clarifies the feed water by removing suspended solids, colloidal particles (partially) as well as bacteria and protozoa, and lowers the turbidity to less than 0.1 NTU. The MF membranes are hollow fibres with nominal pore size of 0.2 μm. The MF permeate flows to the RO unit, which removes salts, very fine colloids, and remaining bacteria and viruses. The RO permeate is irradiated by a 254 nm UV light for disinfection.

The quality of reclaimed water from secondary treated sewage is comparable to the potable water supply as shown in Table 3.10. Although the type of RO membranes are not revealed, most likely they are low-fouling TFC membranes (spiral wound, 20 cm × 100 cm nominal element) with neutral or hydrophilic surface. These membranes are less prone to fouling due to their lower affinity to adsorption of dissolved organics and bacterial growth. According to the case study, the reclaimed water has proven to be a reliable source of feed water for the UPW plant.

Case Study II. Secondary treatment sewage effluent at a paper and fibreboard manufacturing plant is reclaimed by a hybrid membrane plant providing boiler feed water to a 2,640 megawatt (MW) power plant in New South Wales, Australia.[59,60] The membrane filtration system is of the dual-membrane type – MF and RO. The MF membranes are hollow-fibre PP with a nominal pore size of 0.2 μm. The RO membranes are spiral-wound cellulose acetate (CA) polymer based. Pretreated wastewater free of oil and grit is pumped to the MF system at 212 m^3/h (1.3 MGD) as shown in Figure 3.48. The MF permeate is dosed with sodium hypochlorite to

Table 3.10 Typical analyses of membrane treated wastewater and potable water

Parameter	Membrane treated water	Potable water
pH	7.0–7.5	8.0–8.75
Conductivity (μS/cm)	100–180	120–160
TOC (ppm)	0.05–0.15	1.0–1.5
Ions (mg/l)		
Calcium	0.01–0.02	10–15
Magnesium	<0.03	<0.1
Sodium	15–20*	2–3
Boron	0.02–0.08	0.02–0.05
Chloride	8–10	8–10
Sulphate	<0.045	15–20
Phosphate	<0.08	<0.08
Silica	0.6–0.7	3–4

* After caustic dosing
Source: Vishwanath et al.

control biological growth. Sulfuric acid (4%) is added to reduce the pH and provide optimal operating conditions for the CA membranes (pH = 5–6 operating range). Next, an anti-scalant is added upstream of the 5.0 μm cartridge MF (dead-end type) before water flows to the high-pressure RO pump. The RO permeate flows to a degasifier to remove the dissolved carbon dioxide and raise the pH to 7. Degasified water flows to a MB deioniser producing high purity water for power plant steam generators. The RO reject is dosed with ferrous chloride before it flows to a wastewater processing plant prior to disposal.

The MF unit consists of two parallel trains containing 90 modules per train with a total membrane area of 2,700 m^2. The MF system operating conditions are: 90% product water recovery, 212 m^3/h feed water supply, and 4.5 bar g feed pressure. The average membrane flux is 72 $l/m^2.h$ (43 GFD). Membrane fouling is controlled by backflushing the MF hollow fibres every 15–60 min. The MF membranes are cleaned with caustic soda (1%) every 200 h. Membrane integrity is monitored with the aid of an automatic pressure decay test every 24 h. The MF permeate SDI was routinely less than two surpassing the normal level of three required for RO feed water.

The RO unit consists of two parallel trains. Each train is a two-stage (10:6) array. The RO unit is designed to supply 160 m^3/h (1 MGD) of permeate. The single-pass RO unit operating conditions are: 80% product water recovery, 160 m^3/h product water supply, and 15–35 bar g operating pressure. The average membrane flux is 22 $l/m^2 \cdot h$ (13 GFD). The salt rejection was 98% with the permeate TDS equal to 50 mg/l, BOD less than 1 mg/l and turbidity less than 0.1 NTU. The main problems were manganese fouling of MF membranes and organic fouling of RO membranes. Manganese fouling was addressed by using citric acid, and organic fouling was corrected by adjusting the chlorine level in feed water. Annual savings in operating costs were reported to be in the range of $800,000.[59]

Case Study III. One of the water reclamation projects listed in Table 3.9 is located in the West Basin Municipal Water District (WBMWD) in southern California. The facility was built in 1995 to address acute water shortages in southern California. The El Segundo water recycling plant reclaims 4,700 m^3/h (30 MGD) of municipal wastewater from Los Angeles Hyperion wastewater treatment plant producing several different water qualities of water as shown in Figure 3.49.[59] The plant has gone through a series of modifications and expansions since it was completed in 1995, for example prior to the first phase expansion in 1996, the secondary effluent for groundwater restoration was treated in a 790 m^3/h (5 MGD) conventional pretreatment system similar to the one shown in Figure 3.49.

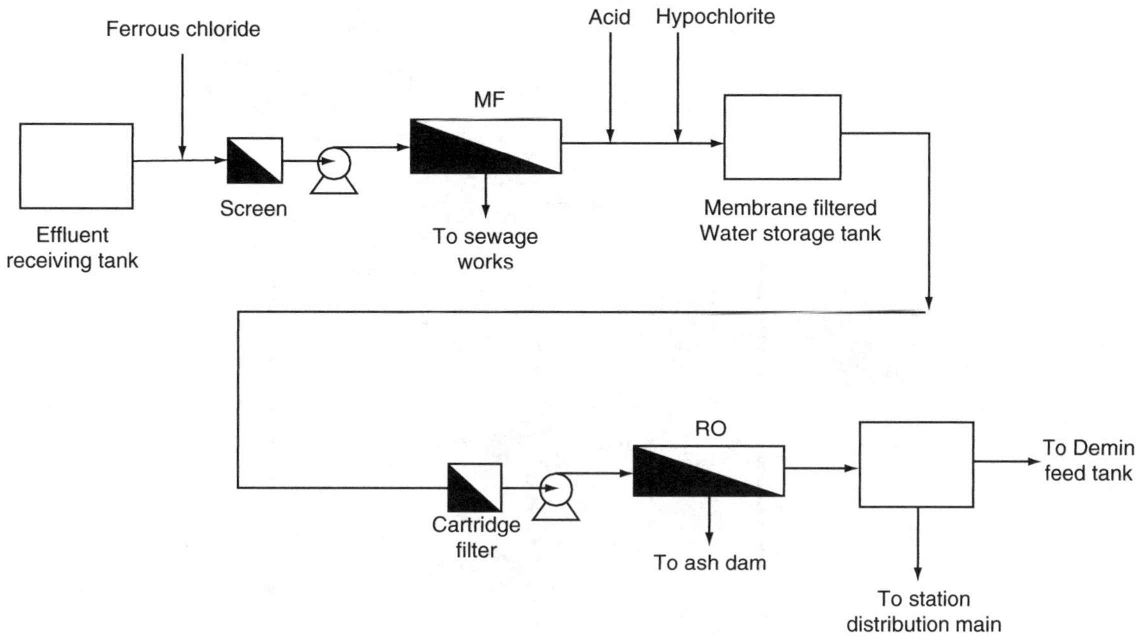

Figure 3.48 Process flow schematic of a hybrid membrane system water reuse treatment plant. Source: Judd and Jefferson

The treatment facility was expanded with MF instead of conventional treatment. The 390 m^3/h (2.5 MGD) MF/RO integrated membrane system is also shown in Figure 3.49. Thus, 7.5 MGD of combined purified water is injected into the ground to prevent seawater intrusion into the groundwater at 50% of the cost of conventional pretreatment. The operating pressure is 15–35 bar g. The average membrane flux is 35 l/m^2.h (21 GFD). The filtered water turbidity is less than 0.1 NTU and SDI is less than three, which exceed minimum RO membrane feed water quality requirements. The hollow-fibre MF membranes are backwashed every 18 min, and chemically cleaned two to four weeks.

The RO system consists of a single-stage array containing 108 pressure vessels with seven CA membrane spiral-wound elements (20 cm × 100 cm nominal) operating at an average flux of 10 GFD. One year after plant came on-line, the CA membranes were replaced with polyamide TFC membranes reducing energy consumption by nearly 50%. The RO system product water recovery is quite high ~85%, and the TFC membranes reduce the TDS from 800 mg/l in the feed water to less than 20 mg/l in the permeate. The membrane cleaning interval typically is 1–2 months.

In 1997, the plant was further expanded to purify tertiary treatment water for the boiler feed IX system at a local oil refinery as shown in Figure 3.49. The MF/RO system purifies tertiary treatment effluent producing 500 m^3/h (3.1 MGD) re-purified water. Tertiary treatment includes ferric chloride addition, coagulation, flocculation, granular media filtration and chlorination. The RO system consists of four trains with 35 pressure vessels, and each vessel houses six TFC membrane spiral-wound elements (20 cm × 100 cm nominal) in series.

According to the WBMWD operating data, tertiary effluent reduces membrane fouling

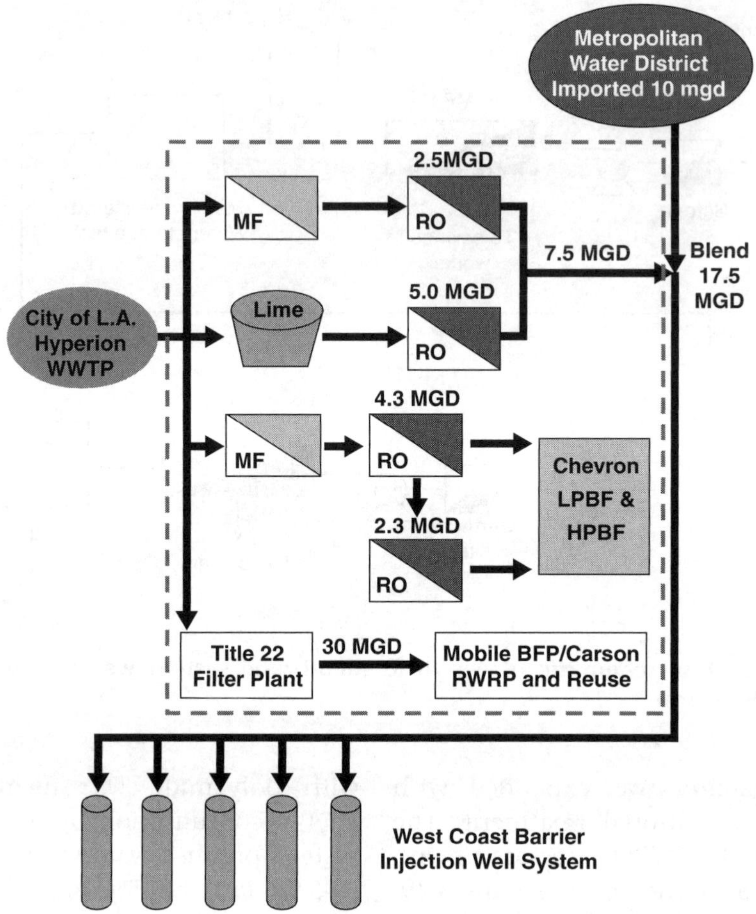

Figure 3.49 City of Los Angeles hybrid membrane water recycling plant flow schematic. Source: Mierzejewski and Guendert

and increases the intervals between chemical cleanings significantly; the MF membrane cleaning interval increased to three months from two to four weeks whereas the RO membrane cleaning interval increased to six-fold in six to nine months. The success of this plant resulted in similar MF/RO plants built at two more local refineries as shown in Figure 3.49. Total RO permeate production using MF/RO integrated membranes is planned for expansion to nearly 1.8×10^5 m^3/day (48 MGD) by 2020.

Case study IV. In another application of membrane filtration for water reuse, a tubular MF membrane unit is used to reclaim and reuse 70% of the RO reject from the make-up RO system that treats industrial wastewater at a General Motors plant in north-east Mexico.[59] The only source of water in this desert region is an aquifer, which is also the only source of drinking water for the local community. The need to conserve water is, therefore, paramount.

In this RO–MF–RO hybrid system, the brine from the main RO system is mixed with media filter backwash and the IX system regeneration (backwash and rinse water) wastewater. The combined wastewater is treated with magnesium chloride in a contact reactor to precipitate silica. Soda ash and lime are added to precipitate calcium hardness as calcium carbonate, and magnesium as magnesium hydroxide. The precipitates and particles are removed in the cross-flow tubular MF membrane unit producing filtered water with a SDI of less than three. The 0.1 μm pore size (nominal) MF membranes are PVDF cast on the inner surface of a porous polyethylene tube. The operating conditions are 1.5–2.5 bar g feed pressure, and 510 l/m²/h (300 GFD) average flux. The MF membranes are backwashed periodically with plant air every 18–30 min, and the chemical cleaning frequency ranges from two to four weeks. Chemical cleaning is a two-step process; cleaning with sodium hypochlorite to remove bacteria followed by hydrochloric acid cleaning to remove hardness scale.

The brine RO system consists of two trains operating in parallel. Each unit is a three-stage (5:2:1) array operating at an average flux of 20 l/m².h (12 GFD). The high quality of the MF filtrate (SDI < 3) made it possible to operate the RO units at 80–90% product water recovery. The brine RO permeate is adjusted for pH with HCl and blended with the primary RO feed water or used as potable water in the automotive plant.

3.5 Advanced membrane filtration

3.5.1 Submerged membrane technology

In this latest membrane technology, MF (or UF) membrane modules are immersed in water that has to be treated. Water permeates the membranes under a small transmembrane pressure (TMP), typically < 1 bar, and is withdrawn from the fibre bore by the permeate pump. The membranes can be flat sheet as shown in Figure 3.50a or hollow fibres as shown in Figure 3.50b. The pore sizes of the membranes vary between 0.1 and 0.2 μm for hollow fibres to 0.4 μm for flat sheet membranes. The MF membranes are polyethylene (PE), hydrophillic PE, polysulphone, PP, or PVDF.[63]

Submerged continuous microfiltration (CMF-S) is an extension of the CMF technology, which was introduced by Memcor®/USFilter in the early 1980s. CMF-S is designed for large wastewater treatment plants. For example, long-term testing of the CMF and CMF-S technologies at Orange County Water District Water Factory 21 (OCWD WF21) has demonstrated the cost-saving benefits of CMF-S for large water treatment plants. The CMF-S process has several advantages over CMF according to a product brochure based on field experience:[64]

- Reduced Footprint – Space requirement for the CMF-S is 50% less than the CMF.
- Reduced Complexity – 80% reduction in peripheral equipment.
- Reduced Costs – 10 to 15% reduction in operating costs and 20% reduction in capital costs.

A typical CMF-S system consists of one or more membrane module cells and the peripheral equipment, including filtrate pumps, blowers, air supply system, clean-in-place (CIP) system and control system. Multiple cells are grouped in 'trains,' typically up to ten cells per train. Each cell in a train shares peripheral equipment and communications system for monitoring and control. Larger systems are split into multiple parallel trains. The submerged membrane modules are attached in a cloverleaf design to a filtrate (clean water) manifold. A process flow sheet of a CMF-S system is shown in Figure 3.51. During the normal filtration cycle, the feed flow enters the bottom of each cell under gravity via a central feed

(a) (b)

Figure 3.50(a) Submerged flat-sheet membrane unit. Source: Shino et al. (b) Submerged hollow-fibre membrane unit. Source: USFilter/Memcor

channel, and passes over and around the membrane. The feed flow surrounds the multiple membrane modules and the hundreds of thousands of individual hollow membrane fibres within the modules. The flow is drawn through the walls of the membrane fibres (outside-in flow) by vacuum to the inside of the fibre (lumen). Filtered water then exits via the top of the module through the clean water manifold.

The PLC (programmable logic control) controlled CMF-S system is cleaned periodically by permeate backwashing based on TMP to maximise the interval of chemical cleanings. At the end of the liquid backwash cycle, air is introduced at the base of the modules to scour the membranes to loosen the dirt and debris from the surface of the membrane fibres. To facilitate scouring and debris removal of the fibres inside the modules, air is introduced within the base of the modules through a series of air connections in the bottom manifold. As the air continues to loosen the dirt on the surface of the membrane fibres, clean, filtered water is back-fed from the inside of the fibres. This operation assists in the flushing of debris from the membrane outside surface.

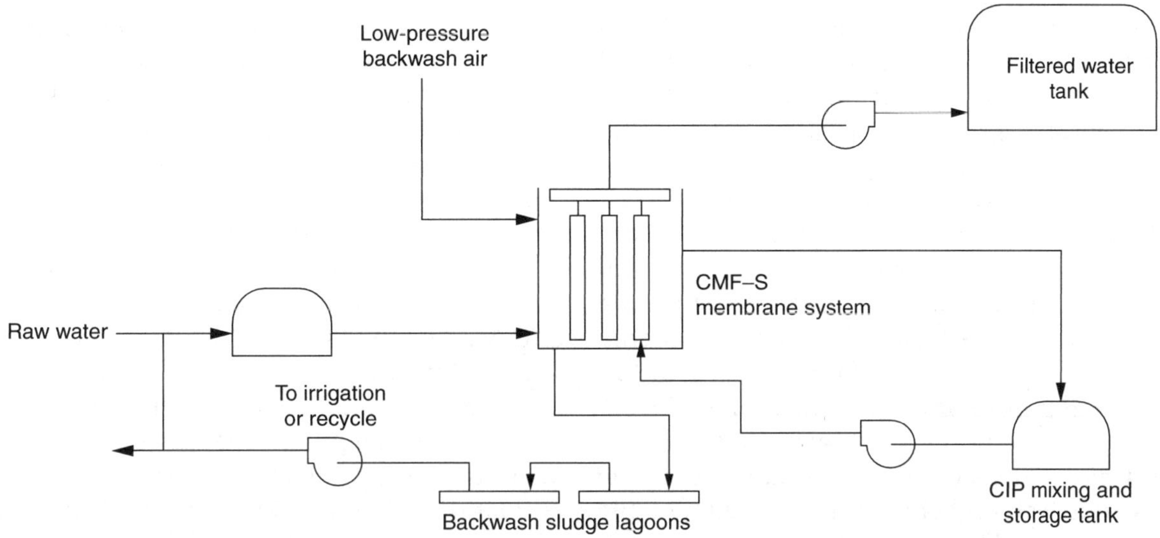

Figure 3.51 Cross-flow microfiltration submerged membrane system process flow sheet. Source: USFilter

The entire backwash sequence takes three minutes.

Several CMF-S systems have been installed by Memcor®/USFilter at several sites in Australia and Canada for drinking water treatment. MF provides a physical barrier to chlorine-resistant organisms such as *Giardia* and *Cryptosporidium*, and produces water with turbidity less than 0.1 NTU. The current size of the CMF-S systems in Australia is around 84 m^3/h (0.53 MGD). The system in British Columbia, Canada has a capacity of 710 m^3/h (4.5 MGD).

3.5.2 CMF-S case studies

I. Bendigo water treatment plant (BWTP). The 12.54×10^4 m^3/day (33 MGD) BWTP has been producing drinking water for nearly 1 million people in central Victoria, Australia since 2002. It is one of the largest if not the largest MF plant in the world. The plant combines submerged microfiltration (CMF-S), ozonation and biological activated carbon (BAC) to treat a variable and difficult raw water. Raw (surface) water is pre-screened, and dosed with lime and carbon dioxide in a contact reactor to control alkalinity and corrosion. Next, water is dosed with a coagulant, liquid aluminium chlorohydrate (ACH) prior to entering the CMF-S plant to remove colour, some organic content, and dissolved metals. The coagulant dosage is typically 5–6 mg/l. The coagulant precipitate is removed by MF. The coagulant/CMF-S process removes up to 15% of the dissolved organic carbon.[64]

The CMF-S system consists of eight cells (six on-line and two on standby) each containing 576 submerged membrane modules. Membrane performance is monitored for turbidity and particle count. A daily pressure decay test is performed to check the integrity of the hollow fibres. The cells are backwashed as described earlier. The MF filtrate is pumped to the ozone/biologically active (BAC) system for disinfection and to reduce the original carbon level in the water, eliminating taste and odour compounds.

II. Orange County Water District (OCWD). OCWD located between Los Angeles and San Diego counties in southern California manages the groundwater basin that supplies about $3.0 \times 10^8\,m^3$ per year potable water to a population of more than 2 million. One of OCWD's best known project is Water Factory 21 (WF21), which protects groundwater from seawater intrusion by injecting up to $5.7 \times 10^4\,m^3$/day (15 MGD) of highly treated reclaimed water blended with deep-well water into four coastal aquifers. More than half of the injected water flows inland and augments potable water supplies.[48] The injected water quality must exceed potable water quality to prevent plugging of the aquifer, and prevent degradation of groundwater quality.

WF21 is an advanced wastewater treatment (AWT) facility that reclaims secondary treatment effluent. In the beginning in 1975, the AWT plant consisted of lime clarification, ammonia stripping, recarbonation, multimedia filtration, granular activated carbon (GAC) filtration and chlorination. RO treatment was added in 1977 to reduce salts and organics in one-third of the flow stream. The $790\,m^3$/h (5 MGD) RO plant consisted of 252 spiral-wound cellulose acetate membranes (20 cm nominal diameter) in a three-stage array (24:12:6). The removal of organic carbon by RO was greater than by GAC, and in 1985, ammonia-stripping towers were removed since RO reduced ammonia and nitrate concentrations by 80%. The WF21 treatment process flow schematic is shown in Figure 3.52.

In 1992, evaluation of MF pretreatment for replacing conventional pretreatment shown in Figure 3.52 (see also Figure 3.47) was initiated with a small pilot unit. The positive results of the pilot testing with Memcor®/USFilter CMF pilot system led to the installation of a $114\,m^3$/h (0.7 MGD) CMF demonstration system. The plant consisting of MF, RO and UV disinfection has been operated since 1997. The plant uses thin-film composite RO membranes producing RO permeate that meets all the requirements of the US National Primary Drinking Water Regulations, and reduces the concentration of TOC to less than 0.1 mg/l.

Based on the success of the CMF demonstration unit, a four module CMF-S pilot unit was commissioned in 1998. After one year of operation, the CMF-S demonstrated increased cleaning intervals while operating at flux equivalent to the CMF pilot and demonstration systems. In early 2000, the evaluation of the CMF-S technology was scaled up to a $36\,m^3$/h (160 gpm), 32 module CMF-S demonstration system. Both systems were operated simultaneously for a year to compare the data. The data was used to design a $950\,m^3$/h (6 MGD) CMF-S system, which was installed in 2004. This system provides injection water during the construction of the $26.5 \times 10^4\,m^3$/day (70 MGD) CMF-S system.

III. Industrial water treatment. In 1997, the first MF/RO plant was commissioned to reclaim secondary treatment effluent at the El Segundo recycling plant of the West Basin Municipal Water District (WBMWD) in southern California for injecting RO purified water into a seawater barrier to control seawater intrusion into groundwater as discussed in section 3.4. The injected water quality must exceed potable water quality to prevent plugging of the aquifer and prevent degradation of groundwater quality. The MF system was a Memcor®/USFilter based CMF system. Subsequently, three more CMF systems were supplied by USFilter to supply re-purified water to three refineries as feed water for their boilers and cooling water. As part of Phase IV expansion, a CMF-S system was supplied by USFilter in 2005 for treating $1{,}840\,m^3$/h (11.6 MGD) secondary treatment effluent.[64]

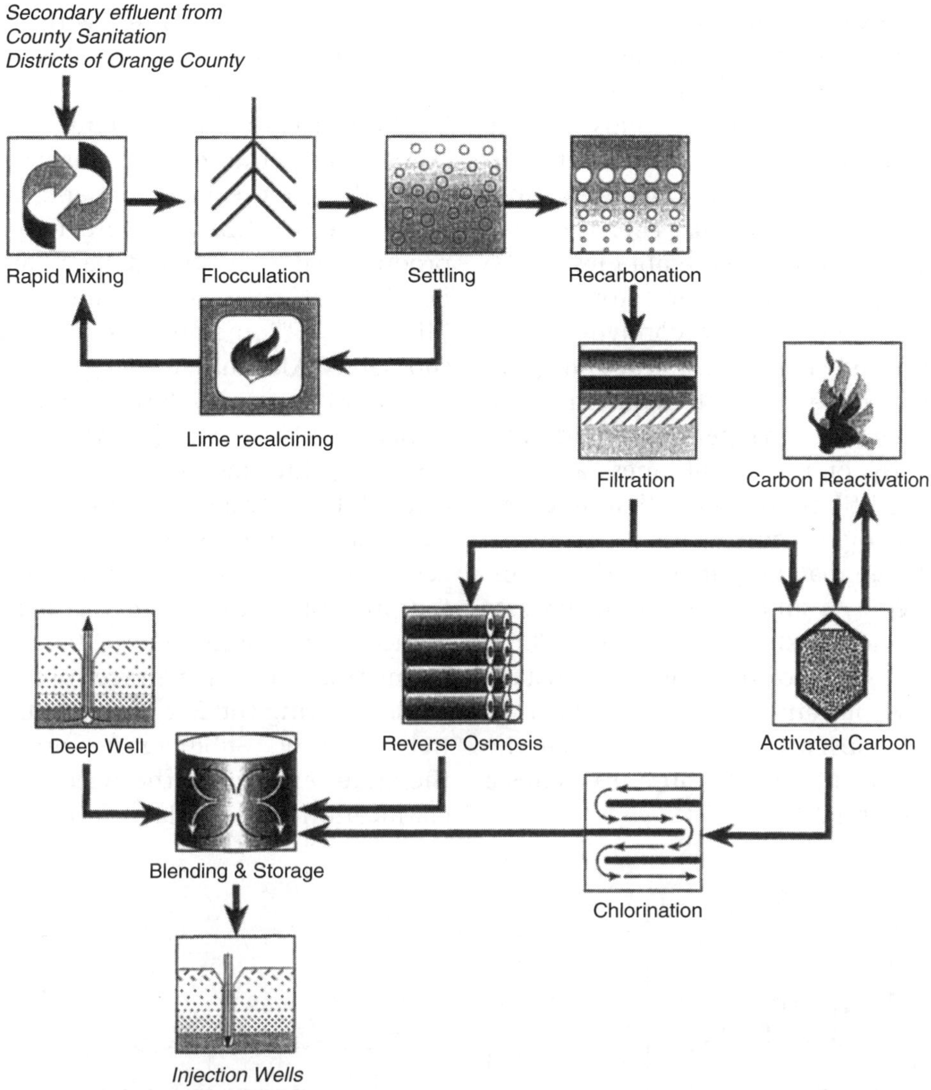

Figure 3.52 RO membrane-based water treatment system at Water Factory 21

Similarly, a 190 m^3/h (1.2 MGD) CMF-S membrane system was installed by USFilter in 2000 at a 1200 MW coal-fired power plant in UK when the well water source that provided feed water for the RO/IX high purity water system was shut down. The CMF-S system treats surface water and cooling tower blow-down water for the RO/IX deionisation system with turbidity less than 0.1 NTU, suspended solids at the detection limit of 1 ppm and SDI less than three. The MF membranes were oxidant-resistant PVDF.

3.5.3 Membrane bioreactor technology

The MBR process combines a bioreactor with a membrane module each with a specific function. Biological degradation of organic pollution is carried out in the bioreactor by adapted microorganisms, and separation of microorganisms in the treated wastewater is done with a membrane module.[63,65] The primary role of the membrane is to provide a barrier against suspended solids in the mixed liquor. The principle microsolute in the liquor is extracellular polymeric substances produced by bacteria. There are two process configurations; the membranes can be placed either outside the bioreactor or submerged in it. The submerged configuration is more economically based on energy consumption: recycle pump is not required since aeration generates a tangential liquid flow in the vicinity of the membranes, and the operating conditions are milder than in external MBR systems because of lower values of TMP and tangential velocities at the membrane surface. A submerged MBR process flow schematic is shown in Figure 3.53. The preferred mode of operation is constant flux – flux is constant with time as the TMP increases vs. constant TMP mode when the flux declines with time – because it ensures a steady throughput.[65]

The first MBR unit with submerged flat membranes was installed in Japan in 1992.[66,67] Since then the technology has been used in three different applications: (a) municipal/domestic wastewater treatment, (b) water reclamation, and (c) industrial wastewater treatment. In order to reuse MBR product water, it should be treated by RO. Besides flat sheet membranes, hollow-fibre MF and UF[68] membranes are used in the submerged MBR units.

The MBR was developed from the conventional activated sludge system by simply installing MF membrane elements in the aerated tank stage. By decoupling the activated sludge process from the settling characteristics in conventional plants, the footprint of a wastewater treatment system is reduced by more than 50%. Submerged membranes prevent passage of contaminants while allowing the free passage of biologically treated water. Submerged membrane MBR, therefore, eliminates the need for secondary clarification and tertiary filtration. Aerators

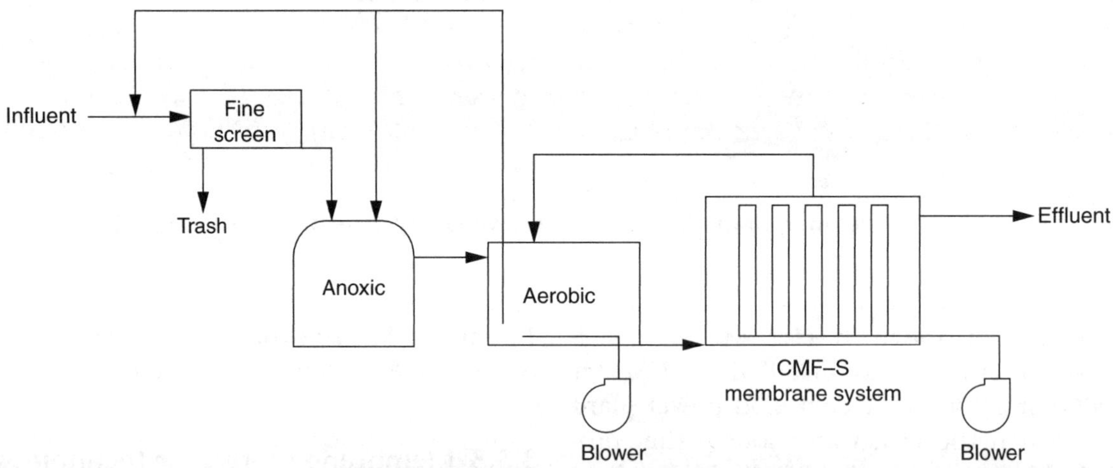

Figure 3.53 Membrane bioreactor (MBR) system process flow schematic. MBR combines biological degradation with membrane separation. Raw municipal water flows to an aerated bioreactor where the organic components are oxidised by the activated sludge. The aqueous sludge then passes through a MF or UF membrane filtration unit, separating water from the sludge. The sludge flows back to the bioreactor while the membrane permeate is discharged or reused. Source: USFilter/Memcor

mounted below the membrane modules or cells release air bubbles, which rise around the membranes creating water circulation and scouring the membrane surface to discourage fouling as shown in Figure 3.54.

Since membrane filtration is unaffected by sludge settleability, the activated sludge process can be optimised for biological activity by operating at very high mixed-liquor suspended solids (MLSS) concentrations (10,000–15,000 mg/l suspended solids), allowing the activated sludge process to treat a higher flow and load in the same tank resulting in more complete biological oxidation.[68,69] Very high mean cell residence times result in higher removals of difficult-to-treat compounds such as endocrine disrupters.

The MBR technology has proven to be cost-effective for various applications, and the number of MBR installations in Asia, Europe and North America now exceeds 1,000.[69] Plants with capacities of 40,000 m^3/day have become operational, and larger plants with capacities in the range of 150,000–300,000 m^3/day are under consideration.

3.5.4 MBR case studies

Two submerged hollow-fibre MBR systems were installed in the Everglades in south Florida in 2004–05 by USFilter for treating 20 gpm (4.5 m^3/h) and 60 gpm (13.6 m^3/h) wastewater. The systems are designed to reduce total phosphorous in the plant effluent to 0.1 mg/l or less. The MBR system replaces an aging secondary wastewater treatment plant. Several case studies using flat sheet submerged membranes are discussed next.[67,68]

- The Porlock plant in UK was commissioned in 1998. It treats 1,900 m^3/day of sewage. The plant consists of four aeration tanks containing six membrane

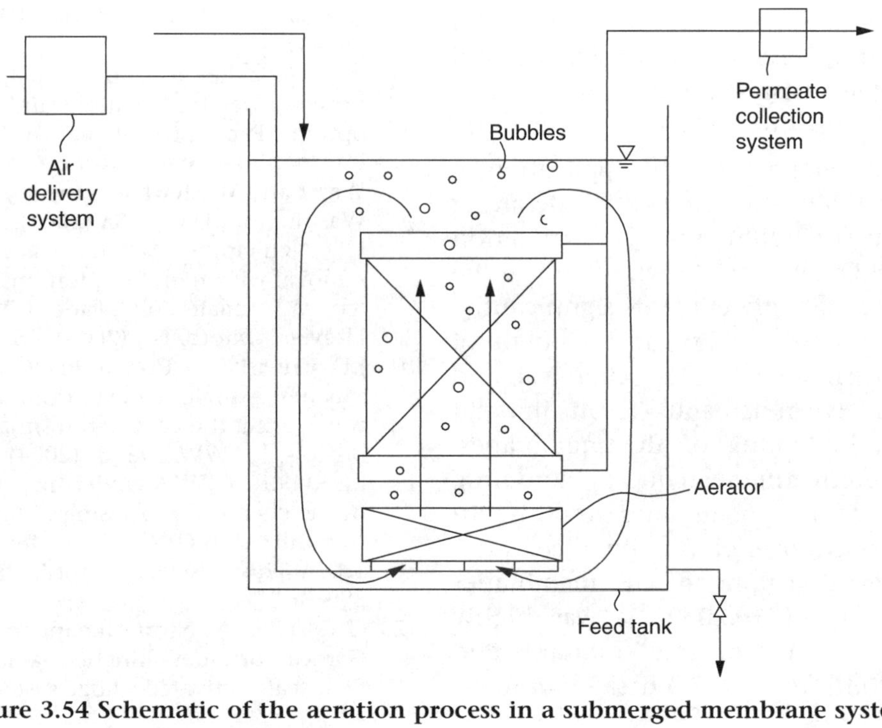

Figure 3.54 Schematic of the aeration process in a submerged membrane system

units. The sewage MLSS varies between 7,000 and 33,000 mg/l.

- The Swange plant in UK was commissioned in 2000. It treats 12,700 m^3/day of sewage. The plant consists of six aeration tanks containing 132 modules for a total of 19,800 membrane panels. During the first three years of operation, 310 panels were replaced, and the membranes were cleaned six times; five times with sodium hypochlorite and once with citric acid, all in situ.
- A milk processing MBR effluent plant in Ireland was started in 1999. The plant consists of 74 membrane treatment units containing 11,100 membranes with a flow of 7100 m^3/day. The MBR effluent quality is total N < 15 mgN/l; P < 0.6 mgP/l; and BOD < 10 mg/l. Due to the nature of the incoming effluent, chemical cleaning is more frequent than municipal effluent treatment plants.
- When treating municipal wastewater primary effluent in San Diego, a pilot MBR unit treated water had a BOD between 1 and 2 mg/l, SS less than 1 mg/l, and turbidity less than 0.1 NTU.

A recent development is the application of an integrated MBR–RO process for desalting mixed tannery effluents for irrigation purposes.[70] RO has been shown to reduce the salt content of tannery effluents significantly. Unfortunately, the high organic content of the tannery effluents – high chemical and biochemical oxygen demand – results in rapid scaling and biofouling of RO membranes. The MBR pretreatment reduces biofouling and scaling of RO membranes resulting in much improved overall performance. MBR using polyether sulphone UF membranes with a MWCO of 100,000 Da was highly effective in removing organic pollutants and suspended solids prior to RO desalination.

References

1 M. Mulder, Basic Principles of Membrane Technology, Kluwer Academic Publishers: Holland (1997).

2 H. Strathmann, Membrane separation processes: current relevance and future opportunities, AIChE J. 47 (5) (2001), 1077–1087.

3 E. Drioli, M. Romano, Progress and new perspectives on integrated membrane operations for sustainable industrial growth, Ind. Eng. Chem. Res. 40 (2001), 1277–1300.

4 S.K. Sikdar, J. Burckle, J. Rogut, Separation methods for environmental technologies, Environ. Progress 20 (1) (2001), 1–11.

5 T.M. Harten, L.M. Vane, D. Szlag, Separations research in the U.S. Environmental Protection Agency. ACS Symp. Series 766 (2001), 222–235.

6 J. Tanninen, L. Kamppinen, M. Nyström, Pretreatment and hybrid processes, in Nanofiltration (eds. A.I. Schafer, A.G. Fane, T.D. Waite), Elsevier: Oxford, UK (2004), 241–262.

7 M. Nyström, J.M.K. Nuortila, M.J. Manttari, Nanofiltration in the pulp and paper industry, in Nanofiltration (eds. A.I. Schafter, A.G. Fane, T.D. Waite), Elsevier: Oxford, UK (2004), 363–393.

8 P.M. Gallagher, St. D.L. Germain, Arsenic in drinking water: regulations and treatment options, Paper presented at the Third NSF International Symposium on Small Drinking Water and Wastewater Systems, 22–25 April, Washington, D.C., USA (2001).

9 L.D. Nghiem, A.I. Schafer, Trace contaminant removal with nanofiltration, in Nanofiltration (eds. A.I. Schafer, A.G. Fane, T.D. Waite), Elsevier: Oxford, UK (2004), 480–520.

10 J.D. Chwirka, C. Colvin, J.D. Gomez, P.A. Mueller, Arsenic removal from drinking water using the coagulation/microfiltration process, J. AWWA 96 (3) (2004), 106–114.

11 M. Kyburz, G.W. Meindersma, Nanofiltration in the chemical processing industry, in Nanofiltration (eds. A.I. Schafer, A.G. Fane, T.D. Waite), Elsevier: Oxford, UK (2004), 329–361.

12 G.T. Tellez, N. Nirmalakhandan, J.L. Gardea-Torresdey, Kinetic evaluation of a field-scale activated sludge system for

removing petroleum hydrocarbons from oilfield-produced water. Environ. Progress 24 (1) 2005, 96–104.
13 P. Remy, H. Muhr, E. Plasari, I. Ouerdiane, Removal of boron from wastewater by precipitation of a sparingly soluble salt, Environ. Progress 24 (1) 2005, 105–110.
14 T. Melin, J. Meier, Wintgens, Nanofiltration in the leachate industry, in Nanofiltration (eds. A.I. Schafer, A.G. Fane, T.D. Waite), Elsevier: Oxford, UK (2004), 395–413.
15 K. Soldenhoff, J. McCulloch, A. Manis, P. Macintosh, Nanofiltration in metal and acid recovery, in Nanofiltration (eds. A.I. Schafer, A.G. Fane, T.D. Waite) Elsevier: Oxford, UK (2004), 459–477.
16 G. Bargemann, M. Timmer, C. Horst, van der, Nanofiltration in the food industry, in Nanofiltration (eds. A.I. Schafer, A.G. Fane, T.D. Waite) Elsevier: Oxford, UK (2004), 305–328.
17 M. Cheryan, Ultrafiltration and Microfiltration Handbook, CRC Press: Boca Raton, FL (1998).
18 S.S. Koseglu, G.J. Guzman, Applications of reverse osmosis technology in the food industry, in Reverse Osmosis (ed. Z. Amjad), Van Nostrand Reinhold: New York, NY (1993), 300–333.
19 Austin, G.T., Schreve's Chemical Process Industries, McGraw-Hill: New York (1984).
20 US Patent, # 6,406,546; # 6,406,547; # 6,406,548.
21 N.W. Pirie, Leaf Protein, Cambridge University Press: London (1978).
22 H.T. Ostrowski-Meissner, Chapter 15 Protein concentrates from pasture herbage and their fractionation into feed- and food-grade products, in Leaf Protein Concentrates (eds. L. Telek, H.D. Graham), AVI Publishing Co.: Westport, CT (1983).
23 O. Kohler, R.H. Edwards, D. de Fremery, Chapter 18 LPC for feeds and foods: The Pro-Xan process, in Leaf Protein Concentrates (eds. L. Telek, H.D. Graham), AVI Publishing Co.: Westport, CT (1983).
24 R. Singh, L.F. Whitney, C.S. Chen, Recovery of alfalfa protein by membrane ultrafiltration, Paper No. 74-6003, Proceedings of the American Society of Agricultural Engineers Annual Meeting, Oklahoma City, OK, June (1974).
25 J.E. Flinn, Green plants as biofactories for drugs, BioPharm Intern. 17 (8) (2004), 42–49.
26 S.S. Koseglu, J.T. Lawhon, E.W.J. Lusas, Membrane processing of crude vegetable oils: pilot plant scale removal of solvent from oil miscellas, J. Amer. Oil. & Chem. Soc. 67 (5) (1990), 281.
27 A.K.S. Gupta, Process for refining crude glyceride oils by membrane filtration. US. Patent # 4,062,882, December 13 (1977).
28 US Patent # 6,833,149.
29 P. Hogan, R.P. Canning, P.A. Peterson, R.A. Johnson, A.S. Michaels, Osmotic distillation, a new option, Chem. Eng. Progress 94 (7) (1998), 49–61.
30 W. Eykamp, Section 22 Alternative separation processes, in Perry's Chemical Engineers' Handbook (eds. D.H. Perry, D. Green), 7th Edition, McGraw-Hill: New York, NY (1999).
31 L.R. Weatherly, Engineering Processes for Bioseparations, Butterworth-Heinemann: Oxford, UK (1994).
32 T. Matsuura, Synthetic Membranes and Membrane Separation Processes, CRC Press: Boca Raton, FL, USA (1994).
33 R. Singh, Will developing countries spur fuel cell surge? Chem. Eng. Progress 95 (3) (1999), 59–66.
34 R. Klassen, A.E. Jansen, Achieving flue gas desulsphurisation with membrane gas absorption, Filtration & Separation, December (2003), 26–28.
35 J.A. Dalan, Things to know about zero liquid discharge, CEP 95 (11) (2000), 71–76.
36 J.A. Cotruvo, Water desalination processes, Water Cond. & Purific., January (2005), 13–17.
37 B.V. der Bruggen, C. Vandecasteele, Distillation vs. membrane filtration: overview of process evolutions in seawater desalination, Desalination 143 (2002), 207–218.
38 National Research Council, Review of the Desalination and Water Purification Technology Roadmap, National Academic Press: Washington, D.C., USA (2004).
39 Antonio von Gottberg, J. McArdie, B. Antrim, D. Halperm, Hollow fibre UF pretreatment for RO brackish water and sweater desalination systems, Ultrapure Water 22 (3) (2005), 23–29.
40 T. Prato, E. Schoepke, L. Etchison, T. O'Brien, B. Hernon, K. Perry, M. Peterson, Production of high purity water from seawater, Ultrapure Water 20 (8) (2003), 20–25.

41 B. Jurenka, C. Owen, Seawater RO designs and costs to meet water quality specifications, Ultrapure Water 22 (1) (2005), 30–37.
42 M. Rahman, Personal Communication, USFilter, January–March 2004.
43 M.F. Lamendola, A. Tua, Desalination of seawater by reverse osmosis, the Malta experience, Des. & Water Reuse 5 (1) (1994), 18–22.
44 S.T. Hsu, K.T. Cheng, J.S. Chiou, Seawater desalination by direct contact membrane distillation, Desalination 143 (2002), 279–287.
45 E. Drioli et al., Desalination 122 (1999), 141–145.
46 C.A. Buckley, C.J. Brouckaert, C.A. Kerr, RO applications in brackish water desalination and in the treatment of industrial effluents, in Reverse Osmosis (ed. Z. Amjad), Van Nostrand Reinhold: New York, NY (1993), 275–299.
47 S.V. Cabibbo, D.B. Guy, A.C. Ammerlaan, A. Ko, R. Singh, Reverse Osmosis Technical Manual. NTIS Publication: Springfield, Virginia, USA (1979).
48 Using Reclaimed Water to Augment Potable Water Resources, Chapter 7, Am. Water Works Assoc., Special Publication: Denver, Colorado, USA (1998), 235–337.
49 S.V. Cabibbo, D.B. Guy, A.C. Ammerlaan, A. Ko, R. Singh, Parametric study on brackish water desalting plants, NWSIA J. July (1979), 17–36.
50 Singh, R., DI water systems for fuel cell power plants, Ultrapure Water 20 (3) (2003), 47–52.
51 G.A. Pittner, The economics of desalination processes, in Reverse Osmosis (ed. Z. Amjad), Van Nostrand Reinhold: New York, NY (1993), 76–103.
52 T. Witham, Reusing nature's most precious resource, WAVE/USFilter 2 (1) (2003), 34–36.
53 F.L. Slejko, Microelectronics water, Ultrapure Water 12 (4) (1995), 4.
54 T. Fulde, Meeting water requirements for 300-mm fab, Ultrapure Water 21(8) (2004), 20–26.
55 W.V. Collentro, Pharmaceutical Water, CRC Press: Boca Raton, FL, USA (1999).
56 H. Moes, R.C. Livingston, Producing affordable high-purity water, BioPharm. Intern. 17 (8) (2004), 28–40.
57 G. Gsell, Water systems utilising multiple effect and vapour compression technologies compared, Pharmaceutical Engineering, March/April (2004), 48.
58 T. Fane, P. Macintosh, G. Leslie, Water reclamation, remediation and cleaner production with nanofiltration, in Nanofiltration (eds. A.I. Schafer, A.G. Fane, T.D. Waite) Elsevier: Oxford, UK (2004), 287–304.
59 M. Mierzejewski, D. Guendert, Industrial applications using microfiltration as RO pretreatment, Ultrapure Water 21 (7) (2004), 29–35.
60 S. Judd, B. Jefferson, Industrial effluent recycling in the power industry, Ultrapure Water 21(4) (2004), 37–42.
61 S. Judd, B. Jefferson, Industrial wastewater recovery and re-use with membranes, Filtration & Separation, September (2003), 38–40.
62 B. Vishwanath, K. Kekre, GuiHe Tao, Reclaimed water: an alternate source for high-purity water production in the semiconductor industry, Ultrapure Water 20 (5) (2003), 41–46.
63 B. Marrot, A.P. Barrios-Martinez, N. Roche, Industrial water treatment in a membrane bioreactor: a review, Environ. Progress 23 (1) (2004), 59–68.
64 Memcor® CMF-S Microfiltration Brochure, USFilter.
65 A. Fane, S. Chang, Membrane bioreactors: design & operational options, Filtration & Separation, June (2002), 26–29.
66 R. Severn, Long term operating experience with submerged plate MBRs, Filtration & Separation, September (2003), 28–31.
67 H. Shino, K. Nishimori, S. Kawakami, T. Uesaka, K. Izumi, Future applications for submerged flat-sheet membranes, Ultrapure Water 21 (8) (2004), 37–41.
68 J. Daly, Advanced wastewater treatment for marine vessels, Filtration & Separation, June 2002, 32–35.
69 G.T. Daigger, B.E. Rittman, S. Adham, G. Andreottola, Are membrane bioreactors ready for widespread application? Env. Sci. & Tech. 39(19) (2005), 399A–406A.
70 W.G. Scholz, P. Rouge, A. Bodalo, U. Leitz, Desalination of mixed tannery effluent with membrane bioreactor and reverse osmosis treatment, Env. Sci. & Tech. 39 (21) (2005), 8505–8511.

Chapter 4

Hybrid membrane system design and operation

'We can't solve problems by using the same kind of thinking we used when we created them'.

Albert Einstein

A hybrid membrane plant encompasses several technologies and processes since a stand-alone membrane system cannot produce high-purity quality product owing to physical and chemical constraints of synthetic membranes, and can foul prematurely. Some of the earliest hybrid membrane plants were designed for dairy plants[1] and for brackish water desalination.[2] Several hybrid membrane systems for various applications are discussed in Chapter 3. In this chapter, we describe the engineering process and operation of a typical hybrid membrane plant for producing high-purity water (HPW). Basically HPW encompasses a wide range of pure-grade water varying in quality: potable (drinking) water, deionised (DI) water for power plants, purified water for beverage and pharmaceuticals industry, and ultrapure water (UPW) for semiconductor manufacturing.

The optimal success of a well-designed plant depends on the ability to operate various systems achieving the highest possible output using minimum energy, and generating as little waste as possible with maximum safety for employees and environment. Plant operation requires a diversity of problem-solving activities, which must occur in an information-rich environment. Current and historic plant data, and characteristics of equipment and information obtained from troubleshooting the failure spots must be available and integrated at points of decision.

The first step is a conceptual design – a process flow diagram (PFD) with mass and energy balances – followed by the design of each unit operation and culminating in well-defined operating conditions.[3] The design of the hybrid membrane plant depicted in Figure 4.1 is based on the raw feed water analysis given in Table 4.1. The key aspect of a hybrid membrane system design is the integration of all processes to produce the specified product under optimal conditions. For membrane systems used in water purification, the feed water quality is almost always unique that poses a great design and operation challenge. Unit operations, system controls, plant operation, troubleshooting and maintenance, process design, and equipment are discussed. These

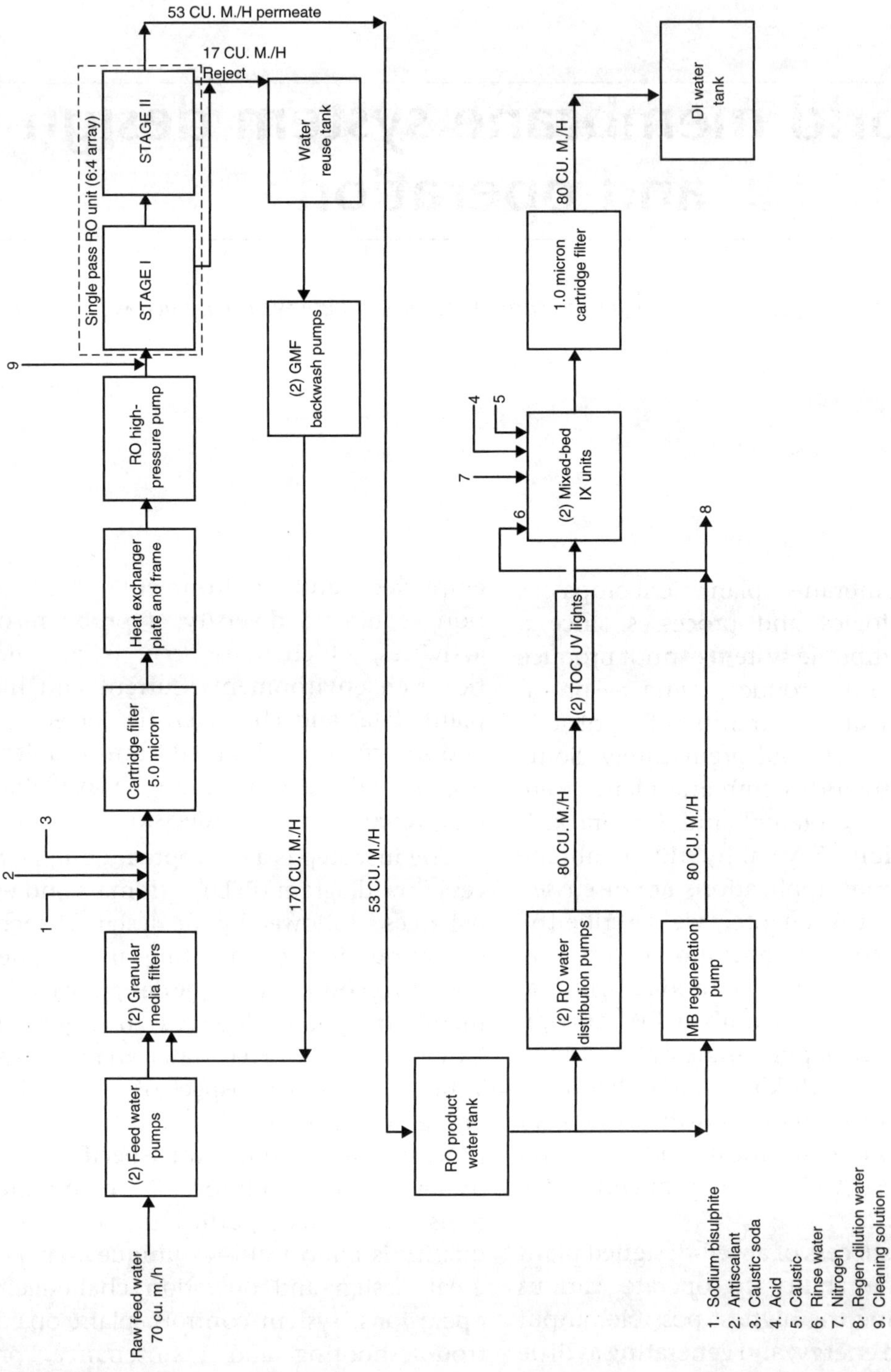

Figure 4.1 Process flow diagram of the hybrid membrane water treatment system

Table 4.1 RO membrane performance projections (Hydranautics)

Calculation created by Rajindar Singh			
Project name	DI Water RO system	Permeate flow	53.00 m^3/h
HP Pump flow	70.7 m^3/h	Raw water flow	70.7 m^3/h
Recommended pump pressure	8.0 bar		
Feed pressure	6.0 bar	Permeate recovery ratio	75.0 %
Feed water temperature	25.0°C (77°F)		
Feed water pH	7.60	Element age	0.0 years
Acid dosage, ppm (100%)	0.0 H_2SO_4	Flux decline % per year	7.0
Acidified feed CO_2	4.2	Salt passage increase, % per year	10.0
Average flux rate	21.6 $1/m^2$-h	Feed type	Surface water

Stage	Perm. Flow m^3/h	Flow/Vessel Feed m^3/h	Flow/Vessel Conc. m^3/h	Flux $l/m^2/h$	Beta	Conc. and Throt. pressures bar	Conc. and Throt. pressures bar	Element type	Elem. No.	Array
1-1	37.5	11.8	5.5	25.5	1.14	4.2	0.0	ESPA2+	36	6 × 6
1-2	15.5	8.3	4.4	15.8	1.11	2.9	0.0	ESPA2+	24	4 × 6

	Raw water		Feed water		Permeate		Concentrate	
Ion	mg/l	meq/l	mg/l	meq/l	mg/l	meq/l	mg/l	meq/l
Ca	7.6	0.4	7.6	0.4	0.03	0.0	30.3	1.5
Mg	3.5	0.3	3.5	0.3	0.01	0.0	14.0	1.1
Na	50.0	2.2	50.0	2.2	0.82	0.0	197.5	8.6
K	0.5	0.0	0.5	0.0	0.01	0.0	2.0	0.1
NH_4	0.0	0.0	0.0	0.0	0.00	0.0	0.0	0.0
Ba	0.100	0.0	0.100	0.0	0.000	0.0	0.399	0.0
Sr	0.000	0.0	0.000	0.0	0.000	0.0	0.000	0.0
CO_3	1.0	0.0	1.0	0.0	0.00	0.0	4.0	0.1
HCO_3	104.0	1.7	104.0	1.7	1.80	0.0	410.6	6.7
SO_4	34.0	0.7	34.0	0.7	0.08	0.0	135.8	2.8
Cl	12.2	0.3	12.2	0.3	0.12	0.0	48.4	1.4
F	0.5	0.0	0.5	0.0	0.01	0.0	2.0	0.1
NO_3	2.8	0.0	2.8	0.0	0.20	0.0	10.6	0.2
B	0.00		0.00		0.00		0.00	
SiO_2	26.0		26.0		0.24		103.3	
TDS	242.2		242.2		3.3		958.8	
pH	7.6		7.6		5.9		8.2	

(Continued)

Table 4.1 Continued

	Raw water	Feed water	Concentrate
$CaSO_4/K_{sp} \times 100$	0%	0%	1%
$SrSO_4/K_{sp} \times 100$	0%	0%	0%
$BaSO_4/K_{sp} \times 100$	304%	304	1631%
SiO_2 saturation	19%	19%	74%
Langelier Saturation Index	−1.05	−1.05	0.67
Stiff and Davis Saturation Index	−0.99	−0.99	0.72
Ionic strength	0.00	0.00	0.01
Osmotic pressure	0.1 bar	0.1 bar	0.5 bar

should familiarise the student of membrane technology about the intricacies of a hybrid membrane system.

4.1 Process description

Hybrid membrane plants (HMP) for water purification consist of membrane (MF–UF–NF) or conventional pretreatment systems, a single-pass or double-pass RO system and, if required, a polishing system.[4–6] The unit operations shown in Figure 4.1 were discussed in Chapter 2. The plant is designed to supply up to 80 m^3/h deionised (DI) water to the point-of-use (POU) with a resistivity of 16.0 MΩ-cm or higher.

The RO unit is the pivotal process since water production by membrane separation decays with time mainly due to fouling and other factors discussed in Chapter 2. The RO system must be run under conditions that minimise decline in flux while maintaining high product water quality. The pretreatment system must be stable and reliable to ensure the RO unit operates continuously without frequent shutdowns for cleaning to restore flux and rejection. A stable RO membrane performance, in turn, is required to ensure the MB polishing system produces water that meets the product water specifications without frequent shutdowns for regeneration of ion-exchange resins.

4.1.1 Pretreatment system

Raw feed water is pumped to the packed-bed granular media filtration (GMF) by the feed water pump at 70 m^3/h and 4.13 bar g, as shown in Figure 4.1. The feed water quality is given in Table 4.1. Water flows through the media filter (one GMF is on standby) downwards through the media bed and out from the bottom at 70 m^3/h. All operating modes – service and wash – and pneumatic valves are controlled by the programmable logic controller (PLC). The pressure drop across the GMF system is monitored by a differential pressure (ΔP) switch. When ΔP exceeds 1 bar (typically every four weeks), the filters are washed to clean the media bed. The media filters should be washed based on differential pressure and not run time.[6]

The wash cycle is initiated manually at the main control panel (MCP), and consists of three steps: backwash, bed settle, and rinse. Once initiated, the PLC controls the wash cycle progressing through each step sequentially – opening or closing the pneumatic valves, and starting or shutting down the appropriate pumps. Only one GMF can be washed at a time.

During the backwash phase, water is transferred from the water reuse tank to the GMF by the backwash pump at 170 m^3/h and 2.76 bar g. Water flows from the bottom upwards through the bed. A high superficial velocity of 25–35 m/h is required to fluidise the media bed and remove dirt from the bed. The backwash phase lasts 15 min, and is followed by a 2-min bed settle phase when no water flows to the filter. This allows the bed media to recover or 'rest' before it is rinsed. During the rinse phase, the standby feed water pump transfers water to the media filter at 70 m^3/h and 2.76 bar g. Water flows downwards through the media bed and out from the bottom to drain. The rinse phase lasts 5 min. Once the GMF has been washed, it is placed in standby mode until the online unit needs to be washed.

The filtered water is monitored for turbidity continuously by a turbidity analyser. The turbidity must be less than 1.0 NTU to prevent fouling of RO membranes by particulate matter. A higher NTU value means the GMF must be washed. The filtered water is injected with chemicals as it flows to the RO system to prevent scaling of membranes by sparingly soluble salts.

Chemical treatment consists of three well-known treatment processes for ensuring reliable operation of RO membranes. First, a 20% sodium bisulphite (SBS) solution is injected by the chemical dosing pump (one pump is on standby). The injection rate is proportional to the RO feed water flow rate, and is controlled by the PLC based on the chlorine concentration monitored by a chlorine analyser downstream of the in-line mixer. Sodium bisulphite like sodium sulphite and sodium metabisulphite is a reducing agent commonly used to dechlorinate RO feed or lower the chlorine concentration to less than 0.05 mg/l in RO plants that use polyamide aromatic membranes. It takes 7.33 mg/l of 20% $NaHSO_3$ (SBS) solution to remove 1 ppm of residual chlorine in water on a stoichiometric basis.

A negatively charged anti-scalant (A/S) is injected by the chemical dosing pump (one pump is on standby) into the pressurised RO feed water line downstream of the SBS injection point. The injection rate or dosage is proportional to the RO feed water flow rate, and is controlled automatically. The A/S prevents sparingly soluble salts of calcium and magnesium from precipitating out and scaling the membrane surface, as discussed in Chapter 2.

Finally, caustic soda (50% NaOH solution) is injected into the RO feed water line by the chemical dosing pump (one pump is on standby) to maintain the pH of RO feed water between 7.5 and 8.5. The injection rate is controlled by the PLC based on the pH value measured downstream of the in-line static mixer. At alkaline pH conditions, dissolved CO_2 gas is converted to bicarbonate ions, thereby, enhancing membrane rejection (lower permeate conductivity). The likelihood of carbonate scaling is also minimised.

The pretreated water flows to the cartridge filter (nominal pore size of 5.0 μm) at 70 m^3/h. The pressure drop (ΔP) across the filter is monitored by a local ΔP indicator/switch. The microfilter operates in a dead-end mode. When ΔP exceeds 1 bar or every 3 months, the plugged cartridges are replaced. The pleated depth filter cartridges are sized based on flow rate and pressure drop, e.g. the rating is 30 l/min per 25-cm long cartridge with a pore size of 5.0 μm.

The RO feed water flows through a parallel plate-and-frame heat exchanger. The PLC maintains the water temperature at 25±1°C by controlling the heat exchanger steam flow. A rule of thumb is that RO membrane flux increases 3% per °C rise in temperature as a result of reduced viscosity. A constant water temperature is necessary to compare RO flux data. Otherwise the flux data must be normalised to evaluate flux with time as discussed in Chapter 2. Heat exchanger sizing is based on the maximum anticipated flow rate and temperature rise. The heat transfer area is a function of heat load (ΔH), overall heat transfer coefficient (U), log mean temperature difference (ΔT_{lm}), and heat exchange interface material.

4.1.2 RO membrane system

The RO membrane separation process reduces the total dissolved solids concentration from 242 mg/l in feed water to less than 4 mg/l in product water corresponding to an average salt rejection of 98%. The RO unit uses thin-film composite (TFC) polyamide membranes in spiral-wound configuration. The single-pass RO system is a two-stage (6:4) membrane array designed to achieve a product water recovery (PWR) of 75% (53 m^3/h permeate / 70 m^3/h feed). The design is based on membrane manufacturer's projections given in Table 4.1. The RO high-pressure pump transfers water to the RO membrane array at 70 m^3/h and 16.7 bar g (total delivery head or TDH). The permeate flows to the RO product water tank at 53 m^3/h, and the reject flows to the water reuse tank at 17 m^3/h.

During start-up the RO skid inlet pneumatic valve opens automatically, and RO pretreated water starts flowing to the RO pump. The membranes are flushed for 60 s at line pressure, and the permeate is diverted to the water reuse tank. When the RO high-pressure pump inlet pressure exceeds the minimum permissible value (see Table 4.5), the RO high-pressure pump starts. Initially, the permeate is diverted to the water reuse tank for a few minutes to allow the product water conductivity to drop below the maximum permissible value. At the end of the quality rinse phase, the RO permeate is diverted to the RO product water storage tank via the service valve. The permeate service and divert valves are controlled manually (automatic valves are also used). RO membrane start-up and post-shutdown flushing are recommended by membrane manufacturers.

The inlet pressure to the membrane array is adjusted manually by a globe valve (when the RO pump speed is controlled by a variable frequency drive, membrane inlet pressure control valve does not require adjustment). The reject water flow rate is controlled manually by adjusting the reject flow control globe valve. Both valves are adjusted to achieve the desired product water recovery (%). The pressure required to achieve the permeate and reject flow rates is set during the initial run. As the membrane performance deteriorates with time, these valves require adjustments.

Product water flow rate higher than the maximum allowable limit is indicative of higher than design recovery. Similarly, reject flow rate lower than the allowable reject flow rate is reflective of higher than design recovery. Both these performance conditions can result in premature scaling by the precipitation of sparingly soluble salts, and must be avoided. Hence, the RO pump operation must be adjusted when the permeate flow rate exceeds the maximum design value or the reject flow rate drops below the minimum design value. When ΔP across the RO unit exceeds the maximum recommended value, the RO system should be shutdown for cleaning the membranes. The RO membranes are cleaned according to the procedure recommended by the membrane manufacturer, as discussed in Chapter 2.

RO permeate flows to the RO product water storage tank at 53 m^3/h. If the tank is full or if the permeate conductivity exceeds 10.0 μS/cm for an extended period of time (minutes), the RO system is shutdown.

The feed water and permeate flow rate data are used by the PLC to calculate % recovery. Similarly, RO feed water and permeate conductivity is monitored continuously, and the data are used by the PLC to calculate salt rejection or simply % rejection.

The RO product water storage tank has a capacity of 76 m^3. The tank is positioned directly downstream of the RO unit, so that the downstream equipment such as mixed-bed (MB) ion exchangers and ultraviolet (UV) units can be operated continuously. With storage, fluctuations and demand for the finished water can be addressed economically. The RO unit can be designed for the average demand, and storage designed to fill the gap between average and maximum.[5] The tank is also provided with level switches, a level transmitter, a low-pressure nitrogen blanket, and a high-efficiency particulate air (HEPA) filter to prevent atmospheric contamination and increase in conductivity due to absorption of carbon dioxide.

Water is transferred from the RO product water storage tank to the TOC destruct UV lamps by the RO water distribution pump (one pump is on standby) at 80 m^3/h. Irradiation with 185 nm ultraviolet light is very effective in destroying residual organic matter (TOC) not removed by the RO membrane. This UV light reduces the organic load on anion resins in the mixed-bed (MB) ion exchangers (deionisers) downstream. The UV units also render more than 90% bacteria ineffective. This is especially important because the MB units operate at neutral pH, and are thus capable of supporting rapid bacterial proliferation.[6]

4.1.3 Mixed-bed ion-exchange polishing

Water flows through the UV lights to the top of the MB deioniser (one deioniser is on standby), down through the resin bed and out from the bottom of the vessel to the final microfilter skid. Each packed-bed MB vessel contains a mixture of 3.26 m^3 strong base anion resin (OH^- form) and 2.18 m^3 strong acid cation resin (H^+ form) in a 60:40 ratio. The bed also contains 0.54 m^3 of inert resin, which forms a separate layer between the two IX resin layers when backwashing thus ensuring there is no mixing of acid and caustic solutions during regeneration of the resins. The ion-exchange process is discussed in Chapter 2.

The MB deionisers are designed to produce demineralised water with resistivity greater than 16.0 MΩ-cm (conductivity <0.063 μS/cm). All operating modes – service, standby, and regeneration – are controlled by the PLC. The PLC opens and closes the pneumatic valves and the corresponding solenoid valves in the local control panel. The process operating conditions are given in Section 4.2.

The deionised product water flows through two 1.0 μm cartridge filters operating in parallel to the DI water storage tank. The cartridge filters remove particles such as resin fines. When ΔP exceeds 0.7 bar, the cartridges are replaced. When the water tank is full or the product water resistivity is less than 15.0 MΩ-cm, water gets diverted to the RO product water storage tank via an automatic valve. This ensures the RO system is in operation continuously without frequent shutdowns and start-ups. It is preferable to replace the final filter cartridges based on time and not differential pressure to ensure there is no bacterial growth since ΔP may not increase substantially unless there is a catastrophic failure of MB internal laterals and distributors.

When the MB product water quality drops, i.e. the resistivity is less than 15.0 MΩ-cm and silica is greater than 30 ppb (μg/l), the MB resins are considered to be exhausted for the purpose of producing high-purity water. The exhausted MB unit is shutdown for regeneration, and the standby MB unit is switched to service mode.

The MB regeneration cycle summarised and described[4,7,8] below was also discussed in Chapter 2:

1. Fill the MB vessel with RO product water slowly through the bottom distributor until the level is 15 cm above the top of the resin bed.
2. Backwash to remove resin fragments and debris (and to separate the resins) for 10–15 min or until no fines are visible in wastewater.
3. After the backwash, allow the resins to settle into three layers – anion, inert, cation – from top to bottom based on density for 5 min.
4. The regeneration efficiency can be improved by warming the bed to 45°C for 30 min for removing silica. Warm water is passed through the anion resin bed and ambient temperature water is passed through the cation resin bed.
5. Commence simultaneous acid and caustic regeneration; 6% HCl solution upflow through the cation bed for 30 min, and 4% NaOH solution at 43–49°C downflow through the anion bed for 60 min. The two streams flow out to wastewater neutralisation tank through the interface collector imbedded in the inert resin layer.
6. Start the slow rinse cycle to displace the caustic and acid solutions from their respective beds with dilution water for 20 min.
7. Repeat the above rinse cycle at a much higher flow rate for 45 min to purge the respective beds of any traces of acid and caustic.
8. Use high-pressure nitrogen to push water in the bed out to drain until the water is at bed level.
9. Mix the resin with nitrogen for 10 min in two steps. In Step I, pressurised N_2 enters the bottom of the MB vessel, mixes the resin, and exits through the top vent.
10. Repeat nitrogen mix. In Step II, N_2 enters the centre of the resin bed through the interface distributor–collector. After 5 min of mixing, water is drained to bed level by opening the vessel drain valve and closing the vent valve. This ensures that the bottom of the bed is well mixed and compacted.
11. After the nitrogen mixing steps, refill the MB vessel with water. Water flows down through the bed from the normal service inlet forcing N_2 through the top of the vessel.
12. Final rinse of the MB resin bed for 5–10 min or until the product water resistivity achieves the specified value.

The regeneration cycle is initiated manually at the MCP. Once initiated, the PLC controls the regeneration cycle progressing through each step sequentially. Only one MB can be regenerated at a time. During acid regeneration, concentrated HCl (35% wt.) is educted from the acid tank, and mixed with dilution water transferred from the RO product water tank to make a 6% acid solution before it flows to the MB deioniser. Similarly, during caustic regeneration, concentrated NaOH (25% wt.) is educted from the caustic tank and mixed with water pumped from the RO product water storage tank to make a 4% caustic soda solution before it flows to the MB deioniser. The dilution water is heated to 43–49°C before it flows to the eductor. The water temperature is controlled by the PLC, which modulates the steam temperature control valve to adjust the steam flow rate. The acid and caustic concentrations (% wt.) of the regenerating solutions are measured downstream of the eductors by their respective conductivity sensors. The PLC adjusts each chemical flow control valve based on the measured concentration value.

Mixing of anion and cation resins is critical for achieving the design resistivity and silica during quality rinse. Cross contamination during regeneration can also affect the rinse time when the separation is poor. Under such circumstances, the IX resins are exhausted to achieve complete separation.

4.2 Process design and controls

Whether designing a new system or plant, modifying an existing one, or operating the plant, understanding the interface between process engineering and system controls is important. The process flow diagram in Figure 4.1 and the operating conditions given in this section form the basis for designing the unit operations. While system operation and performance are determined by the specifications of components, the automated control system is required to run the plant at system design conditions. Process specifications are discussed below. Equipment requirements are described in Section 4.5.

4.2.1 Granular media filtration (GMF)

GMF design criteria

The design of media filters is based on three criteria: (a) linear velocity, (b) freeboard for backwashing, and (c) operating pressure.[4,9] During normal service, linear velocity should be 10–20 $m^3/h.m^2$ cross-sectional area. Higher velocity affects filtration adversely, and can result in channeling in the media bed. The vessel diameter is based on the normal service feed flow rate to achieve the recommended linear velocity. The media bed consists of 0.6–0.8 mm anthracite (45 cm deep); 0.45 mm filter sand (35 cm deep); 0.25 garnet (5 cm deep) from top (coarser) to bottom (finer) with a depth of 0.9–1 m. The media bed depth is typically based on a maximum pressure drop of 1 bar at the design flow rate. The vessel straight shell height, in turn is based on bed depth and a bed expansion of 50–60% during backwashing. The backwash linear velocity should be 25–35 $m^3/h.m^2$ at 15–27°C. The design of vessel internal distributors and laterals is based on a minimum backpressure of 0.3 bar to achieve a uniform distribution of water across the bed cross section during the service run.

GMF operating conditions	
Service flow rate	70 m^3/h
Operating pressure	4.5 bar g
Operating temperature	12°C
Bed pressure drop	<1 bar
Backwash flow rate	170 m^3/h
Rinse flow rate	70 m^3/h
Valve sequencing	See Table 4.2

4.2.2 RO membrane system

RO design criteria

The RO unit design is based on meeting the specified product water quality and flux. RO unit service run is based on permeate conductivity (% rejection), productivity (% recovery), time, and feed pressure, which are affected by membrane scaling and fouling. The performance criteria for cleaning the membranes are discussed in Section 4.3.

The computer-generated design of the single-pass, two-stage RO membrane array (6:4) is shown in Figure 4.2. The stream numbers in the figure are: 1 is raw feed

Table 4.2 Granular media filter valve operation

Process cycle	Open valves*	Prime mover
GMF-061		
Normal service:		
Service water in/out	OV-061-1/OV-061-3	P-041 or P-042
Standby	All closed	off
Washing:		
Backwash in/out	OV-061-2/OV-061-4	P-211 or P-212
Bed settle	All closed	off
Fast rinse	OV-061-1/OV-061-5	P-041 or P-042
GMF-062		
Normal service:		
Service water in/out	OV-062-1/OV-062-3	P-041 or P-042
Standby	All closed	off
Washing:		
Backwash in/out	OV-062-2/OV-062-4	P-211 or P-212
Bed settle	All closed	off
Fast rinse	OV-062-1/OV-062-5	P-041 or P-042

* See Figure 4.4 for valve location. OV-xxx ... valves are pneumatic with manual override

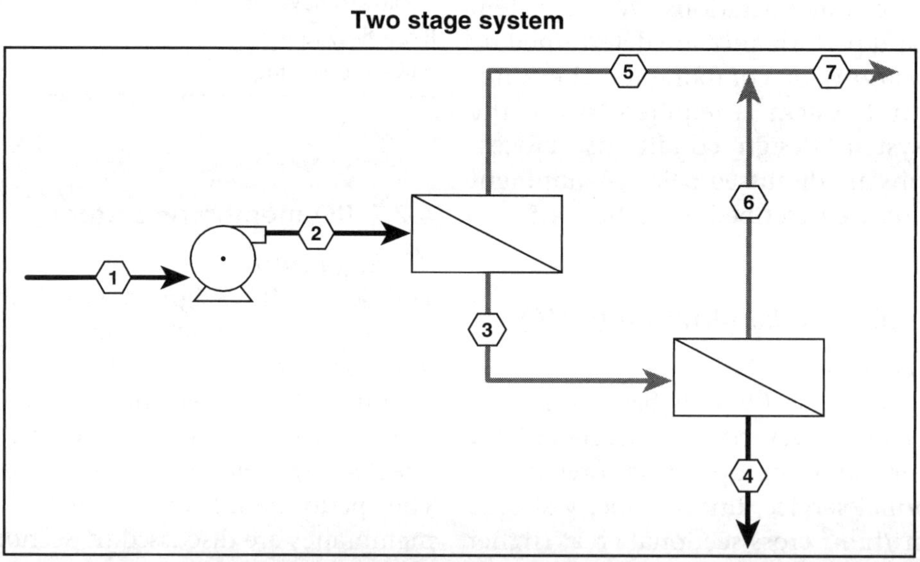

		1	2	3	4	5	6	7
Flow	(m^3/h)	70.7	70.7	33.2	17.7	37.5	15.5	53.0
Pressure	(bar)	0.0	6.0	4.2	2.9	0.0	0.0	0.0
TDS	(ppm)	242.2	242.2	513.2	958.8	2.2	6.1	3.3

Figure 4.2 RO membrane projection process flow schematic diagram

water, 2 is first-stage feed, 3 is first-stage reject/ second-stage feed, 4 is system reject, 5 is first stage permeate, 6 is second stage permeate, and 7 is system permeate. The RO membranes are thin-film composite polyamide wrapped in spiral-wound modules, 20 cm diameter × 100 cm long. Additional information is provided in Table 4.1. The β factor is within range, and the LSI is negative as required.

RO operating conditions	
Product water recovery	~75 %
Product water flow rate	53 m^3/h
Product water conductivity	<10.0 μS/cm
Average salt rejection	~99.0%
Feed water flow rate	70 m^3/h
Operating pressure	~15 bar g
Operating temperature	25°C
RO array ΔP	<4.1 bar
Cleaning solution flow rate/stage	0.2 m^3/h
Cleaning solution pressure	3 bar g
Cleaning solution temperature	<30°C
Cleaning solution pH	4.0–9.0
RO feed water SDI	<3.0
RO feed water LSI	<0
Valve sequencing	See Table 4.3

The RO systems are designed with the following general conditions for safe and reliable operation of an RO unit[2]:

- The initial minimum % rejection is based on chloride ions, and applies to each membrane element. The % rejection may be lower depending on the array. Further, membrane deterioration may cause individual element salt passage to double within three years.
- Always operate the RO unit at the lowest system pressure that produces the design flow rate and % rejection.
- Product water back-pressure greater than the system pressure at any time will damage the RO membranes.

4.2.3 Mixed-bed ion-exchange system

MB design criteria

The deioniser vessels are 2.1 m diameter × 3 m straight shell height. Each MB vessel contains 3.26 m^3 strong base anion resin (OH^- form), 2.18 m^3 strong acid cation resin (H^+ form), and 0.54 m^3 inert resin. This translates into a freeboard of 72% for bed expansion when the bed depth is 1.7 m. The service linear velocity of 25 $m^3/h/m^2$ (10 gpm/ft^2) is within the resin manufacturer's range of 25–35 $m^3/h/m^2$ (10–15 gpm/ft^2). The backwash linear velocity is 3 $m^3/h/m^2$. Based on resin manufacturer's hydraulic expansion curves, the bed expansion is about 60% when the water temperature is 16°C. A general procedure for designing ion exchangers is given in Chapter 5.

MB operating conditions	
Product water flow rate	80 m^3/h
DI water resistivity	≥16.0 MΩ-cm
Operating pressure	5.9 bar g
Operating temperature	Ambient
Service run time	250 h
Maximum bed ΔP	2.1 bar
Backwash flow rate	45 m^3/h
HCl flow rate	5 m^3/h at 6% conc.
NaOH flow rate	7 m^3/h at 4% conc.
NaOH temperature	43–49°C
Slow rinse flow rate:	
Cation bed	4 m^3/h
Anion bed	6 m^3/h
Fast rinse flow rate:	
Cation bed	24 m^3/h
Anion bed	24 m^3/h
Final rinse flow rate	80 m^3/h
Nitrogen supply	425 N m^3/h at 2.1 bar g
Valve sequencing	See Table 4.4

Table 4.3 RO unit valve operation

Process condition	Open valves*	Prime mover
Service feed/ product/reject	OV-161-1/ MV-161-3/ MV-160-2	P-161
Membrane array feed pressure control	MV-161-1 (open 30%)	N/A
Reject flow control	MV-161-2 (open 50%)	N/A

*See Figure 4.7 for valve location. OV-161-1 is pneumatic with manual override

4.2.4 System controls

Generally, the purpose of a control system is to supervise, control, monitor, schedule, document, and record process parameters that are vital to plant operation. The control system of any plant or system is related to the process design of the plant and ultimately its operation, and must be able to cope with various uncertainties. In other words, the control system is closely tied to process operability. Process operability, in turn, involves the coordinated effort of both the operator and the automatic control system. The automatic controls system is programmable logic controller (PLC)-based and includes the main control panel (MCP) and local control panels (LCP). PLC systems are preferred to discrete control system because of the flexibility in establishing set points and providing multiple functions.

Instrumentation and control systems are provided for monitoring and controlling the major process variables related to each equipment and unit operation, as shown in the process flow sheets (PFSs) in Figures 4.3–4.13.

Table 4.4 Mixed-beds valve operation

Process cycle	Open valves*	Prime mover
MB-281		
Normal service:		
Service water (In/Out)	OV-281-19; OV-281-11/ OV-281-13	P-231 or P-232
Standby	All closed	off
Regeneration:		
Backwash (In/Out)	OV-281-18; OV-281-12/ OV-281-14	P-233
Acid regeneration (In/Out)	OV-330-2; OV-281-16/ OV-281-23	P-233 and eductor
Caustic regeneration (In/Out)	OV-340-2; OV-281-17/ OV-281-23	P-233 and eductor
Nitrogen mixing (In/Out)	OV-281-20; OV-281-12/ OV-281-22; OV-281-15	In-house nitrogen supply
Rinse (In/Out)	OV-281-18; OV-281-11/ OV-281-15	P-233
MB-282		
Normal service:		
Service water (In/Out)	OV-282-19; OV-282-11/ OV-282-13	P-231 or P-232
Standby	All closed	off
Regeneration:		
Backwash (In/Out)	OV-282-18; OV-282-12/ OV-282-14	P-233
Acid regeneration (In/Out)	OV-330-2; OV-282-16/ OV-282-23	P-233 and eductor
Caustic regeneration (In/Out)	OV-340-2; OV-282-17/ OV-282-23	P-233 and eductor
Nitrogen mixing (In/Out)	OV-282-20; OV-282-12/ OV-282-22; OV-282-15	In-house nitrogen supply
Rinse (In/Out)	OV-282-18; OV-282-11/ OV-282-15	P-233

*See Figures 4.11–4.13. OV-xxx ... valves are pneumatic with manual override

Instrumentation includes from the primary element or sensor monitoring a controlled variable to a modulating control valve performing the throttling action. The pumps, media filters, and the mixed-bed deionisers can be controlled automatically or manually, for example to control the flow rate or storage tank's water level. However, the RO feed pressure and reject flow control valves are controlled only manually. All pump motors are controlled at either the LCP or the MCP. The pneumatic valves can be controlled manually by opening the override button for each corresponding solenoid valve in the local panel.

Analogue data signals (4–20 mA) from the instruments are sent to the PLC input module located in the MCP, and to a digital display. These displays provide operators with necessary data. All analogue signals for control such as pressure and flow for process water temperature control are obtained from the PLC. Alarms are produced by data received by the PLC from the instruments. Alarms are only produced when the process, the instrument is monitoring, is active. Most alarms often have several seconds delay associated with them to ensure the process has stabilised before activating the alarm.

The PLC responds to uncertainties and alarm conditions for each equipment or unit operation, as defined in Table 4.5. The process description in Section 4.1 and process conditions defined in Section 4.2 provide the basis for operating the DI water plant equipment shown in Figures 4.3–4.13.

4.3 System operation

Initial start-up is the key to operating any plant or system efficiently and successfully because start-ups are often plagued with problems. Hence, strict criteria must be adhered to, defined either by the customer specifications or the plant manufacturer's standards. During start-up, the field engineers ensure all equipment is installed correctly, the vessels and the piping has been hydro-tested, chemical tanks are full and the chemical pumps are calibrated, wiring is complete, the centrifugal pumps have been checked for rotation, all instruments are calibrated, the UV lamps and microfilter cartridges have been loaded, the RO membranes have been installed in the pressure vessels, and the packed bed vessels have been filled with the specified media. In addition, during initial start-up the filter media is washed, the RO membranes are flushed of all preservatives, and the mixed-bed resins are regenerated and thoroughly mixed.[5,6,10,11]

4.3.1 Start-up requirements

Criteria for acceptance of equipment should be reviewed first. After the equipment is installed, standard preparation procedures listed below should be followed to ensure a safe and reliable operation.[5,6,10,11]

1. Using the process flow sheets (PFSs), equipment and instrument lists, check each system for piping, valves, and interconnecting piping.
2. Check all electrical connections.
3. Check all instruments and accessories.
4. Check all vents are operational.
5. Check lubrication on all pumps.
6. Check all centrifugal pumps for alignment and rotation.
7. Turn on instrument air and process nitrogen.
8. Check that steam is available for the heat exchangers.
9. Set pressure relief valves as required.
10. Check automatic valves operation.
11. Hydro-test the vessels and piping, and check for leaks.

Table 4.5 Process alarms

Process condition	Controller action	Set-point
Raw water high turbidity	Alarm	> 1.0 NTU
Raw water pump low pressure	Alarm; pump off	0.7 bar g
Raw water pH range	Alarm	6.0–9.0
Granular media filters high ΔP	Alarm	1 bar
Bisulphite tank low level	Alarm; bisulphite pump off	15%
Anti-scalant tank low level	Alarm; anti-scalant pump off	15%
NaOH tank low level	Alarm; NaOH pump off	15%
RO feed water high pH	Alarm	9
RO feed water low pH	Alarm	6
RO feed water high chlorine	Alarm	0.1
Cartridge filter high ΔP	Alarm	1 bar
RO feed water high temp.	Alarm; Temp. control valve closed	29°C
RO feed water low temp.	Alarm; Temp. control valve open	18°C
RO pump low inlet pressure	Alarm; RO pump off	0.7 bar g
RO pump high outlet pressure	Alarm; RO pump off	18 bar g
RO low feed pressure	Alarm;	12 bar g
RO low reject pressure	Alarm;	6.9 bar g
RO high feed pressure	Alarm; RO pump off	19 bar g
RO high reject pressure	Alarm	18 bar g
RO high product pressure	Alarm	3.5 bar g
RO array high ΔP	Alarm	5.5 bar
Tank T-200, low level	Alarm; Transfer pumps off	15%
Tank T-200, high level	Alarm;	95%
Tank T-220, low level	Alarm; Transfer pumps off	15%
Tank T-220, high level	Alarm; RO pump off	95%
RO permeate high conductivity	Alarm; permeate divert valve open	10.0 μS/cm
RO permeate low rejection	Alarm; permeate divert valve open	<92%
RO permeate high recovery	Alarm	>80%
MB product low resistivity	Alarm; DI tank divert valve open	≤15.0 MΩ-cm
MB product high silica	Alarm; DI tank divert valve open	20 ppb
Cartridge filter high ΔP	Alarm	0.7 bar
DI water low resistivity	Alarm	≤15.0 MΩ-cm
Acid tank low level	Alarm	15%
Caustic tank low level	Alarm	15%
Regen. HCl low concen.	Alarm	6%
Regen. HCl high concen.	Alarm	8%
Regen. NaOH soln. low concen.	Alarm	3%
Regen. NaOH soln. high concen.	Alarm	5%
Regen. NaOH soln. high temp.	Alarm; Temp. control valve closed	29°C

12. Drain vessels and visually inspect strainers and internals.
13. Rinse and drain all storage tanks.
14. Purchase all required chemicals.
15. Ensure the HEPA filters have been installed on the purified water storage tanks.
16. Load the filter media and ion-exchange resins.
17. Set and check temperature controller settings of the two heat exchangers.
18. Load RO membranes as per the membrane manufacturer's instructions.
19. Check the Silt Density Index (SDI) of raw water. It should be <4. Check the turbidity of raw water. It should be <1.0 NTU.

Typically each unit operation (UOP) shown in Figures 4.3–4.11 is started in sequence to ensure there are no bottlenecks during start-up. It also ensures any trouble spots are rectified before proceeding to the unit operation, equipment, or instrument. The initial start-up procedure is often used as a hands-on-training exercise for the plant operators.

Granular media filters
The filters are washed to remove media fines prior to initial start-up as well as during the normal service run when required, as described in Section 4.1. The wash cycle consists of backwash, bed settle, and rinse steps. The operating conditions and valve sequencing are given in Section 4.2. It is required that the system is shutdown whenever there is a leakage.

RO membrane unit
All new RO elements are usually shipped with chemical preservatives (e.g. sodium meta bisulphite) to protect them from biological fouling. The RO membranes must be flushed to displace the preservative before using any RO product water as per the membrane manufacturer's instructions for flushing the membranes. The membranes are flushed for 2–3 h at line pressure (RO high-pressure pump is off). The system operating conditions and valve sequencing are given in Section 4.2. It is required that the system is shutdown whenever there is a leakage.

Mixed-bed deionisers
The operating conditions and valve sequencing are given in Section 4.2. The ion-exchange resins are regenerated during initial start-up to wash the resin beds and to restore the resins to their desired operating conditions as well as during the normal service run when required as described in Section 4.1 and outlined in Table 4.6. The strong base anion resins are regenerated to the hydroxyl form (OH^-), and the strong base cation resins to the hydrogen (H^+) form. It is required that the system is shutdown whenever there is a leakage.

If the IX resin is not fully exhausted prior to regeneration, it is difficult to separate the anion and cation resins completely. Complete separation is required to prevent cross contamination and to produce ultra-high-purity water with resistivity greater than 18.0 MΩ-cm. The procedure described below called 'caustic kill' is employed to convert the heavier cation-exchange resin sites to the sodium form resulting in even higher density than the lighter anion resins in the hydroxyl form. The procedure ensures complete separation of resins during backwashing:[8]

1. Flow a 4% NaOH solution through the entire bed for 15–20 min at the standard flow rate required for regeneration.
2. Monitor the resistivity of the water flowing to the drain. Continue until the MB effluent resistivity is 0.01 MΩ-cm indicating cation resin exhaustion.
3. Mix the resins thoroughly to ensure further exhaustion.

Table 4.6 Mixed-bed regeneration cycle operation

Step	Cycle	Flow rate (m^3/h)	Duration (min)	Remarks
1	Pause for resin exhaustion	n/a	n/a	All valves closed
2	Caustic sol. for exhaustion	7.3	30	4% NaOH at 0.9 m^3/h
3	Bed rinse	65.5	155	P-233 is on
4	Pause for backwash	n/a	n/a	All valves closed
5	Backwash I	35	8	P-233 is on
6	Backwash II	26.1	20	P-233 is on
7	Settle the bed	n/a	10	All valves closed
8	Bed pre-heat	10.2	14	Water at 43°C
9	Caustic regeneration/acid regeneration	7.3/4.8	54	4% NaOH downflow/6% HCl upflow
10	Caustic regeneration/dilution water block	7.3/3.6	27	4% NaOH downflow/dilution water upflow
11	Slow hot rinse/anion bed and cation bed	6.4/3.6	43	Anion bed downflow/cation bed upflow. Dilution water at 43°C
12	Slow cold rinse/anion bed and cation bed	6.4/3.6	15	Anion bed downflow/cation bed upflow
13	Fast rinse I/anion bed and cation bed	24.3/24.3	1	Anion bed downflow/cation bed upflow, 50/50
14	Fast rinse II/anion bed and cation bed	24.3/24.3	38	Anion bed downflow/cation bed upflow, 50/50
15	Blow-down I	n/a	1	Bed drain
16	Blow-down II	n/a	4	Drain to 20 cm above bed level
17	Re-mix I	n/a	30	Nitrogen at 119 m^3/h
18	Re-mix II	n/a	1	Nitrogen at 425 m^3/h
19	Bed compaction	n/a	5	All valves closed
20	Vessel re-fill	18.2	4	Vent valve open
21	Pause for final rinse	n/a		
22	Final rinse	79.5	100	Rinse until resistivity is $\geq$16.0 MΩ-cm. P-233 is on
23	Quality rinse	79.5	50	P-233 is on, if necessary

4. Rinse the bed for 4–5 min.
5. Backwash the bed to separate the resins.
6. Regenerate the resins as described above.

Miscellaneous equipment

Other equipment including the microfilters and the ultraviolet (UV) lights must be rinsed before installing the specified cartridges and the UV lamps.

4.3.2 System operation

Once all the system components have been fine-tuned, the plant is ready for operating in its normal service mode. In modern hybrid membrane plants, system operation has become less dependent on the operator and more dependent on the automatic systems in place. However, human intervention is required during start-up. During normal operation, operators are required for starting the systems, for performance monitoring and data collection, for cleaning membranes, and sometimes for manually initiating media filters wash cycle and mixed-beds regeneration cycle, replacing membranes and filter cartridges, routine maintenance of equipment and instruments as well as for correcting problems that may occur.

Service operation checklist

Follow the checklist given below to guide the engineer through the service run preparation process.[5,6,10,11] In addition, the equipment maintenance records provided in Section 4.4 should be reviewed prior to start-up.

1. Measure the SDI value of RO feed water. It should be less than three. The turbidity must be less than 1.0 NTU.
2. Review the process operating parameters given in Section 4.2.
3. Activate all instruments including analysers, process monitors, level and pressure switches, and pressure and temperature transmitters.
4. Check that the anti-scalant (A/S), sodium bisulphite (SBS) and caustic soda (NaOH) tanks are full.
5. Ensure that the regeneration skids HCl and NaOH tanks are full.
6. Verify if the UV lamps have been installed.
7. Verify if the microfilter cartridges have been installed.
8. Turn the nitrogen supply for instruments, control valves, mixing resins and for blanketing RO product water storage tank.
9. Open the inlet and outlet valves of all the pumps as well as the storage tanks inlet and outlet valves.
10. Close the drain valves of all vessels. Open the vent valves of cartridge filter housings.
11. Open all service manual valves on all skids including the media filters, cartridge filters, heat exchangers, UV lights, RO unit, and mixed-beds.
12. Set the RO water tank nitrogen blanket pressure at 3 in. w.c.
13. Turn the steam supply on for the RO feed water heat exchanger.
14. Review valve sequencing for the multimedia filters given in Table 4.2.
15. Review valve sequencing for the RO unit given in Table 4.3.
16. Review valve sequencing for the mixed-beds given in Table 4.4.
17. Turn the main control panel (MCP) power switch on.
18. Verify if the alarm set-points given in Table 4.5 have been entered in the PLC.
19. Power the following equipment in automatic mode at the MCP:
    (a) Feed water pump, P-041 or P-042
    (b) Sodium bisulphite pump, CP-131A or CP-131B
    (c) Anti-scalant pump, CP-132A or CP-132B

(d) Caustic soda pump, CP-133A or CP-133B
(e) RO high-pressure pump, P-161
(f) RO water distribution pump, P-231 or P-232, and
(g) UV lights, UV-231/ UV-232.

20. Use thc process flow sheets in Figures 4.3–4.13 to guide you when operating each equipment.

Service run

1. Ensure that the equipment has been screened for operation as defined above.
2. Prepare the daily data log such as Table 4.7 for data entry.
3. Press the service run pushbuttons for the GMF, RO, and MB units at the MCP.
4. Water will start flowing to the media filter skid at 70 m^3/h and 4 bar g. Monitor the flow rate using FQIT-060. Monitor the pressure drop across the filter using DPIS-060. It should be less than 1 bar.
5. The media filter, GMF-061 is shutdown when the differential pressure across the media bed exceeds 1 bar, and the standby filter, GMF-062 is brought online.
6. Turn the backwash pump, P-211 switch to the auto position. Initiate the backwash cycle for GMF-061 at the MCP. Once initiated, the PLC controls the cycle, as described in Section 4.2.
7. Monitor RO feed water chlorine using analyser AIT-130. It should be less than 0.1 ppm.
8. Monitor RO feed water pH using analyser pHIC-130. It should be between 8 and 8.5.
9. Check the pressure drop across the cartridge filter, F-141 using DPIS-140. Replace the cartridges when ΔP is >1 bar or 3 months whichever is sooner.
10. Monitor RO feed water temperature. It should be between 24 and 26°C.
11. The RO unit has several start-up conditions that must be satisfied:
    (a) During the 60-s pre-service flush cycle, the RO pump is off while the RO membrane brine channels are flushed with water at line pressure (<2 bar g).
    (b) The pre-service flush cycle is followed by a 5-min quality rinse cycle when the RO permeate is diverted to the water reuse tank, T-200 via the permeate divert valve, MV-161-4.
    (c) After 5 min, the RO pump, P-161 starts provided the inlet pressure exceeds the minimum permissible value as monitored by PSL-161.
12. Monitor the RO product water conductivity using conductivity monitor CIT-161. It should drop below 10.0 μS/cm.
13. Adjust the reject flow control valve, MV-161-2 until the reject flow rate is 17 m^3/h when the RO permeate flow rate is 53 m^3/h as monitored by FQIT-161B/FQIT-161A, respectively.

14*. It may be necessary to throttle the RO pump discharge valve, MV-161-1 (30% open nominal) while adjusting the reject flow control valve, MV-161-2 (50% open nominal) to achieve the desired flow rate. When making adjustments, be sure to avoid excessive pressure and flow.

15. When the product water conductivity is ≤10.0 μS/cm, open the permeate service valve, MV-161-3, and close the divert valve, MV-161-4. Water will start

*The RO unit contains only one user controllable function, that being the control of the final reject flow rate. The reject flow control valve (50% open) is used in conjunction with the RO feed pressure control valve (25–30% open). The reject flow valve and the RO feed pressure valve are throttled to achieve the proper operating pressure necessary to achieve the desired product water recovery. These control valves may require fine-tuning to achieve the desired productivity. In addition, as the membrane performance declines with time due to changes in the material properties of the polymer, the control valve settings would require adjustments, when higher feed pressures would be required.

Table 4.7 A typical daily data collection form

Parameter	S	M	T	W	T	F	S
	(date)	(date)	(date)	(date)	(date)	(date)	(date)
Feed water data:							
Flow rate (m^3/h)							
Granular media filters ΔP (bar)							
Turbidity (NTU)							
Total hardness (mg/l)							
SiO_2 (ppm as SiO_2)							
Total dissolved solids (mg/l)							
pH							
RO unit data:							
Cartridge filter ΔP (bar)							
RO feed water SDI							
Feed temperature (°C)							
Feed pH							
Feed chlorine (ppm)/ORP (mV)							
Feed conductivity (μS/cm)							
Feed flow rate (m^3/h)							
RO pump inlet pressure (bar g)							
1st stage operating pressure (bar g)							
2nd stage operating pressure (bar g)							
Permeate pressure (bar g)							
Membrane array ΔP (bar)							
Permeate flow rate (m^3/h)							
Reject flow rate (m^3/h)							
Product water recovery (%)							
Normalised productivity (m^3/h)							
1st stage permeate conductivity (μS/cm)							
2nd stage permeate conductivity (μS/cm)							
Product water conductivity (μS/cm)							
Salt rejection (%)							
MB deioniser data:							
Product water resistivity (MΩ-cm)							
Product water silica (ppb)							
Mixed-bed ΔP (bar)							
Cartridge filter ΔP (bar)							
DI water resistivity (MΩ-cm)							

Table 4.8 RO unit performance analysis

Item	Diagnosis	Resolution
% Rejection – Low	• Product water recovery higher than projected design recovery	• Reduce recovery and adjust flow rates to match RO projections
	• Inadequate operation pressure	• Increase RO operating pressure to match design projections
	• Change in feed water chemistry	• Check feed water chemistry. Run a new RO projection
	• Membranes may be fouled	• Determine the nature of the foulant(s) and clean the membranes
	• RO pretreatment equipment malfunction	• Check feed water chemistry
	• Exposure to chlorine	• Check sodium bisulphite injection pump; replace damaged elements
	• Mechanical leaks in membrane elements	• Check product water conductivity at each housing to isolate the leaking seals
% Recovery – Low	• RO operating pressure is too low	• Readjust membrane feed pressure
	• Permeate back-pressure high	• Ensure there are no restrictions in the permeate line
	• Membrane elements fouled by sparingly soluble salts; mineral scale	• Determine the nature of the foulant(s) and clean the membranes
	• RO feed temperature is colder	• Adjust the temperature to 25 ± 2°C
Permeate flux – High	• Chemical attack/chlorine exposure	• Replace damaged elements
	• Mechanical leaks in membrane elements	• Check the product conductivity at each housing to isolate the leaking seals

flowing to the RO product water tank, T-220 at 53 m^3/h.

16. Enter RO performance data in Table 4.7, and use the data along with information given in Table 4.8 to evaluate the performance.
17. Clean the membranes, as described in Chapter 2, to restore the performance. Cleaning is implemented when any of the conditions discussed in Section 4.3.3 are true.
18. The RO water distribution pump, P-231 or P-232 transfers water to the mixed-bed polishing system at 80 m^3/h and 5 bar g.
19. Water flows through two TOC destruct ultraviolet (UV) lamps, UV-241/UV-242. The UV lamps switch on automatically when the RO water distribution pumps are turned on, and water starts flowing through the UV units. Monitor UV lamp intensity and temperature.
20. Water flows to the mixed-bed MB-281 at 80 m^3/h. Monitor the pressure drop across the vessel. The ΔP should be <1 bar.
21. Monitor the product water resistivity using analyser RIT-281. It should be >16.0 MΩ-cm.
22. Monitor the product water silica. It should be <20 ppb.
23. Monitor the pressure drop across the final filters, F-291/F-292 using DPIS-290. Replace the cartridges when ΔP is >0.7 bar.
24. Monitor the system product water resistivity using analyser RIT-290. It should be ≥16.0 MΩ-cm.

25. Enter MB performance data in Table 4.7.
26. Evaluate the MB unit performance based on the criteria given in Table 4.9.
27. The mixed-bed, MB-281 is shutdown when the product water quality deteriorates. The standby mixed-bed, MB-282 is brought on-line simultaneously.

Table 4.9 Mixed-bed performance analysis

Item	Diagnosis	Resolution
Resin capacity – Low	• Increase in total dissolved solids (TDS) concentration in feed water	• If no change, go to next step
	• Loss of resin	• Measure bed depth with resin in fully exhausted state. If no change, go to next step
	• Faulty regeneration conditions	• Verify, (a) amount of regeneration solution used monthly is constant, (b) regeneration flow rates, and (c) regeneration concentrations
	• Inaccurate measurement of throughput	• Check integrator/totaliser flow meters
	• Faulty end-point measurement	• Check analytical instruments for measuring resistivity and silica
	• None of the above	• Analyse resins. Take a core sample from each quadrant of the resin bed. Combine the samples (1 l) and send for analysis
Product water quality – Inferior	• Higher than specified sodium leakage	• Check regeneration sequence at off-site regeneration facility
	• Silica leakage due to oxidative loss of functionality of anion resin	• Verify if there is chlorine in water
	• Silica leakage due to thermal loss of functionality of anion resin	• Check caustic solution temperature <54°C; verify the PLC controller set-point
	• Higher silica leakage due to incomplete NaOH regeneration	• Check NaOH regeneration solution concentration and flow rate
	• Fouled resins	• Check ΔP across the bed
Bed ΔP – High	• Inlet distributor blocked by finer resin beads due to a high backwash flow rate	• Reduce backwash flow rate. Verify water temperature is not colder than design temperature
		• Check resin bed is not compacted
	• Bed distributors/laterals blocked	• Check resin bed laterals distributors
	• Inaccurate flow meter readings	• Check flow meters regularly to ensure correct flow rates
	• Vessel walls lining	• Check the quality of rubber lining by spark testing
	• Resin breakdown due to physical (osmotic and mechanical attrition) and thermal shock effects	• Remove fines by back washing. Change operating conditions

28. Regenerate the MB unit based on the procedure given in Table 4.6. Initiate the regeneration cycle for MB-281 at the MCP, as discussed earlier in this section. Once initiated, the PLC controls the cycle.

4.3.3 Performance evaluation

Data collection and analysis is critical for monitoring system performance. As a minimum requirement the data must be entered in Table 4.7. Mathematical relationships for converting field data to performance data are given in Chapter 2. The normalised data must be plotted to study the performance trends, for example RO membrane rejection vs. time and RO productivity vs. time. The key performance parameters in Tables 4.8 and 4.9 should be used in conjunction with normalised data to evaluate system output. The data along with RO systems design guidelines provided in Chapter 2, and RO projections given in Table 4.1 should be used to understand and troubleshoot the performance over time.

Most membrane problems occur due to a failure to record and analyse data. RO product flow data must be normalised to the same feed water temperature since RO membrane flux increases 3% per degree rise in temperature (°C). Data trending helps in diagnosing the system. Without data trending it is difficult to analyse when the RO unit needs to be shutdown for cleaning the membranes. Membrane manufacturer's recommend that membranes should be cleaned when any of the following conditions are true:

- Permeate flow has dropped to 10–15% below the rated flow at normal pressure.
- Temperature-corrected feed pressure has increased 10–15% to maintain the rated permeate flow.
- Permeate quality has decreased 10–15%; salt passage has increased 10–15%.
- Applied pressure has increased about 10–15%.
- The differential pressure across an RO unit has increased noticeably (>15%).

Once the membranes are cleaned, the operating conditions at the time of restart must be noted. This is the new base data for comparing the performance before the next shutdown. Membrane flux is never restored 100% after cleaning due to aging of the membranes, and especially if the membranes are fouled, as shown in Figures 2.34 and 2.35. For a well-designed RO pretreatment system, there is a gradual decline in flux with time. The slope of flux vs. time curve is a function of membrane compaction, restoration of flux after cleaning of mineral scales, and the severity of fouling. Cleaning may not restore the flux if the fouling is irreversible.[2,9]

4.4 System diagnosis and maintenance

4.4.1 Diagnostics

Considerable troubleshooting is often necessary during plant start-up. During normal service, troubleshooting is critical for understanding why a certain component is malfunctioning, and what can be done to fix the problem for example by changing either the operating conditions or the equipment. Early detection of potential problems is possible through consistent record keeping and performance analysis. Typical symptoms and recommended procedures are discussed. For alarm-related conditions, the information provided in the alarm chart in Table 4.5 can serve as a valuable guide. Once the problem has been detected, the cause of the problem must be identified and follow-up action taken according to the recommended remedial action.

According to an old adage, "prevention is better than cure". Maintenance and calibration of instruments are important since accurate readings are required to detect a problem. Proper corrective action must be taken as soon as possible; otherwise other equipment may be affected adversely. Preventive maintenance procedures are discussed in Section 4.4.12. In addition, the manufacturer's recommendations should be followed.

4.4.2 Process water pumps

General troubleshooting guidelines are provided for centrifugal pump equipment and process conditions. The pumps can be single-stage or multi-stage, horizontal, vertical, or submersible. Process-related problems depend on process design conditions for a particular system.

4.4.3 Process water storage tanks

General troubleshooting guidelines are provided for process water storage tanks. Such tanks act as equilibrium tanks for receiving water from upstream equipment and supplying water to downstream equipment or other equipment in the plant. The tanks are always supplied with level switches and or level transmitters to start or shut-off upstream/downstream equipment, e.g. shutdown the process transfer pumps to protect the pumps from running dry and getting damaged or preventing tank overflow.

Problem (4.4.2)	Solution
I. Equipment related:	
Pump leaks at the pump head connection	• Inspect and replace the pump head seal
Pump leaks at the pump shaft	• Inspect and replace the pump shaft mechanical seal
Excessive weep rate	• Examine the pump high-speed shaft assembly
Pump capacity is below design	• Check pump suction line for blockage, including valves • Check pump discharge line for blockage, including valves • Check pump impeller for damage
The performance has dropped below the pump curve	• Examine the pump high-speed shaft assembly
Pump performance is low or the pump is noisy	• Check pump suction and discharge lines for blockage, including closed valves • Check the system for entrapped air in the pump suction line • Check for damage to the pump impeller
II. Process related:	
Pump does not run	• Verify that the pump is energised • Ensure that the inlet valve is not closed • Check the low-pressure switch set-point. Increase the feed pressure • Ensure that the feed water tank is at low-level set-point Fill the tank
Pump shuts down	• The discharge pressure is too high. Check the high-pressure switch set-point • Process water receiving tank is full • Downstream equipment is not calling for water

Problem (4.4.3)	Solution
Low-level alarm	• Check the status of tank water transfer pump to ensure that the pump was not in manual mode • Check the low-level alarm setting in the PLC • Check that the tank inlet water valves are open
High-level alarm	• Check that the high alarm set-point corresponds to the high-high tank level • Check that the tank inlet valves are closed, and the divert valves are open • Check that the tank feed water supply pumps/systems are not operating (should turn off automatically due to a process inter-lock • Check that the feed pumps are not on

4.4.4 Chemical feed systems

General troubleshooting guidelines are provided for chemical dosage systems including tanks, pumps, and tank mixers. Not all problems may be relevant to a particular system. Problems may depend on system design and process conditions.

4.4.5 Granular media filters

General troubleshooting guidelines are provided for media filters. These guidelines are also applicable to other media filter systems such as activated carbon filters. Problems may depend on system design and process conditions.

4.4.6 Dead-end cartridge microfilters

General troubleshooting guidelines are provided for cartridge microfilters. These filters vary in size rating, generally varying in pore size from $< 0.1\,\mu m$ for final polishing filters to $> 5.0\,\mu m$ filters for RO pre-filters. Sometimes cartridge filters with pore sizes of $10.0\,\mu m$ are used in lieu of multimedia filters including backwashable cartridge filters. The pore size rating is usually nominal for $1.0\,\mu m$ or higher cartridges and absolute for less than $1.0\,\mu m$ cartridges. Absolute cartridges usually

Problem (4.4.4)	Solution
Chemical pumps fail to operate	• Check the start button is not in the auto position • Ensure the water flow rate exceeds the minimum permissible value • Check whether the solution level in the tank is below the low-level set-point value
Pump dosage rate is low	• Verify that the electronic pump controller is functioning at the desired rate. Verify pump speed and stroke length • Check the operation of the flow switch and the set-point value
Tank mixer fails to operate	• Check the start button is not in the auto position • Check whether the solution level in the tank is below the low-level set-point value
Low-level alarm	• Check the low-level alarm setting in the PLC • Check whether the solution level in the tank is below the low-level set-point value
Level switches fail to alarm or alarm prematurely	• Check the operation of the level switches
High chlorine (or ORP) alarm	• Verify chlorine (or ORP) analyser/controller set-point value • Check sodium bisulphite pump flow setting and calibration
High pH alarm	• Verify pH analyser/controller set-point value • Check NaOH pump flow setting and calibration

Problem (4.4.5)	Solution
No supply of water	• Check that the feed water pumps are operative, and the downstream equipment is calling for process water
The filter does not run in service mode	• Ensure that the downstream equipment is calling for process water • Ensure that the filter start switch is on • Verify the filter is not on in the wash cycle
Filter effluent turbidity is greater than 1.0 NTU	• Recalibrate the turbidity analyser • Initiate the wash cycle • Verify wash cycle flow rates match the design condition
Excessive ΔP acros the media bed	• Check that the ΔP switch is not out of calibration. • Initiate a two-step backwash sequence; start at one-half the flow rate gradually increasing to the design flow rate • Verify that the bed has not shrunk or compressed • Make provisions for air scouring to break loose the particles trapped in the upper layers of the bed

Problem (4.4.6)	Solution
High ΔP alarm	• Check the differential pressure switch set-point • Change the filter elements
Cartridge filter elements foul prematurely	• Check the suspended solids removal performance (turbidity) of upstream media filter • Check the composition of raw water for any changes • Measure the SDI value of feed water • Media or resin fines in the cartridge filter indicates that a vessel distributor/lateral has failed

have a net negative charge to ensure >99% rejection of particles both by ionic repulsion and pore blockage.

4.4.7 Heat exchangers

General troubleshooting guidelines are provided for heat exchanger performance. Process-related problems depend on operating and process design conditions, and may not be unique to the system deployed, e.g. steam heated vs. chilled water heat exchangers. Procedures for cleaning heat exchangers are provided by the manufacturers. Heat exchangers may be plate-and-frame or shell-and-tube type.

Problem (4.4.7)	Solution
Loss of temperature control	• Check the temperature set-point value in the PLC • Check the operation of temperature transmitter • Check the performance of steam pressure regulating valve • Check the performance of temperature control modulating valve; check the operation of the positioner • Check for loss of instrument air pressure.
High-temperature alarm	• Verify that the controller set-point is below the high-temperature alarm value • Check the set-point of the temperature switch
Low-temperature alarm	• Verify that the controller set-point is above the low-temperature alarm value. Check the set-point of the temperature switch

4.4.8 RO membrane system

General guidelines are provided for troubleshooting RO system performance. Process-related problems depend on operating and process design conditions. Performance analysis information in Table 4.8 provides additional information for solving process and system problems.

Problem (4.4.8)	Solution
Feed/Permeate high conductivity	• Check the conductivity cell sensor. Check the conductivity analyser calibration • Check RO feed water composition
Low-rejection alarm	• Check the conductivity cells. Check the calibration of conductivity analysers • Check the conductivity of permeate from each pressure vessel manually, and verify feed bypass due to damaged O-rings or membranes
Product water recovery is high	• Verify the permeate and reject flow rates are within the design specifications • Check the calibration of flow monitors • Adjust reject flow control valve and/or pump speed
Feed high-pressure alarm	• Check the alarm set-point of the high-pressure switch • Check the ΔP across the RO unit; if it has increased significantly, higher feed pressure would be required to maintain normal productivity
Permeate high-pressure alarm	• Check the set-point of the high-pressure switch • Verify if there is any back-pressure in the permeate line
High ΔP alarm	• Check the historical performance data to ascertain if the increase is gradual or sudden. A gradual increase indicates fouling; a sudden increase indicates a faulty pressure indicator or a catastrophic membrane failure

4.4.9 Mixed-bed ion exchangers

General guidelines are provided for troubleshooting mixed-bed ion-exchange systems. Process-related problems depend on operating and process design conditions. Performance analysis in Table 4.9 provides additional information for solving process and system problems.

Problem (4.4.9)	Solution
Product water high silica concentration	• Re-mix the resin during start-up • Regenerate the resin bed. Check caustic regeneration cycle temperature meets the design condition • Monitor rinse water quality
Low-resistivity alarm	• Re-mix the resin during start-up • Regenerate the resin bed • Monitor rinse water quality
Mixed-bed fails to 'rinse up' after regeneration	• Verify the standard regeneration procedure, given in Table 4.6, and as follows: (a) Resin separation during the backwash step. Ensure there is a sharp resin separation as seen through the middle sight-glass (b) Check caustic dilution water flow rate, concentration, and duration. Also check the acid dilution water flow rate required for blocking the caustic flow (c) Check the acid and caustic dilution water flow rates and solution concentrations during the caustic addition and acid addition steps (d) Ensure that the water level after the bed drain-down is no more than 3–6 cm above the bed level (e) Confirm proper 'air' mix flow rate and duration

	• Re-mix the resins
Resin bed high ΔP	• As discussed in Table 4.9

4.4.10 MB regeneration systems

General troubleshooting guidelines are provided for troubleshooting mixed-bed ion-exchange systems. Use the guidelines in Section 4.4.9 when solving system problems. Process-related problems depend on operating and process design conditions, and may not be unique to the system deployed.

Problem (4.4.10)	Solution
Chemical pumps/ dilution water pumps do not operate in automatic mode	• Check the start button is not in the auto position • Check if the storage tank level reading at the local level indicator/controller is below the low-level set-point value
Chemical tank low-level alarm	• Check the low-level alarm setting in the PLC to ensure that it is set correctly
Chemical tank high-level alarm	• Check that the alarm set-point is set for the maximum operating capacity of the tank
Level switches fail to alarm or alarm prematurely	• Check the operation of level switches
High-temperature alarm – caustic skid dilution water heat exchanger	• Check caustic tank heater temperature alarm set-point
High or low concentration – regeneration solution concentration	• Take a sample and check the specific gravity of the dilute solution. Use specific gravity to check and re-calibrate the conductivity meter; readjust the set-point • Adjust the dilution water and acid (or caustic) flow rates to obtain the desired concentration by adjusting the flow rate with valves on the skids

4.4.11 Ultraviolet lamps

General troubleshooting guidelines are provided for ultraviolet (UV) units. A low radiation intensity indicator is provided with all UV units. The lamp intensity should be greater than 80%.

Problem (4.4.11)	Solution
Low UV intensity	• Check if the UV lamp needs to be replaced • Check the operation of the UV intensity monitor and probe
Low-temperature alarm	• Verify if the set-point is above the low-temperature alarm value. Check the set-point of low-temperature switch
High-temperature alarm	• Verify if the set-point is below the high-temperature alarm value. Check the set-point of high-temperature switch

4.4.12 Preventive maintenance

Maintenance of equipment and components is critical for safe and efficient operation of a manufacturing plant. Equipment manufacturers usually supply maintenance procedures with the equipment. A recommended maintenance schedule is given in Tables 4.10 and 4.11. Adherence to the schedule and recording of information is necessary to satisfy potential warranty claims and for taking corrective action to protect the equipment from any potential damage. This schedule along with the equipment manufacturer's recommendations should be used to service or replace the components prior to component breakdown. It is important to maintain a log of regular maintenance service such as calibration of monitors, analysers, flow

Table 4.10 Equipment maintenance schedule

Item	Maintenance frequency
1. General	
Inspect skids for alarm conditions, leaks, damages	D
Review operational data	D
Do control panel lamp tests	D
2. Transfer pump	
Replace mechanical seals	A
Replace head seals	A
3. High-pressure RO pump	
Lubricate the air motor located on pump suction	Q
Replace the pump lubricant and lubricant filter	Q
Replace seals and O-rings in the high-speed shaft assembly	A
4. Chemical dosage pump	
Inspect and clean the chemical containment area	W
Inspect and clean any chemical weeping from the pump injection heads	M
Replace the diaphragm	SA
5. RO unit	
Membrane cleaning	R
Inspect the operation and settings of the pressure and flow control valves	Q
6. Mixed-bed deioniser	
Check silica level for breakthrough	D
Check resistivity for breakthrough	D
Resin replacement	R
7. Heat exchanger	
Inspect the controller temperature set-point	M
Calibration of temperature transmitter	Q
Calibration of temperature switch	Q
Inspect the operation and setting of the nitrogen system temperature control valves	Q
8. Water storage tank	
Level transmitter and level switch calibration	Q
Inspect the tank ladder, railing and associated attachment hardware for loose nuts and bolts	Q
Inspect the tank tie-down hardware for loose bolts and nuts, or stress failure	Q
Inspect the operation and setting of the nitrogen system pressure control valve	Q
Inspect the operation and setting of the nitrogen system pressure relief valve	Q
Inspect and replace the HEPA filter	Q
Inspect the inside of the tank for damage and/or untimely surface wear or build-up of foreign matter on the tank bottom or walls	SA
9. Chemical solution tank	
Level transmitter and level switch calibration	Q
Clean the inner surface of tank	SA
Re-fill the tank	R
10. Dead-end cartridge filter	
Replace the cartridge filter housing head O-ring	A
Replace RO pre-filter cartridge elements	R
Replace final cartridge elements	SA/A
Autopsy the cartridge element	A

(Continued)

Table 4.10 Continued

Item	Maintenance frequency
11. UV lamp	
Check the intensity and lamp indicator settings	D
Replace UV lamps	SA/R
Replace UV sleeves	SA/R
Calibrate the intensity meter	SA
12. Switches	
Temperature switches calibration check	Q
Pressure switches calibration check	Q
Check the operation of automatic level switch	Q
In the case of float switches, check their operation by manually depressing the level float	Q
Differential pressure switches calibration check	Q
13. Pneumatic valve	
Check the operation of solenoid valve	Q
Inspect the operation of automatic valve	Q
14. Miscellaneous	
Check pumps/motors for excessive vibration, noise, and over-heating. Check lubrication level	W
Check equipment mounts especially pumps/motor	A
Backwash media filters to prevent biological build-up	W
Inspect ion-exchange vessel internals and lining	A
Repaint surfaces as necessary	A
Electrical relays; check and clean contacts	M

D: Daily; W: Weekly; M: Monthly; Q: Quarterly; SA: Semi-annually; A: Annually; R: As required

meters, switches, transmitters, and gauges. Maintaining records of pump components, membrane cleaning, regeneration of ion-exchange resins as well as replacement of membrane elements and cartridge filters is especially important.

4.5 Process equipment

Design, operation, and maintenance of a system involve selection of process equipment based on sound and proven specifications. Specifications are the workhorse of any new plant. They provide design and performance requirements discussed in Sections 4.1 and 4.2.

4.5.1 Equipment selection

Engineers select and evaluate process equipment. The selected equipment must be technically sound, reliable, safe and easy to operate, serviceable, economically attractive, and environmentally sound, e.g. minimal noise pollution, and no mechanical and fire hazards. All pneumatic valves except for the modulating valves should be air-to-open/spring-to-close valve type for safe shutdown unless the valve is a divert drain valve. Similarly, a heat exchanger should have enough surface area, and the process water tank has excess capacity to provide a margin for continuous operation. Some of the equipment used in RO plants operates under severe conditions, e.g. 30 bar g at

Table 4.11 Instrumentation maintenance schedule

Item	Maintenance frequency
1. pH analyser	
Check pH indicating transmitter controller set-point	M
Standardise pH controller	M
Replace pH element	SA
2. Differential pressure monitor	
Check the calibration of ΔP indicating transmitter	Q/SA
3. Chlorine/ORP analyser	
Check the consumable chemicals	M
Calibrate the analyser	M
4. Flow indicator/totaliser	
Calibrate flow indicators and transmitter	Q/SA
Check the calibration of flow controller	Q
5. Temperature transmitter	
Inspect the set-points of temperature controller	M
Inspect and check the calibration of	Q
6. Pressure transmitter	
Check the calibration of pressure indicating transmitter	Q
Check the calibration of pressure controller	Q
7. Level indicator and transmitter	
Calibrate level transmitter	Q
Calibrate ultrasonic level transmitter	Q
8. Conductivity analyser	
Check the calibration of the conductivity monitor	SA
Clean the sensor	Q
9. Polishing analysers	
Check the calibration of resistivity monitor	SA
Clean the sensors	Q
Calibrate silica analyser	M
Replace silica analyser reagent	M
Calibrate particle count analyser	
10. Turbidity analyser	
Clean the turbidity cell	M
Check the calibration of the analyser	Q

3,000 rpm and 480 VAC/3 phase/50 Hz for operating the RO feed water pumps. The major components specifications are discussed in Section 4.5.2.

The process equipment should be designed with extra capacity, for example a pump must be sized large enough to overcome the pressure drop in the control valve or an orifice while delivering water or other fluid at peak flow and pressure. Some of the key features of centrifugal pumps and control valves are discussed in Chapter 5. The pump motor should be sized large enough for the pump's maximum impeller even when the pump comes with a smaller impeller initially; it is cheaper to replace the impeller than the motor in the future. The electric motor should have enough power to be non-overloading to ensure the motor would have enough power when the control valve is fully open with the pump following its pump curve. Similarly, a heat exchanger should have enough surface area, and the process water tank should have excess capacity to provide a margin for continuous operation.

RO high-pressure centrifugal pumps are typically multi-stage, vertical, or horizontal depending on the capacity. Increasingly, the pump motors are using variable-frequency drives (VFD) to maintain constant flow rate and pressure. VFDs allow the operation of pumps at pre-selected speeds; increasing or decreasing the flow and pressure characteristics for a given impeller size and motor power.

The RO systems are often designed around 75% product water recovery (corresponding to a feed/brine concentration factor of four) to minimise the potential for scaling by sparingly soluble salts that get concentrated in the reject channel as permeate is removed from the membrane module. In the case of seawater desalination, product water recovery is held to less than 50% (concentration factor = 2) due to the very high osmotic pressure that in turn affects the size of the RO feed pump and membrane pressure vessel ratings. Operation at reduced recovery implies more membrane surface area resulting in higher initial capital cost, which is compensated by lower operation costs due to fewer shutdowns to clean the membranes and longer life. These aspects of RO design are discussed in detail in Chapter 2.

4.5.2 Equipment specifications

The major components of the hybrid membrane plant are described below. Refer to the figure numbers to identify the equipment. A list of some of the leading process equipment manufacturers is given in Table 4.12 (detailed information available on their web sites).

Feed water pumps

The feed water pumps shown in Figure 4.3 are 316L stainless steel (SS) horizontal centrifugal pumps rated for a flow rate of 70.5 m^3/h at 2.8 bar g each. The totally enclosed fan cooled (TEFC) motors are rated for 18.7 kW, 3000 rpm, 460V/3 phase/50 Hz. The motor service factor is 1.15.

Granular media filters

The media filters shown in Figure 4.4 are rubber-lined (5 mm thick) carbon steel (ASME coded-stamped) vessels, 2.4 m diameter (D) × 1.83 m straight shell height (SSH) and rated for a working pressure of 6.9 bar g. The hydraulic test pressure rating is 10 bar g. The internal distribution system consists of a PVC upper distribution system, and a header and lateral bottom collection system. Each vessel is provided with two 8 cm × 30 cm sight glasses and a 40 cm × 50 cm elliptical manway. All face piping is PVC, and accessories include pneumatic butterfly valves, isolation ball valves, pressure gauges, differential pressure gauges, and paddle wheel flow elements. Each media filter contains 4.25 m^3 of granular media of varying porosity and density from top to bottom: anthracite (top), silica sand (centre), and garnet (bottom), i.e. the lightest and most coarsely graded material on top, the heaviest and the most finely graded material on the bottom, as shown in Figure 2.4.

Chemical feed system

The chemical feed system consists of three chemical feed systems shown in Figure 4.5. The sodium bisulphite (SBS) system consists of a 570 l feed tank and two injection pumps, rated for 18.9 l/h at 5.5 bar g. The anti-scalant (A/S) system consists of a 570 l feed tank and injection pumps, rated for 18.9 l/h at 5.5 bar g. The caustic soda (NaOH) system consists of a 570 l feed tank and injection pumps rated for 170 l/h at 5.5 bar g. The PVC injection pumps are positive displacement type with PVDF

Table 4.12 Partial list of equipment manufacturers

Component	Manufacturer
RO membranes (spiral-wound TFC)	Desal, Dow/FilmTec, Fluid Systems, Hydranautics, IonPure, Osmonics, Saehan, Toray, TriSep
Microfilters (dead-end)	CUNO, Filterite, Gelman, Millipore, Optimum Filter, Pall, Pentec, Shelco, USFilter
Ion-exchange resins	Dow, Purolite, ResinTech, Rohm & Haas, Sybron, USFilter
Centrifugal pumps	Durco, Fristam, Fybroc, G&S, Goulds, Grundfos, Triclover, Worthington
Diaphragm pumps	LMI, Milton Roy
Positive displacement pumps	Milton Roy, Pulsafeeder
Submersible pumps	Pumps Unlimited, Tonkaflo, T&T
Variable-frequency drives	Allen-Bradley, Siemens, Yasakawa
Heat exchangers	Alfa-Laval, ITT, Trantor, Yula
Ultraviolet lamps	Aquafine, Trojan, Wedeco Ideal Horizon
Loop controllers	Foxboro
PLC	Allen Bradley, Honeywell, Siemens, Toyota
Pressure gauges/switches	Ashcroft, United Electric
Process monitors/transmitters	Endress-Hauser, Foxboro, GF Signet, Thornton, Yokogawa
Temperature gauges/switches	Ashcroft
Chemical monitors	GLI, Orbisphere, Rosemount, Thornton
Chemical analysers	Anatel, HACH, Scientific Instruments
Plastic valves**	Asahi, George-Fisher, Hayward, Plastomatic
Ball valves	Apollo, Jamesbury, Swagelok
Butterfly valves	Bray, Keystone
Check valves	Apollo, Check-All, Hayward
Control valves	Fisher, Foxboro, Neles-Jamesbury, Stonel
Diaphragm valves	ITT
Needle valves	Marsh, Marquest, Nupro
Pressure-regulating valves	Anderson Greenwood, Jordan Mark, Plasto-Matic, Protecto-Seal
Relief valves	Crosby, Kunkle, Protecto-Seal

** Ball, butterfly, check, diaphragm, relief, others; CPVC, PVC PP, PVDF

diaphragms. Each tank is provided with a level switch. All motors are rated for 220V/1 phase/50 Hz. The caustic tank includes a mixer.

Cartridge filter

The RO pre-filter shown in Figure 4.6 contains 24 polypropylene (PP) cartridges, 76 cm long (L), rated for 5.0 μm (nominal pore size). The cartridge separation efficiency is greater than 85% rejection for particles >5.0 μm in size. The filter vessel is 304 SS, ASME code construction, and rated for a design pressure of 10 bar g.

Feed water heat exchanger

The feed water heat exchanger systems shown in Figure 4.6 is used for heating 70.5 m^3/h RO feed water to 25°C with low-pressure steam. The heat exchanger is a 304 SS plate-and-frame, counter current type sized for 100% total flow. The rated design pressure is 6.9 bar g. The steam flow rate is 9103 lb/h at 1 bar g.

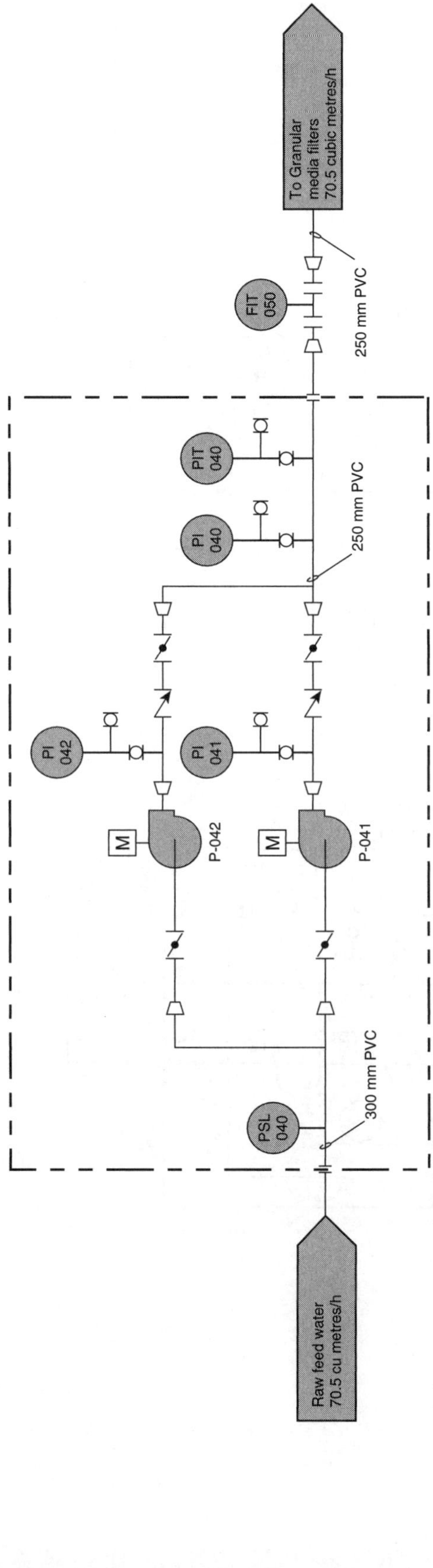

Figure 4.3 Feed water pumps process flow diagram

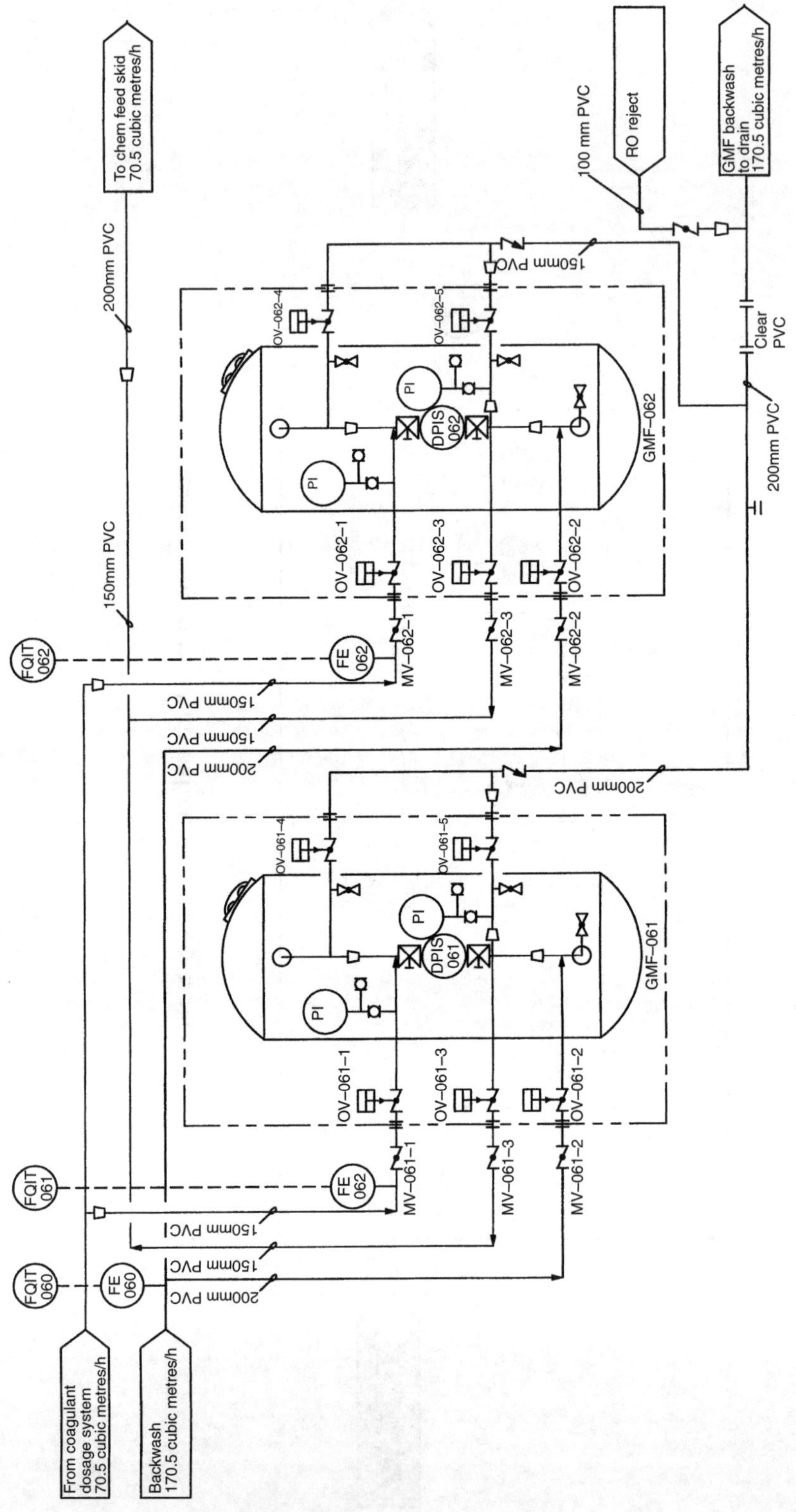

Figure 4.4 Process flow diagram of granular media filters

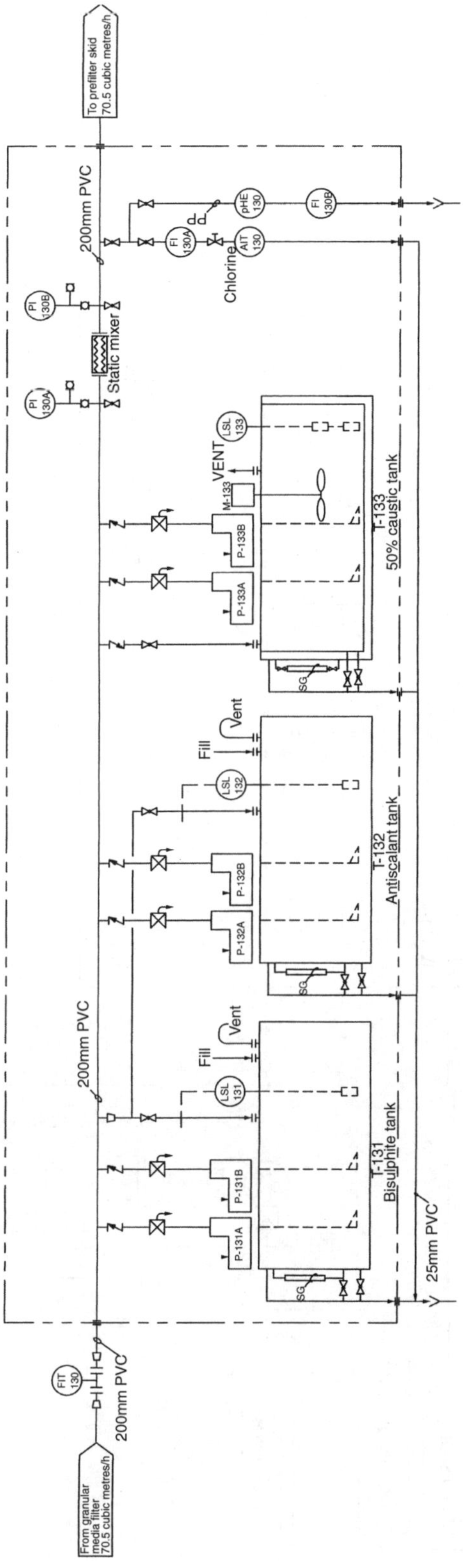

Figure 4.5 Chemical feed systems process flow diagram

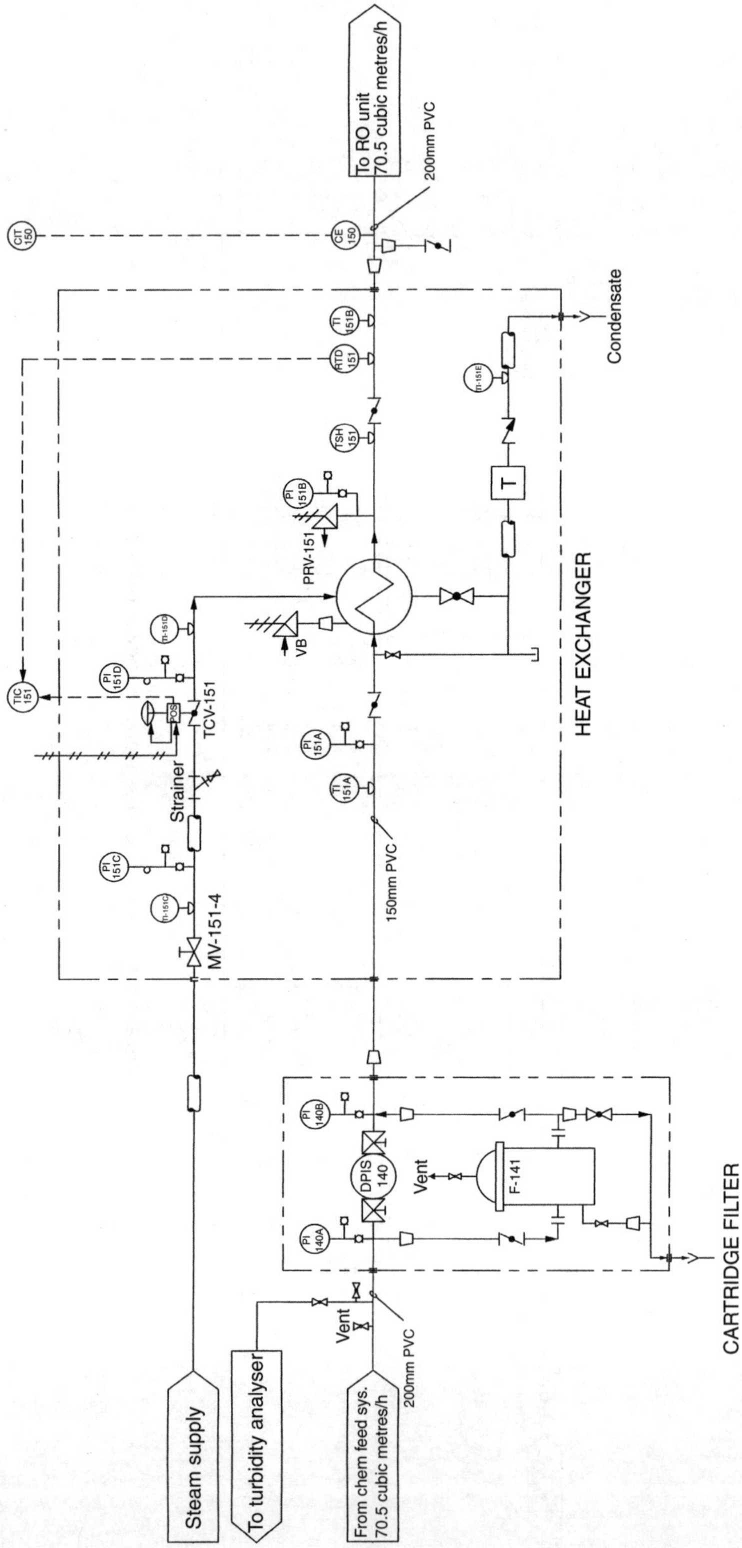

Figure 4.6 Process flow diagram of 5.0 μm cartridge filter and heat exchanger skid

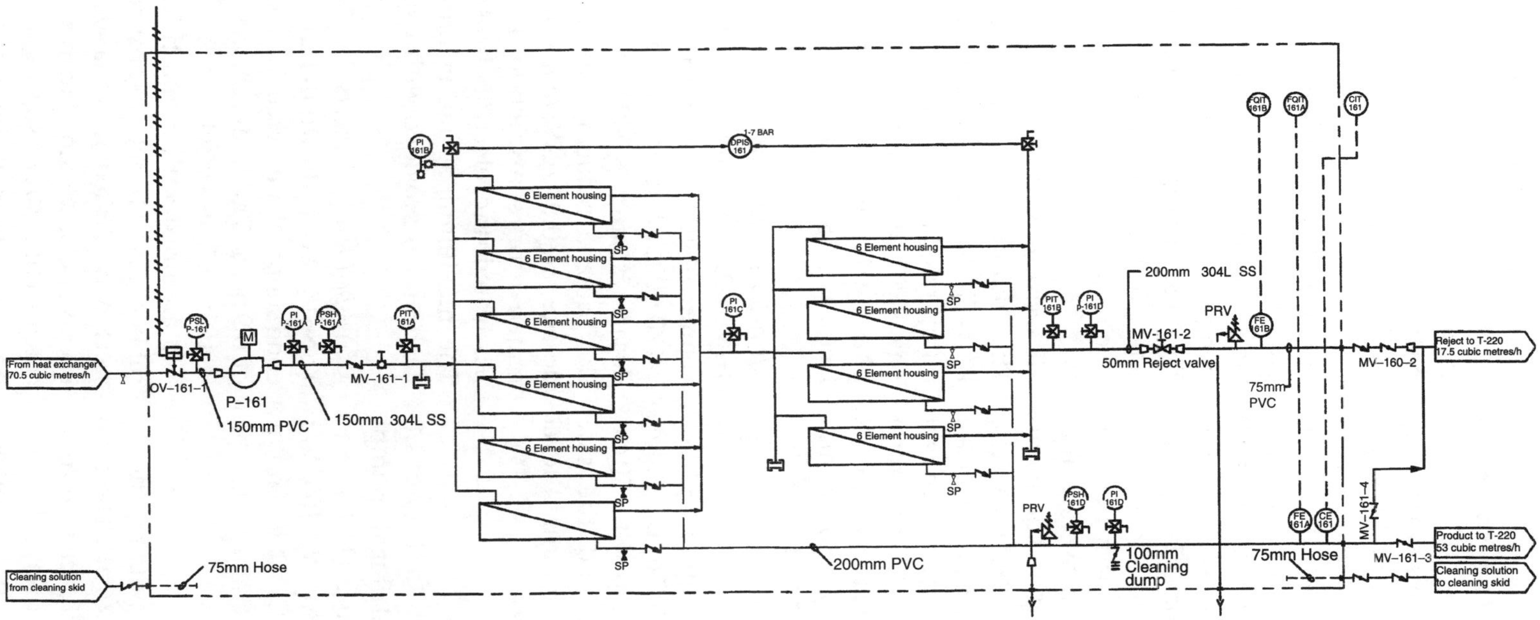

Figure 4.7 Single-pass two-stage RO system process flow diagram

RO unit

The single-pass RO unit systems shown in Figure 4.7 is a 6:4 staged array (10 pressure vessels) designed for a product water recovery of 75% using spiral-wound, thin-film composite polyamide, full-fit RO membranes with an initial salt rejection of >99.0%. Each pressure vessel (module) contains six elements connected in series for a total of 60 elements. The membrane elements are housed in FRP pressure vessels rated for 40 bar g. Each membrane element is 20 cm.D × 100 cm L long. The elements are rated for a nominal flow rate of 34 m^3/day with an average flux of 22 l/m^2.h based on well water. The membrane projections are given in Table 4.1. The RO high-pressure pump, P-161 (Figure 4.7) is a 304L SS submersible, multi-stage centrifugal pump rated for 70.5 m^3/h at 17 bar g total delivery head (TDH). The TEFC motor is rated for 45 kW, 460 V/3 phase/50 Hz, and 3,000 rpm with a service factor of 1.15.

Water reuse tank

The water reuse tank shown in Figure 4.8 has a capacity of 76 m^3. The tank is 3.7 m D × 7.3 m SSH; FRP, flat-bottom, dome-top with 50 cm man-ways in the top dome and the side shell. It is provided with side-mounted level switches and a level transmitter.

GMF backwash pumps

The backwash pumps shown in Figure 4.8 are 316L SS horizontal centrifugal pumps rated for 170 m^3/h at 2.8 bar g (TDH). The TEFC motors are rated for 30 kW, 460 V/3 phase/50 Hz, and 1,750 rpm. The motor service factor is 1.15.

RO product water tank

The RO product water tank systems shown in Figure 4.9 has a capacity of 76 m^3. The tank is 3.7 m D × 7.3 m SSH; FRP, flat-bottom, dome-top with 50 cm. man-ways in the top dome and the side shell. It is provided with side-mounted level switches and a level transmitter, a pressure relief valve/vacuum breaker, a nitrogen blanketing system and a high-efficiency particulate air (HEPA) filter.

RO water distribution pumps

The distribution pumps shown in Figure 4.9 are 316L SS vertical, multi-stage centrifugal pumps rated for 70.5 m^3/h at 4.8 bar g (TDH). The TEFC motors are rated for 33 kW, 460 V/3 phase/50 Hz, and 3,000 rpm. The motor service factor is 1.15.

MB regeneration pump

The dilution water pump shown in Figure 4.9 is 316L SS vertical, multi-stage centrifugal pump rated for 70.5 m^3/h at 4.8 bar g (TDH). The TEFC motor is rated for 33 kW, 460 V/3 phase/50 Hz, and 3,000 rpm. The motor service factor is 1.15.

Ultraviolet light lamps

The TOC destruct UV units shown in Figure 4.10 are rated for a dosage >120,000 μW s/cm^2 at 185 nm. The rated dosage is after 8,000 h of operation (Note, the bacteria reducing lamps are rated for a dosage >30,000 μW s/cm^2 at 254 nm). The material of construction is 316L SS with passivated and electropolished wetted parts. The maximum operating pressure of 8 bar g. Each unit is provided with two light traps.

Mixed-bed ion exchangers

The MB vessels shown in Figure 4.11 are ASME code stamped and rated for a working pressure of 6.9 bar g. The hydraulic test pressure rating is 10 bar g.The vessels are 2.1 m D × 3 m SSH, and are lined with 5-mm thick unpigmented rubber. Each vessel is constructed of carbon steel with dished heads and a false-bottom. Each vessel is provided with a 60 cm man-way, three 8 cm × 30 cm sight glasses and a 15 cm port for removing the resins. Each MB vessel

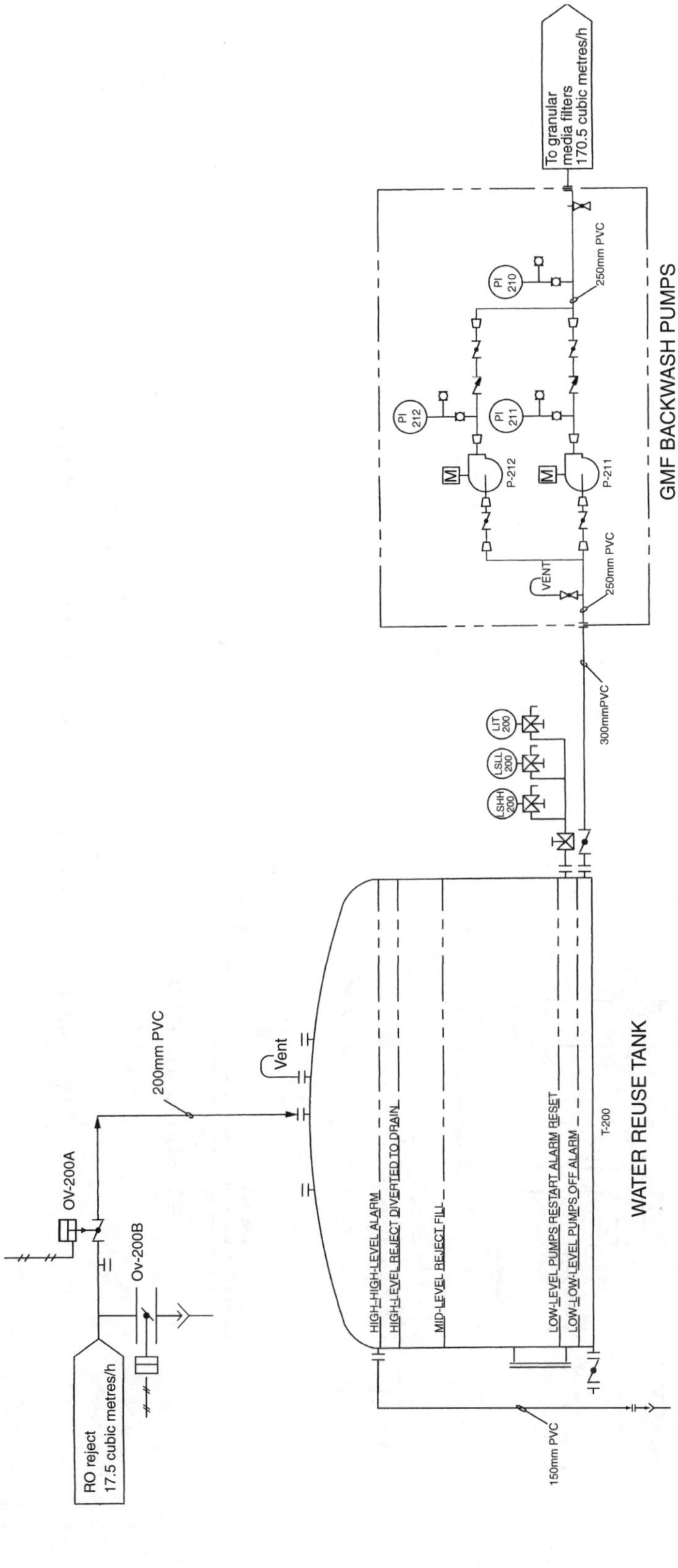

Figure 4.8 Water reuse tank system process flow diagram

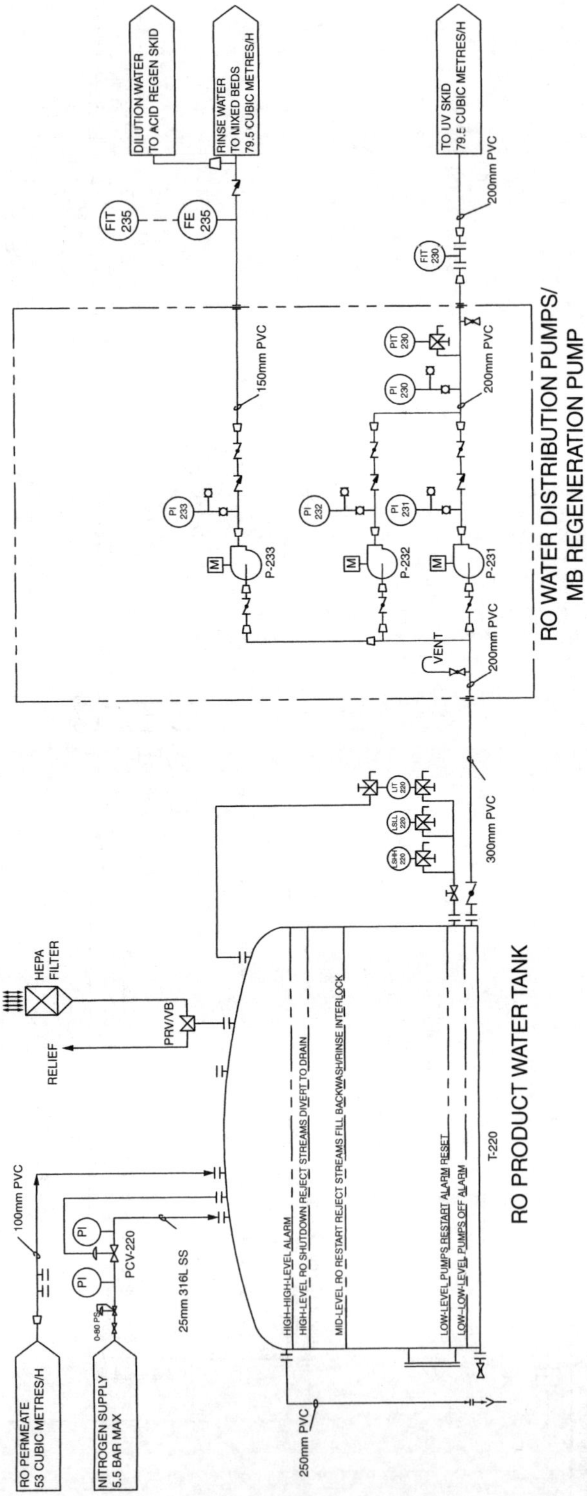

Figure 4.9 RO product water tank system process flow diagram

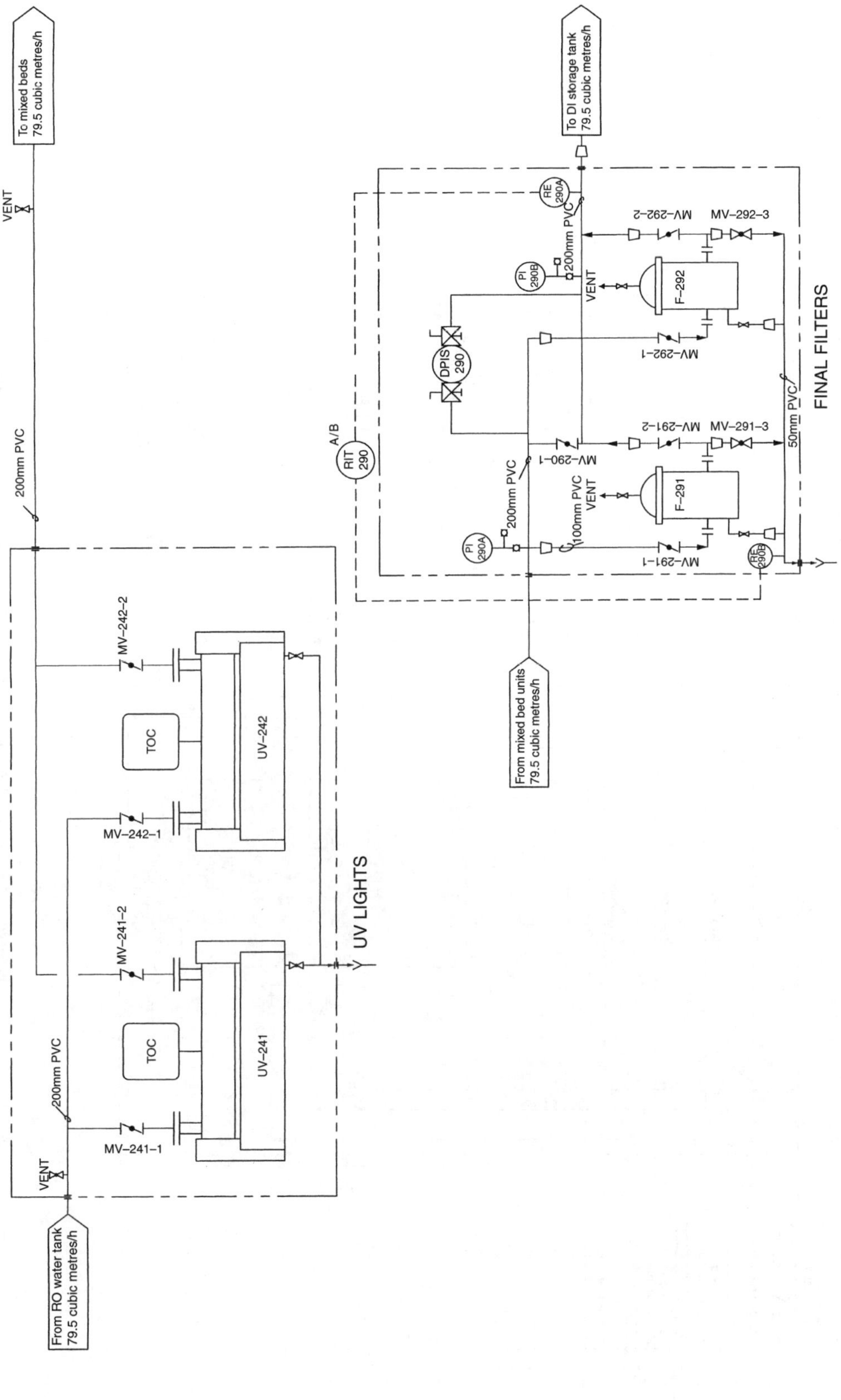

Figure 4.10 Process flow diagram of UV light (185 nm) and 1.0 μm cartridge filter skid

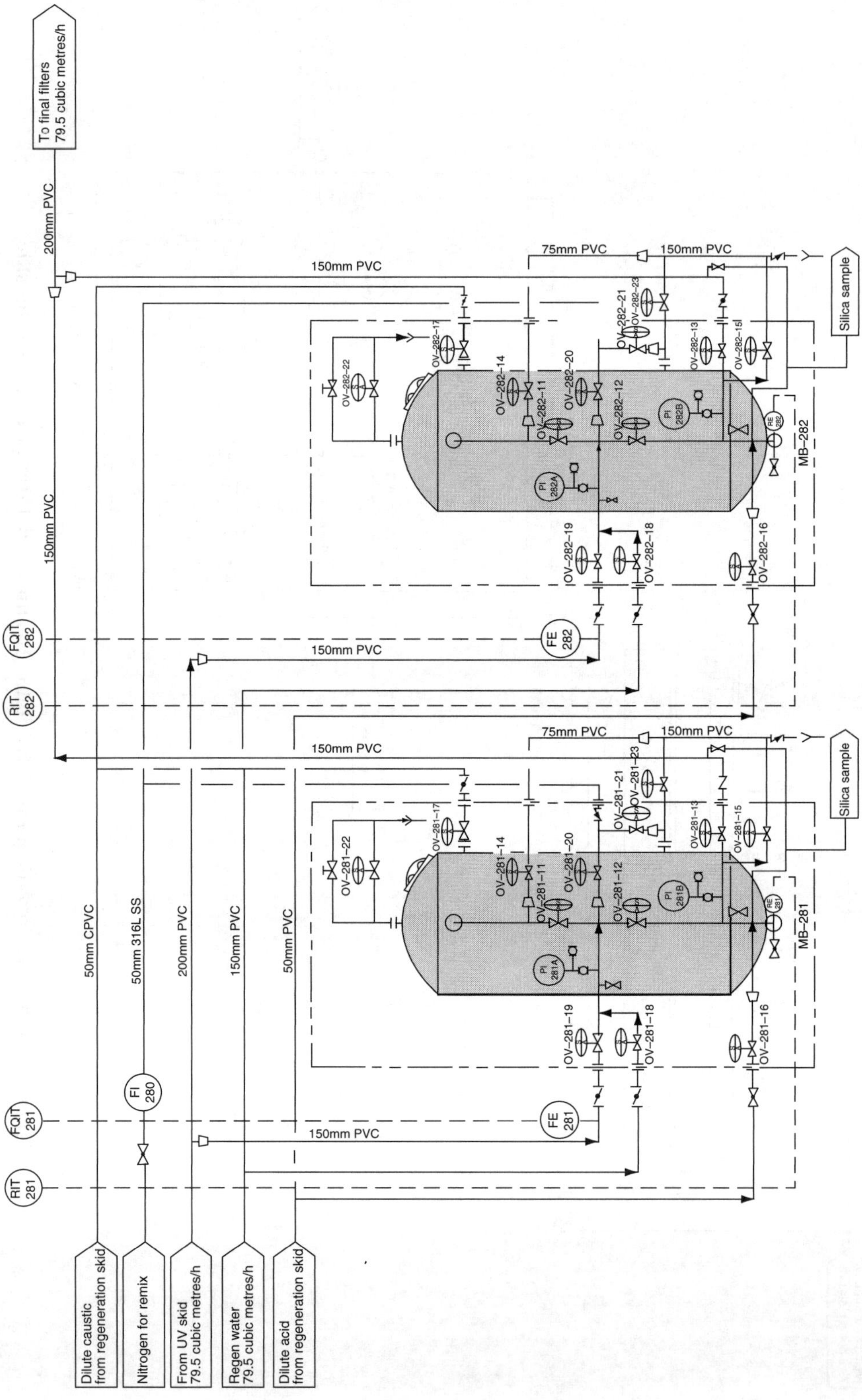

Figure 4.11 Mixed-bed ion exchangers process flow diagram

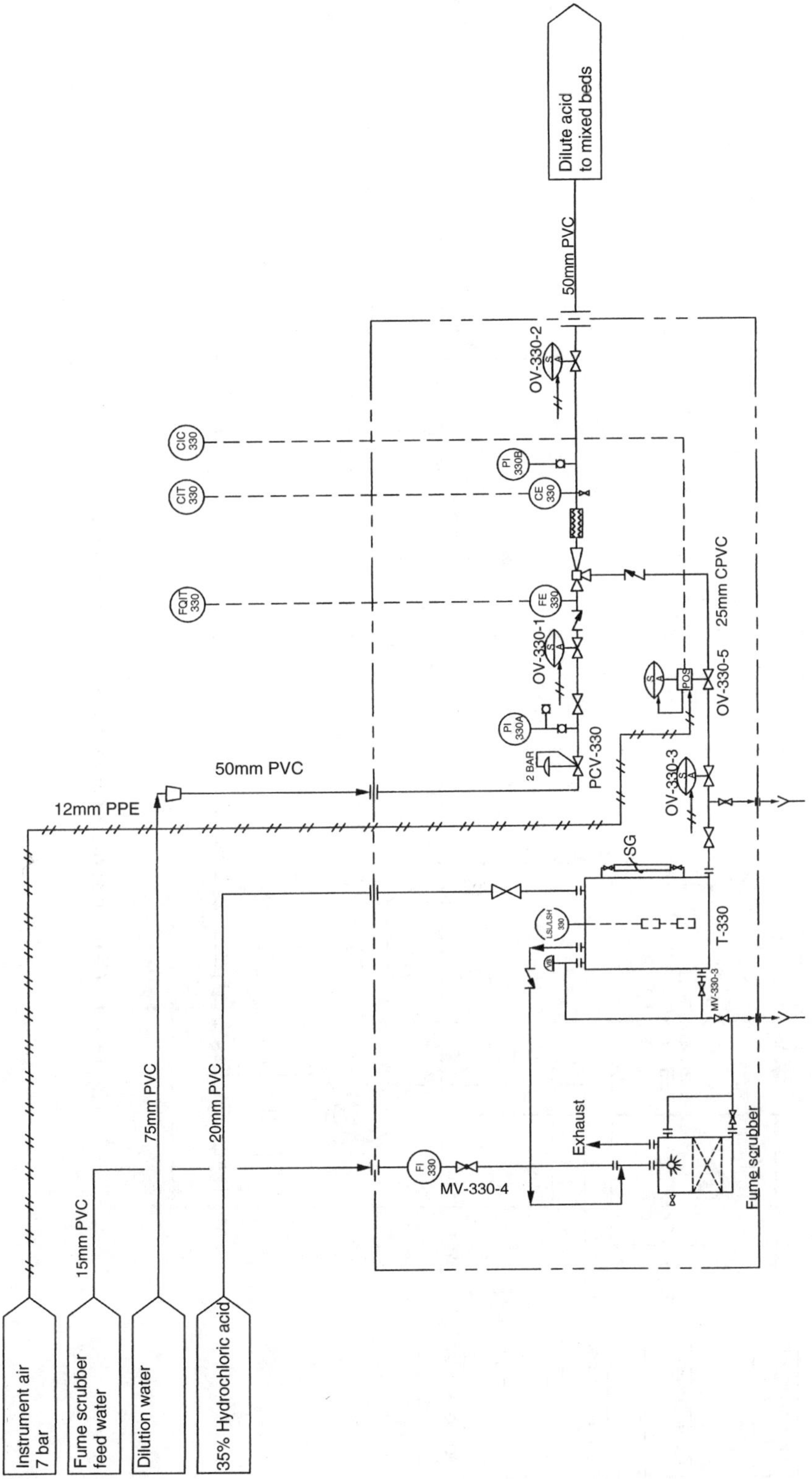

Figure 4.12 Acid regeneration system process flow diagram

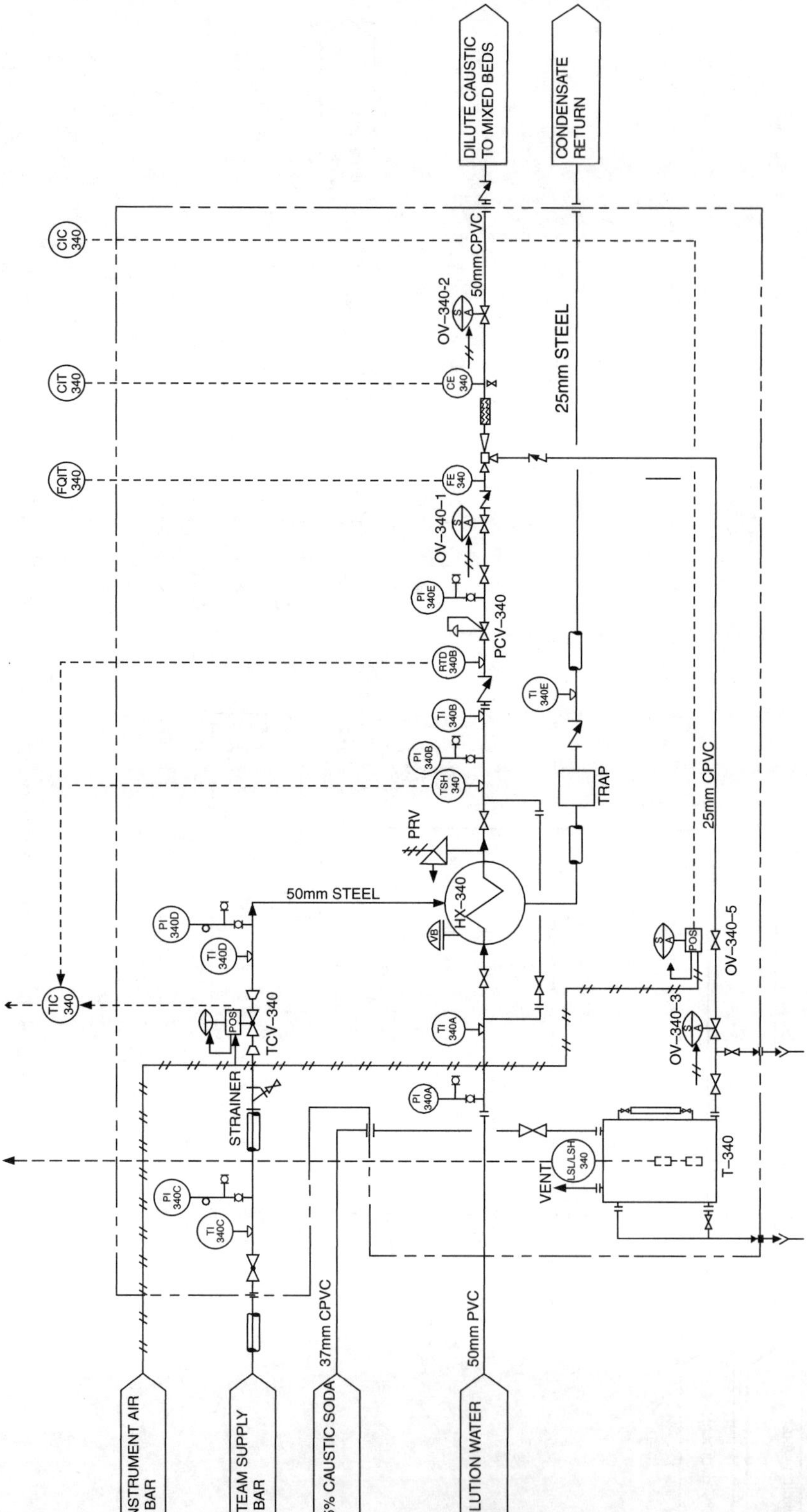

Figure 4.13 Caustic soda regeneration system process flow diagram

contains 3.26 m^3 strong base anion resin (OH^- form), 2.18 m^3 strong acid cation resin (H^+ form) and 0.54 m^3 inert resin. The internal piping consists of a 316L SS upper distributor, a 316L SS bottom/acid distributor with wedge-wire screens, and 316L SS interface and caustic distributors. All service water external piping and valves are PVC (or PVDF). Acid and caustic solution piping is CPVC. The nitrogen line piping is 316L SS.

Cartridge filters

The microfilters Figure 4.10 contain 4.19 polysulphone/nylon 66 cartridges, 76 cm long, rated for 1.0 μm (absolute) each. The cartridge separation efficiency is >99% rejection for particles ≥1.0 μm in size. The filter housing is 316L SS, ASME code construction, and rated for a design pressure of 10 bar g. The cartridges are sized for 15 l/m per 25 cm length of the cartridge. The piping is PVC or PVDF.

MB regeneration skids

The MB regeneration skids shown in Figures 4.12 and 4.13 include hydrochloric acid dilution and caustic soda dilution systems, respectively. Each skid contains a 570 l FRP tank, a CPVC eductor for mixing water with the chemical, a conductivity (% concentration) analyser and pneumatic valves. Each tank is provided with an ultrasonic level switch. The acid skid is provided with a scrubber. The skid piping is CPVC. The caustic regeneration skid heat exchanger shown in Figure 4.13 is used for heating 9.5 m^3 dilution water to 43°C. The heat exchanger is a 304 SS plate-and-frame, counter-current type sized for 100% total flow. The rated design pressure is 6.9 bar g. The steam rate is 956 lb/h at 1 bar g.

Control panels

The main control panel (MCP) houses the entire system control package including the system programmable logic controller (PLC), graphic panel lights, annunciator, selector switches as well as monitors. The remote input/output (I/O) or local control panels (LCP) are provided on each skid for interacting with the MCP via the PLC. All panels are NEMA 4X enclosures.

4.6 Concluding remarks

A hybrid membrane plant is discussed in detail to familiarise the reader with the complexities of a relatively simple membrane-based water treatment plant. The design, equipment specification and operation of process plants depend on the quality and quantity of feed water and product water. Often plants are modified or upgraded to meet newer demands. The initial or preliminary design is based on a simple line diagram incorporating mass balances followed by designing the plant based on feasibility and cost. Increasingly RO membrane plants integrated with MF, UF, and/or NF are being built to supply potable water and treat wastewater.[12,13] Currently, more than 1 million m^3/day of water is being treated with integrated membrane plants. Furthermore, integrated membrane systems are more competitive than conventional processes. Hybrid membrane systems permit low energy consumption, higher quality of final products, and low environmental impact. Several hybrid membrane process options for various applications are discussed in Chapter 3.

References

1 M. Cheryan, Ultrafiltration and Microfiltration Handbook, CRC Press: Boca Raton, FL (1998).
2 S.V. Cabibbo, D.B. Guy, A.C. Ammerlaan, A. Ko, R. Singh, Reverse Osmosis Technical Manual, NTIS Publication: Springfield, Virginia, USA (1979).

3 J.M. Douglas, Conceptual Design of Chemical Processes, McGraw-Hill: New York, NY (1988).
4 S.D. Strauss, Water treatment, Power, June (1993), 1–52.
5 G.A. Pittner, High purity water production using reverse osmosis, in Reverse Osmosis, (ed. Z. Amjad), Van Nostrand Reinhold: New York, NY (1993), 334–363.
6 W.V. Collentro, Pharmaceutical Water, CRC Press: Boca Raton, FL, USA (1999).
7 S.B. Applebaum, Demineralization by Ion Exchange, Academic Press: New York, NY (1968).
8 D.L. Owens, Practical Principles of Ion Exchange Water Treatment, Tall Oaks Pub.: Littleton, CO (1995).
9 R. Bradley, Design considerations for reverse osmosis systems, in Reverse Osmosis (ed. Z. Amjad) Van Nostrand Reinhold: New York, NY (1993), 104–137.
10 D.R. Woods, Process design and engineering practice, Prentice-Hall: Upper Saddle River, NJ (1995).
11 C. Branan, Rules of Thumb for Chemical Engineers, Gulf Publ. Co. (1998).
12 E. Drioli, M. Romano, Progress and new perspectives on integrated membrane operations for sustainable industrial growth, Ind. Eng. Chem. Res. 40 (2001), 1277–1300.
13 A.I. Schafer, A.G. Fane, T.D. Waite, eds., Nanofiltration, Elsevier: Oxford, UK (2004).

Chapter 5

Appendix: Engineering data and notes

'Anyone who believes that exponential growth in a finite world can go on forever is either a madman or an economist'.

Kenneth Boulding (Economist)

5.1 Glossary/terminology

Absorption. It involves the dissolution (partitioning) of a contaminant from one phase into another without chemical reaction, e.g. absorption of oxygen gas in water or the absorption of DDT in hexane, an organic solvent.

Acidity, water. All waters – natural, domestic wastewater, and also frequently, industrial wastewater – with pH less than 8.5 contain acidity. Acidity in natural waters is caused mainly by dissolved carbon dioxide that produces a weak acid, carbonic acid. Natural waters are buffered by a CO_2 – bicarbonate system. See Carbon dioxide.

Activated diffusion. The process by which a penetrant molecule moves through a membrane from one transient vacancy to another with an energy interaction between the penetrant and the membrane.

Adsorption. The process by which ions or molecules present in one phase condense and concentrate on the surface of another phase (surface phenomena), e.g. adhesion of the molecules of gases, dissolved substances, or liquids to the surfaces of solids or liquids with which they are in contact. There are three general types of adsorption: physical (activated carbon), chemical, and exchange (ion exchange).

Aeration. Operation by which air is introduced into water and/or to release dissolved gases from water.

Agglomerate. To gather fine particulate matter together into a larger mass.

Alkalinity. It is a measure of the buffer capacity to resist drop in pH when acid is added, i.e. it is the measure of the chemical constituents of water that react with hydrogen ions, and thus resist a decrease in pH. The major portion of alkalinity is due to hydroxide, carbonate, and bicarbonate. Other contributors include salts of weak acids such as borates, silicates, phosphates, organic acids such as humic acid. Alkalinity can be expressed as 'M' alk. to a methyl orange titration end-point ($pH = 4.2$) or 'P' alk. to a phenolphthalein end-point ($pH \geq 8.2$).

Amphoteric. Compounds such as amino acids that form salts with acids or bases. The hydroxides of aluminium, zinc, chromium, and a few other elements are soluble in both acids and bases. Hence it is used to accomplish separations in water treatment.

Anion. A negatively charged ion resulting from the dissociation of a salt, acid, or an alkali in an aqueous solution.

Anion permeable membrane. Membrane containing a fixed positive polar group that will permit passage of anions and reject cations.

Anisotropic (asymmetric). Defines a particular type of ultrastructure of microporous membranes. The surface of the membrane where separation occurs is more dense than the rest of the membrane body. The pore diameter increases in a direction perpendicular to the membrane surface, with the pore opening near the separation surface being smaller than the pore opening on the bottom of the membrane. This 'skin' layer is typically present in polymeric membranes made by the phase-inversion process. Some asymmetry is also present in many inorganic membranes.

Aqueous chemistry. Also called solution chemistry, aquatic chemistry, water chemistry, electrolyte chemistry – any chemical system that involves water and dissolved salts.

ATD (Anti-Telescoping Device). A circular ring with spokes fitted to the ends of a spiral module to prevent the membrane from pushing itself outwards under a high-pressure drop.

Attrition. Breakage or wear of ion-exchange resins.

Back-pressure. Pressure exerted against flow.

Backwash. Reverse (upward) flow of water through a packed bed to remove foreign material from the bed, and clean and expand the bed.

Backpulse. Brief reversal of flow from permeate to feed side of the membrane.

Base. The hydroxyl form of a cation or a compound that can neutralise an acid.

Bicarbonate alkalinity. The bicarbonate ion concentration expressed in 'ppm' as $CaCO_3$ equivalents. To convert ppm of bicarbonate to ppm of $CaCO_3$ equivalents multiply by 1.64.

Bioreactor. A vessel for conducting biological reactions.

BOD (Biochemical Oxygen Demand). A measure of the amount of oxygen required to support growth of bacteria during the breakdown of organic compounds in a sample. Usually expressed in parts per million (ppm). It is a bioassay procedure that involves the measurement of oxygen consumed by living organisms (mainly bacteria) while utilising the organic matter present in waste under conditions that simulate natural conditions.

Boundary layer. The region near a sold surface where fluid motion is affected by the surface. The boundary layer is a major resistance to transport (e.g. by heat, mass, or momentum). In order to improve transport, the thickness of this boundary layer must be reduced.

Brackish water. Generally water having dissolved solids content of 1,000–10,000 mg/l.

Buffers. Substances in solution that resist change in pH as acids or bases are added to or formed in the solution. Buffer solutions usually contain mixtures of weak acids and their salts (conjugate bases) or weak bases and their salts (conjugate acids).

Calcium carbonate equivalents. The impurities of naturally occurring waters are frequently expressed as $CaCO_3$ equivalents (Table 5.6). To convert the ppm (mg/l) of a given constituent to ppm expressed as $CaCO_3$, multiply 'ppm' by the ratio of the molecular weight (MW) of $CaCO_3$ to that of the constituent. For example, to convert ppm of carbonate to ppm of $CaCO_3$ equivalents multiply by 1.67. The MW of $CaCO_3$ is 100.

Capacity. Operating – the fraction of total capacity used in an ion-exchange operation. Total – The ultimate or maximum ion-exchange capacity of a resin.
Carbonate alkalinity. The carbonate ion concentration expressed in 'ppm' as $CaCO_3$ equivalents.
Carbon dioxide. CO_2 gas is a natural component of natural waters. It is also generated by biological oxidation of organic matter, especially in polluted water. Carbon dioxide is very soluble in water above pH 5 and forms bicarbonate and carbonate buffers by exposure to atmosphere. In the case of surface waters, it can enter water from the atmosphere when its concentration in water is less than that in equilibrium with CO_2 in the atmosphere in accordance with Henry's law. Free CO_2 is dissolved carbon dioxide gas in water.
Carbonate hardness. Carbonate hardness is chemically equivalent to the bicarbonate plus carbonate hardness (formerly called 'temporary hardness'). It is expressed as ppm $CaCO_3$.
Catalysts. See Enzymes.
Cation. A positively charged ion resulting from the dissociation of molecules in solution.
Cation permeable membrane. Membrane containing a fixed negative polar group that will permit passage of cations and reject anions.
cfu. Colony forming units, a measure of the number of microorganisms in a sample that are capable of forming colonies on a plate.
Channelling. Liquid flow through a path of least resistance in a packed bed.
Chelating agent. Organic compounds having the ability to form soluble complexes with ions in an aqueous solution. The solubility of metal ions is increased by chelating agents. See EDTA. Natural chelates include haemoglobin (contains iron), vitamin B-12 (containing cobalt), and chlorophyll (containing iron). Natural organic substances, e.g. humic are capable of chelating.
Chemical potential. Defined in terms of Gibbs free energy, it is a measure of how the Gibbs function for a system changes when a specified amount of component is added to the system.
Chlorination. Addition of chlorine in the form of free chlorine or sodium hypochlorite. Chlorine, hypochlorous acid, and hypochlorite ions are called 'free chlorine residuals' whereas the chloramines are referred to as 'combined chlorine residuals.'
***Cryptosporidium*.** A microbial pathogen found in surface water. Ingestion can cause cryptosporidiosis, an abdominal infection.
Coagulant. A chemical (e.g. alum of iron salts) used to form a floc; used in the clarification of turbid or coloured water, and as an aid in removing colloids. See also Zeta Potential.
COD (Chemical Oxygen Demand). A chemical measure of the amount of oxygen required to breakdown organic matter in a sample expressed in parts per million (ppm). It does not differentiate stable from unstable organic matter. COD values are greater than BOD values.
Colloids. Particles (0.001–1.0 μm) that form a suspension or emulsion, cause turbidity in water. All colloidal particles are electrically charged (+ve or –ve), have very large surface areas, and are too stable to be removed by gravitational settling (can be removed by UF in their original state). However, if they are destabilised by charge neutralisation or allowed to form aggregates with coagulating agents, they can be removed by filtration. Hydrophilic colloids such as protein compounds have minimum solubility at their iso-electric point (pH = 4.0–6.5).

Concentrate. It is that portion of the feed solution retained on the high-pressure side of the membrane. As a result, the retained components are usually more concentrated than they were in the original feed solution. The terms concentrate, reject, and retentate are used interchangeably.

Concentration polarisation. Accumulation of solutes within the boundary layer above the membrane surface. Solutes that do not pass through the membrane accumulate on the membrane surface causing either an increased resistance to solvent transport or an increase in local osmotic pressure (either of which decreases flux), and possibly a change in the separation characteristics of the membrane.

Concentration ratio or concentration factor. The ratio of the initial feed volume (or weight or flow rate) to the concentrate volume (or weight or flow rate).

Contact angle. Acute for hydrophilic, and obtuse for hydrophobic at solvent–solid interface.

Cross-contamination. Inter-mixing of one resin with another of opposite charge.

Cross-linking. The 'setting up' of chemical links between two or more polymer chains.

Cross-flow. The flow of fluid across a membrane surface, parallel to its surface or tangential flow. These modules have one inlet (feed) and two outlets (concentrate and permeate). This is in contrast to dead-end modules.

DE. Diatomaceous earth, which is the fossilised skeletal remains of microscopic prehistoric plants; used as a filter aid in conventional filtration.

Dead end. Mode of operation where there is only one feedstream and one outlet stream (the filtrate of permeate). Refers to modules that do not have a means for cross-flow, e.g. stirred cells. Cross-flow modules could be operated in the dead-end mode by shutting off the reject outlet of the module.

Deionisation. See Demineralisation.

Demineralisation. Any process used to remove mineral salts from water; commonly restricted to an ion-exchange process.

Desalination. The process of removing salts from water by reverse osmosis or distillation.

Detergents. Organic materials that have the property of being 'surface active' in aqueous solution. See Surfactants.

Diafiltration. A mode of purification where water is added to the feed tank or the retentate loop at the same rate as permeate flow. The concentration of the permeable solutes decrease in proportion to the volumes diluted and their individual rejections.

Diffusivity/diffusion coefficient. A measure of a solute's diffusive mobility in a solvent (cm^2/s).

Dissolved oxygen. All atmospheric gases are soluble in water to some degree. Oxygen and nitrogen are poorly soluble, and since they do not react with water chemically, their solubility is directly proportional to their partial pressures. The solubility of atmospheric oxygen in fresh water ranges from 14.6 mg/l at 0°C to 7 mg/l at 35°C at 1 bar atmosphere pressure. The equilibrium concentration of oxygen in water at 20°C at 1 bar air is 9.2 mg/l. See Henry's law. The low solubility of oxygen restricts the purification capacity of natural waters, necessitating treatment of wastes to remove pollutants prior to discharge. The solubility of oxygen is also less in saline water than in clean water.

Divinylbenzene. A di-functional monomer used to cross-link polymers.

EDTA. Ethylene diamine tetra acetic acid; the Na^+ salt is the usual form of this chelating agent.

Eductor. An in-line static mixing device that, by flow of water through it creates a vacuum, thereby, drawing a solution into the water stream.

Electric double layer. Most substances acquire a surface electric charge when brought into contact with a polar (e.g. aqueous) medium possibly due to ionisation, ion adsorption, and ion dissolution. Ions of opposite charge (counter-ions) are attracted towards the surface, and ions of like charge (co-ions) are repelled away from the surface. This together with the mixing tendency of thermal motion, leads to the formation of an electric double layer made of the charged surface and a neutralising excess of counter-ions over co-ions distributed in a diffuse manner in the polar medium. The theory of the electric 'double layer' deals with this distribution of ions and the magnitude of the electric potentials, which occur in the area of the charged surface. See also 'Zeta Potential.'

Electroneutrality. When the positive charges equal the negative charges.

End point (breakthrough). A pre-selected value or measurement used to determine when the service cycle is completed or that a breakthrough has occurred.

Enzymes, catalysts. Temperature sensitive, organic catalysts produced by living cells. Enzymes are proteinaceous catalysts that permit biochemical reactions at room temperature. Catalysts change the rate of a chemical reaction by lowering the activation energy of the reactions.

Equivalent. Equal valence is the weight in grams of a substance that combines with or displaces one gram of hydrogen; usually obtained by dividing the molecular weight by the valence.

Exchange sites. Reactive groups on an ion-exchange resin.

Exothermic. Giving off heat in a chemical reaction. Reverse of endothermic.

Feed-and-bleed. A continuous mode of UF or MF operation where the feed is pumped into the recirculation loop of the membrane system at the same rate as the sum of the concentrate flow and permeate flux out of that loop. The feed concentration within the recirculation loop is determined by the concentration factor.

Filter aid. A material added to a liquid to be filtered or on to the filter medium to improve the separation of solids from the liquid.

Fines. Particles smaller than 50 mesh in size.

Floc. A gelatinous precipitate formed by a coagulant used to suspended material.

Flocculants. Materials that can form gelatinous clouds of precipitate to enclose fine particles of suspended dirt.

Fluoropolymers. See Thermoplastics.

Flow rate. The volumetric rate of flow of fluid parallel to the membrane surface. This is expressed in terms of volume/time (e.g. l/min or gallon/min). Flow rate is velocity (v) × cross-sectional area of the feed channel. This is the major determinant of the state of turbulence in a membrane element.

Flux. Membrane permeate throughput (flow rate per unit area) usually expressed as volume per unit membrane area per unit time, e.g. $l/m^2/h$, $m^3/m^2/h$, or $gallons/ft^2/day$.

FMA. Free mineral acidity; the sum of the mineral acids.

Fouling. Phenomenon in which the membrane adsorbs or interacts in some manner with solutes in feed resulting in a decrease in membrane performance (lowering of the flux and/or increase in rejection of solutes). This is usually irreversible and time-dependent, which distinguishes it from concentration polarisation.

Freeboard. The space above a packed bed in a vessel that allows for bed expansion during backwashing.

Functional groups. See Exchange sites.

Gel layer. Extreme case of concentration polarisation, describing a physical state of solids accumulating on the membrane surface.

Gellular. A jelly-like appearance with no apparent physical pores, e.g. gellular resins.

Glass transition temperature (T_g). The temperature where the polymer transits from a rubbery state to a glassy state. The thermal expansion coefficient in the rubbery state is two to three times greater than that in the glassy state because of greater molecular (chain) mobility. T_g varies between $-120°C$ and $+130°C$ depending on the type of polymer. Rubbery polymers such as elastomers have –ve T_g, i.e. T_g is well below their use temperature whereas common plastics such as polyvinyl chloride (PVC) polymers have +ve T_g, i.e. T_g is well above their use temperature. However, if used above its T_g, PVC would display the usual rubber-like behaviour.

Hardness. It is caused by multivalent metallic cations, e.g. the combination of calcium and magnesium in water expressed as equivalents of $CaCO_3$. The main cations causing hardness in water and the major anions associated with them are: Ca^{2+}, Mg^{2+}, Sr^{2+}, Fe^{2+}, Mn^{2+}, HCO_3^-, SO_4^{2-}, Cl^-, NO_3^- and SiO_3. It can also include iron and other multivalent cations.

Henry's law. The law states that the weight of any gas that dissolves in a given volume of a liquid at constant temperature is directly proportional to the pressure the gas exerts above the liquid. The Henry's law constant (H) reflects the relationship between gas phase over a solution and dissolved species. Hence, Henry's law can be used to calculate the amount of gas present at saturation at any temperature.

Housing. A pressure vessel in which a membrane element or elements are placed.

Humic substances. High molecular weight polymeric substances, phenolic in character that are formed by the decomposition of plant materials that releases coloured substances of a highly refractive nature. The macromolecular structure is composed largely of aromatic building blocks that have been heavily substituted with carboxylic acid (—COOH) and alcoholic (—OH) functional groups, as shown in Figure 5.1. It has an amphiphilic character due to the hydrophilicity of the —COOH and —OH functional groups, and the hydrophobicity of the aromatic and aliphatic groups in the core of the molecule.

Hydrolysis. The addition of water to split a chemical bond, e.g. the splitting of salt and water into its ions and formation of a weak acid or base, or both.

Hydrophilic. A material that is compatible with or preferentially interacts with water ('water loving'). See Contact angle.

Hydrophobic. A material that is incompatible with and cannot interact with water ('water hating'). These materials preferentially interact with oils and other hydrophobic components. See Contact angle.

Hydroxide alkalinity. The carbonate ion concentration expressed in 'ppm' as $CaCO_3$ equivalents. To convert ppm of hydroxide to ppm of $CaCO_3$ equivalents multiply by 5.88.

Hyperfiltration. See Reverse osmosis.

Inert media. A resin with no reactive groups.

Ion exchange. The interchange of one ion in solution with another ion on an insoluble material.

Iso-electric point. Substances such as proteins acquire their charge mainly through the ionisation of carboxyl and amino groups to give COO^- and NH_3^+ ions. The ionisation of these groups and the net molecular charge depends strongly on the pH, of the solution. At low pH, a protein molecule is positively charged, and at high pH, it is negatively charged. The pH at which the net charge is zero is called the 'iso-electric' point.

Isotropic. Refers to the ultrastructure of a membrane, where the pores are essentially uniform in size from the top to the bottom of the membrane body, and the body is more or less of a uniform density.

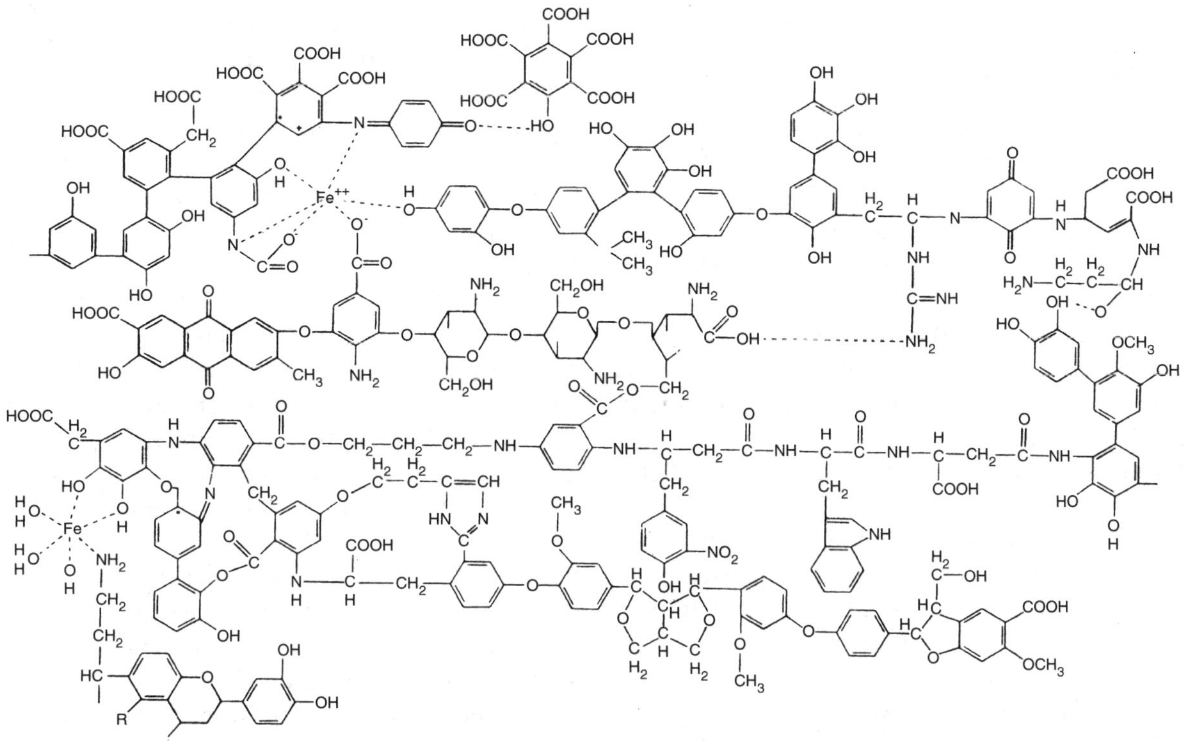

Figure 5.1 A suggested structure of humic acid. Humic materials are highly surface active; basis for much of their fouling characteristics in membrane systems. Source: Harland

Jackson Turbidity Units (JTU). A measure of turbidity or suspended particles by optical obstruction of light rays passed through a water sample. See NTU.

Langelier Saturation Index (LSI). A means of expressing the degree of saturation of water as related to calcium carbonate solubility. It quantifies calcium carbonate scaling potential. The LSI value must be negative to ensure that the scaling potential is zero. Common practice is to feed sufficient acid or an antiscalant to reduce the LSI in the concentrate stream to less than one.

Leakage. Leakage or slippage is the amount of an ion or ions in the effluent from an ion exchange bed during the service cycle.

Macroporous. A material with large pores, e.g. macroporous resins for removing organics.

Mass Transfer Coefficient (*k*). A measure of a solute's mobility due to forced or natural convection in the system. Analogous to a heat transfer coefficient, it is measured as the ratio of the mass flux to the driving force. When using the 'film theory' to model mass transfer, k is also defined as D/δ, where D is solute diffusivity and δ is the thickness of the concentration boundary layer.

Membrane element. Refers to a membrane module without the pressure vessel housing.

Meq/l. Milli-equivalents per litre.

Membrane (bio) reactor. A system for simultaneous reaction and separation with a membrane. Could be conducted in the same physical enclosure or in separate enclosures if the reaction vessel and membrane unit are coupled in a semi-closed-loop configuration.

Mesh. The number of strands in a linear inch of filter material. Higher numbers mean finer particles are retained.

Micrometre (micron). Indication of size of microscopic particles; 1.0 μm = 0.0001 cm.

Microfiltration (MF). Dead-end or cross-flow membrane separation process using microporous membranes with a pore size range of 0.1–10 μm.

Mil. One-thousandth of an inch. 1 mil = 0.0254 mm. A common way of expressing the thickness of spacers in flat-sheet membrane elements.

Molality. One mole of solute per 1,000 g of solvent.

MWCO. Molecular weight cut-off, a term used to describe the potential separating capabilities of a UF membrane. Molecular weight of a theoretical solute with 90% rejection by the membrane in Daltons.

Nanofiltration (NF). Also called 'loose RO.' NF operates at lower pressure than RO but at a higher flux. A process that rejects multivalent ions and organic compounds greater than 200 molecular weight. Rejection of monovalent ions is <70%.

Neutral. Neither acidic nor basic, i.e. pH = 7.0.

Nitrification. An aerobic respiration process in which ammonia is converted into nitrate in the presence of oxygen.

NMWCO. Nominal molecular weight cut-off.

Non-carbonate hardness. The amount of hardness that is in excess of carbonate hardness, i.e. hardness in water caused by chlorides, sulphates, and nitrates of calcium and magnesium (formerly called 'permanent hardness'). See Hardness.

Non-reactive silica. Polymerised colloidal silica.

NTU. Nephelometric turbidity units. A measure of turbidity in water by scattering of light from particles. Replaces JTU.

Oleophilic. Same as hydrophobic.

Osmosis. It is the movement of a solvent through a membrane that is impermeable to a solute, from the more dilute to the more concentrated solution. Water passes through the membrane in both directions but it flows more rapidly in the direction of the salt solution resulting in a difference in hydrostatic pressure (Figure 1.10). The tendency of the solvent to flow can be stopped by applying pressure to the salt solution. The excess pressure that must be applied to the solution to produce equilibrium is known as 'osmotic pressure'. The pressure exerted in excess of the osmotic pressure 'reverses' the flow of the solvent, and is called 'reverse osmosis'.

Oxidation. A chemical reaction in which an element or ion is increased in positive valence, losing electrons to an oxidising agent. Oxidants such as chlorine and ozone are added to water and wastewater to bring about the desired inorganic and organic transformations as well as to disinfect.

Oxidation–reduction reactions. Many reactions in wastewater treatment including organic oxidation, methane fermentation, nitrification, and denitrification are mediated by bacteria. They are important in the solubilisation of iron and manganese.

Permeate. It is that portion of the feed solution which passes through the membrane.

pH. A measure of the hydrogen ion concentration in solution. pH of water determines the percentage of inorganic carbon that is in the form of carbon dioxide, bicarbonate, or carbonate. It also determines the extent to which carbon dioxide is in the permeate water, or whether calcium carbonate is likely to precipitate.

Poiseuille's law. Describes the behaviour of liquids flowing through capillary tubes in relation to their viscosity [$V = (\pi Pr^4)/(8\,\mu l)$].

PPM. An acronym for 'parts per million' (mg/l). Similarly, PPB = μg/l.

PWR. Product water recovery defined as: permeate flow/ feed water flow (%).
Pyrogen. Any substance capable of producing a fever when injected into mammals. Produced by bacteria and not necessarily inactivated by conventional heat treatment methods.
Reduction. A chemical reaction in which an element or compound gains electrons, being reduced in positive valence.
Regenerant solution. A chemical used to restore the ion-exchange resin back to its desired or original form.
Rejection. A measure of how well a membrane retains or allows passage of a solute. When based on concentrations in the bulk of the permeate and retentate streams, it is the 'apparent' rejection. The 'intrinsic' rejection is based on concentrations at the membrane surface. Sometimes used interchangeably with retention factor.
Reject stream. Also called 'retentate,' is the concentrated feed stream that does not flow through the membrane.
Reverse osmosis (RO). The reverse of natural 'osmosis' achieved by external application of sufficient reverse pressure to cause the solvent to flow in its unnatural direction, i.e. from the more concentrated solution to the dilute solution. See Osmosis.
Reynolds number (Re). A measure of the state of turbulence in a fluid system. It is the ratio of inertia effects to viscous effects. Reynolds number is dimensionless. Flow systems with Re values <2,200 are considered to be laminar flow, while Re > 4,000 are considered to be fully turbulent (Table 5.1).
Salt splitting. To replace the cation (or anion) of a neutral salt such as sodium chloride (NaCl) with hydrogen (or hydroxyl) ion.
Schmidt number (Sc). A measure of the ratio of momentum transfer to mass transfer.
Silt Density Index (SDI). It is an empirical measurement (ASTM Standard D-4189-82, 1987) used for measuring the colloidal fouling potential of RO/NF feed water. It is not a direct measure of the particle concentration, which is more properly measured by turbidity. The SDI value is based on the time required to filter a volume of feed water through a 0.45 μm filter pad at a feed pressure of 2.1 bar g. The SDI value is calculated from the formula:

$$SDI = P_{30}/T_t = 100(1 - T_i/T_f)/T_t$$

where,

P_{30} = % plugging; P_{30} should not exceed 75%
T_t = total test time, usually 15 min
T_i = initial time in seconds required to obtain a sample
T_f = time required to obtain a sample after 15 min

A value <1 is indicative of a system that will run for several years without problems; a system with a value in the range of 3–5 is considered a 'problem' and requires regular cleaning; and a value >5 indicates a system with severe fouling potential. Most spiral-wound RO/NF membrane systems require a SDI < 4.
Sequester. To form a stable water soluble complex.

Table 5.1 Reynolds number for various flow systems

Flow system	Re[1]
Flow past wing of a Jumbo jet	5×10^7
Spin-down of a stirred cup of tea	5×10^3
Finger moved through water	3×10^2
Golden syrup draining from a spoon	2×10^{-3}
Swimming spermatozoan	1×10^{-3}
Spiral-wound membrane elements[2]	$<1 \times 10^3$

Source: Acheson, From Calculus to Chaos, Oxford, 1997
[1] $Re < 2.1 \times 10^3$ is laminar; $Re > 4 \times 10^3$ is turbulent. The transition from laminar to turbulent flow in porous channels and tubes, e.g. tubular membranes, $Re < 2 \times 10^3$
[2] Flow is considered turbulent because of mesh spacers

Sherwood Number (Sh). A dimensionless measure of the ratio of convective mass transfer to molecular mass transfer. If the mass transfer coefficient k is defined in terms of the film theory, then Sh is a measure of the ratio of hydraulic diameter to the thickness of the boundary layer.

SHMP. Sodium hexametaphosphate, a sequestering agent used for complexing calcium sulphate, thereby, preventing $CaSO_4$ from precipitating on the membrane surface.

SI units. International System of Units was established in 1973–74 to develop a uniform method of reporting results worldwide. The basic units of the SI system are: metre (m), kilogram (kg), second (s), ampere (A), kelvin (K), mole (mol), and candela (cd). Acceptable units outside the SI system are: litre (l), grams per litre (g/l), moles per litre (M), day (d), hour (h), and minute (min).

Solubility product. Solubility Product Constant (K_{sp}) reflects the relationship between dissolved species and precipitated species. K_{sp} is commonly used in solubility calculations to determine the precipitation potential of mineral salts during RO/NF processing. Certain combinations of cations and anions form sparingly soluble salts in water, and scaling of a reverse osmosis membrane may occur when the salts are concentrated within the RO/NF element beyond their solubility limits. (Table 5.7).

Solute. The solid (e.g. salt) which is dissolved in water (solvent) to form a solution.

Spacers. A mesh-like material used in flat-sheet modules (e.g. plate, spirals, pleated sheet) to separate successive layers of membranes. Spacers control the feed channel dimensions in these modules. Serve as turbulence promoters in spiral-wound membrane elements.

Space velocity. Flow per unit volume of media ($m^3/h/m^3$ or gpm/ft^3).

Superficial linear velocity. Flow per unit area ($m^3/h/m^2$ or gpm/ft^2).

Surfactant. 'Surface active agent' is a compound that alters the surface tension of a liquid in which it is dissolved. All surfactants have large polar molecules. One end of the molecule is soluble in water (due to carboxyl, sulphate, hydroxyl, or sulphonate groups) and the other end is readily soluble in oils (organic groups). Synthetic surfactants are of three types: anionic, cationic, and non-ionic.

Total Dissolved Solids (TDS), in mg/l or ppm is a measure of the total weight of impurities found in water supply. The measure permits a quick, rough estimate of RO/NF permeate quality and feed pressure requirements.

Thermoplastic. Polymers such as polyethylene (PE) and polypropylene (PP) can be processed and fabricated in the molten state. When actually used they are in the crystalline state so that the desired physical properties can be exploited. The replacement of all the hydrogen atoms from the linear PE with fluorine atoms results in a *fluoropolymer* such as polytetrafluoroethylene (PTFE) with a much higher melting temperature.

TOC. Total organic carbon is a measure of the organic carbonaceous material in water. Under normal circumstances, TOC concentrations are less than 15 ppm, ranging in most water supplies from two to six ppm. Under unusual circumstances or in wastewater streams, the TOC level can be 50 ppm or higher. Organic compounds with a molecular weight greater than 200 are easily rejected by RO and NF membranes. Only 40% compounds of lower molecular weight such as trihalomethanes are removed. TOC is not a key parameter in the design of RO systems. It can, however, be a source of organic or bacterial fouling of the membrane.

Trace contaminants. Contaminants in groundwater or drinking water that are carcinogenic when present in excess of minute concentrations as set by the US government maximum contaminant levels. Mostly halogenated hydrocarbons (typically 1–100 μg/l or ppb) and heavy metals (typically 0.1–1.0 mg/l or ppm).

Transmembrane pressure (TMP). The driving force for flux. In cross-flow systems, it is measured as the average of the feed inlet and concentrate (reject) outlet pressures minus the permeate back pressure.

Turbidity is a measure of the extent to which light is scattered by particulate matter in the water. It is caused by a wide variety of suspended materials ranging in size from colloidal to coarse dispersions. Municipal water supplies normally have a turbidity level of less than one NTU indicating a content of less than 0.5 ppm of suspended solids. The turbidity should be less than 1 NTU to avoid particulate fouling of RO/NF membrane elements.

Turbulence promoter. Devices that are inserted into the feed/retentate channel of a module to improve mixing characteristics, increase turbulence, and thus improve flux. Typical turbulence promoters are spacers, spiral wires, static Kenics®-type mixers, spheres, or balls.

Ultrafiltration (UF). Cross-flow membrane separation process using membranes with a pore size range of about 0.005–0.1 μm corresponding to MWCO of 1,000–500,000 Da.

Valence. The number of positive or negative charges of an ion. It is measured by the number of atoms or ions of hydrogen it takes to combine with or be replaced by an element or radical.

Vapour pressure. The pressure exerted by a vapour in equilibrium with a liquid or solid.

Velocity. The linear rate of flow of fluid parallel to the membrane, expressed in units of length/time (e.g. m/s). This is calculated as flow rate/cross-sectional area of feed channel.

Void volume. The space between particles of a settled packed-bed, or 'interstitial volume.'

Water hammer. Instantaneous surges of water pressure caused by sudden interruptions in water flow in pipe or tank systems.

Zeolite. A mineral composed of hydrated silicates of aluminium and sodium or calcium. The term has been used, sometimes improperly, to describe softening by synthetic IX resins.

Zeta potential (ζ). It is the magnitude of the charge at the surface of shear – the boundary surface between the fixed ion layer and the solution serves as a shear plane when the particle undergoes movement relative to the solution. $\zeta = 4\pi\delta q/D$ where q is the charge at the shear surface, δ is the thickness of the diffuse layer, and D is the dielectric constant of the liquid. Zeta potential gives an indication of the effectiveness of added electrolytes in lowering the energy barrier between colloids (i.e. reducing surface potential and electrical double-layer thickness), and thus can be used to select the optimum conditions for coagulation. Most colloids in natural water sources have a negative charge that helps to repel each other and to keep them suspended in solution. Overall the colloidal water has a negative ζ. The negative charge can be neutralised by the addition of cationic coagulants such as aluminium sulphate (alum) and ferric chloride. The coagulants are added until ζ reading is zero (neutral). The colloids without charges do not repel each other and coagulate into larger particulate groups, which can be easily filtered out by filtration. See also 'Electric double layer'.

5.2 Membrane polymer performance

In this section, we demonstrate how membrane morphology and, in turn performance are altered by changing the chemical structure of the membrane polymer.

The permeability of synthetic membranes is dependent on both the chemical and physical structure of the membrane. High permeability may be realised by (a) increase in porosity, (b) decrease in thickness, (c) increase in operating pressure, and (d) increase of hydrophilicity. Water flux of polyamide (PA) membranes can be enhanced by the introduction of highly hydrophilic pendant carboxylic (—COOH) or sulphonic acid (—SO_3H) groups. Nitration of polymers also imparts hydrophilicity to the polymer. However, it was found that the nitrated PA membranes had the opposite effect; the membrane shrank slightly and there was a reduction in flux[1] when thin-film composite (TFC) PA membranes were nitrated to produce a chlorine–resistant membrane (Section 5.3).

The drop in flux can be shown quantitatively by the drop in the value of the membrane constant, A. For a perfectly semi-permeable membrane with feed side mixing or flow conditions such that the boundary layer is insignificant,

$$A = J/(\Delta P - \pi) \tag{1}$$

where A is the membrane constant (g/cm^2.s.atm); J the permeate flux (g/cm^2.s.); ΔP the pressure drop across the membrane; and π the osmotic pressure of the salt solution (atm). The membrane constant is based on the membrane properties alone.[2] The value of A decreased for the TFC membrane from 1.92×10^{-5} to 0.47×10^{-5} g/cm^2.s.atm or fourfolds after nitration, most likely due to qualitative morphological changes in the membrane discussed next. Any shrinkage in membrane would lead to a decrease in porosity (ε), and an increase in skin-layer thickness (λ). These in turn would lead to a high tortuosity (τ). In other words, the term ($\lambda\tau/\varepsilon$) defined here as a 'morphology constant' would increase. This is further evident (J decreases as $\lambda\tau/\varepsilon$ increases) if we consider the equation for the special case of Poiseuille flow through membrane pores, which have the ideal critical diameter (2t) for desalination given by

$$J = (\varepsilon\pi P\gamma^4)/(8\eta\lambda\tau) \tag{2}$$

where t is the thickness of a monomolecular layer of pure water on the membrane surface based on the preferential-sorption capillary flow model of Sourirajan, J the permeability of water, P the effective pressure, γ the effective pore radius, and η the average viscosity.[3] Typical values for homogenous PA membranes are reported in the literature as $\lambda = 0.1$–$0.3\ \mu$m, $\tau = 2.5$ and $\varepsilon = 0.005$–0.052.[4]

It was also reported that membrane permeability decreased while membrane selectivity increased in the case of nitrated polysulphone (PS) membranes[5] due to increasing packing density. Since the packing density is a common measure of polymer 'free volume,' an increase in packing density was believed to be due to an increase in polymeric interchain attraction caused by the polar nature of the nitro (—NO_2) group, and/or an increase in chain–packing efficiency. Further, the addition of nitro groups to the backbone of the PS polymer most likely decreases polymer permeability by stiffening the polymer matrix, thereby restricting chain torsional mobility. The stiffening of the PS polymer by nitration was due to an increase in the glass transition temperature (T_g) of the highly nitrated PS polymer. Hence, a lower flux after nitration of the PA membrane was most likely due to morphological changes, playing a far more dominant role than hydrophilic factors. Similar behaviour was also noted when a polysulphone UF membrane was nitrated by a gas–phase technique.[6] The UF membrane shrank transforming into a NF membrane.

5.3 Chlorination of polyamide (PA) membranes

In order to form suitable membranes, the membrane polymer must possess glass transition temperatures (T_g), and melting temperatures (T_m) that are well above the use temperature. Thin–film composite (TFC) polyamide (PA) membranes in spiral-wound elements are the work-horse of RO and NF systems, as discussed in Chapter 1. These membranes have remarkable properties for separations in aqueous streams; high flux, high rejection (RO membranes), low fouling, satisfactory thermal properties, and tolerance to operating conditions over a wide pH range. In the last few years, TFC PA membranes have been developed that are sanitisable with hot water at 80°C, an essential requirement for applications in the pharmaceutical and beverage industries.

As remarkable as the performances of these membranes are, all aromatic PA membranes have one severe drawback; their susceptibility to degradation by oxidants such as chlorine. The removal of chlorine prior to membrane separation by activated carbon, sodium sulphite, or sodium bisulphite addition is, therefore, mandatory. Dechlorination by $NaHSO_3$ in stoichiometric excess can, however, be ineffective in seawater feed because dissolved oxygen in seawater reacts with the chemical. Further, the absence of chlorine can lead to biofouling that is often irreversible.

Aromatic PA polymers used for RO membranes are essentially random copolymers consisting of repeat units shown in Figure 5.2. The amide ring is sensitive to attack by chlorine because it is an electron-rich region.[1,7] Hence, deleterious ring, chlorination occurs only on the electron-rich phenylenediamine rings and not on the electron-deficient phthalate rings, as shown in Figure 5.3. The bulkiness of the halogen

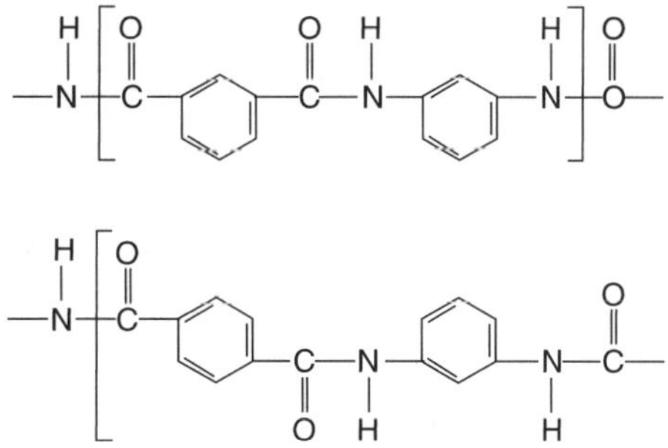

Figure 5.2 Aromatic polyamide polymer structure. Source: Singh[1]

substituents on the amide ring results in changes in hydrogen-bonding modes from inter-molecular to intra-molecular. This results in chain deformation and alteration of the average inter-chain displacement within the membrane skin structure. The damage to the skin layer of TFC composite membranes is severe rendering the membrane useless with rejections dropping to as low as 20%.[8–10] A reaction scheme for chain cleavage is shown in Figure 5.4.

The effect of oxidative chain cleavage on PA polymer can be investigated by exposing the polymer to a sodium hypochlorite solution over a period of time and measuring the viscosity of the dissolved polymer after chlorination. The severe drop in viscosity after several exposure times confirmed the breakdown of the polymer.[7] Why viscosity? Solution viscosity is a well-recognised measure of the molecular weight of a polymer. According to the Mark–Houwink equation, for example, the relationship between the intrinsic viscosity [η], and the average molecular weight, M is expressed as:

$$[\eta] = K \cdot M^{X} \tag{3}$$

where K is a proportionality constant.[3] Hence the breakdown of the PA polymer due

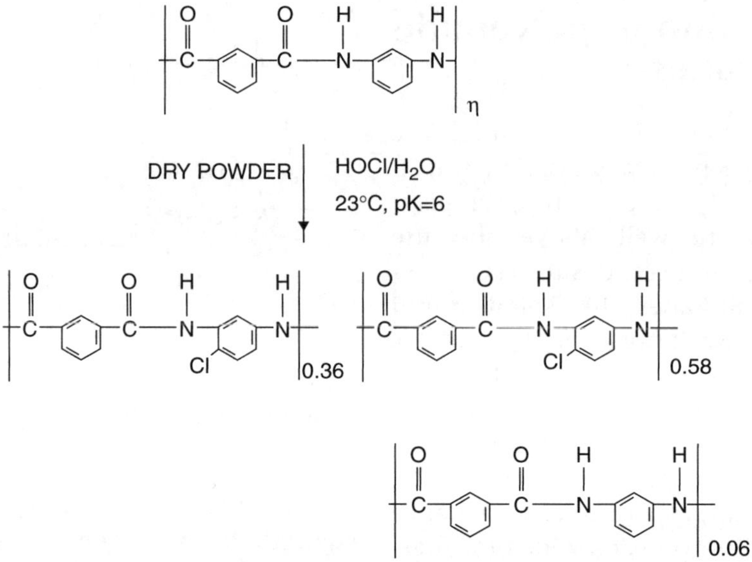

Figure 5.3 Polyamide ring chorination based upon chlorine analysis and NMR evidence. Source: Singh[1]

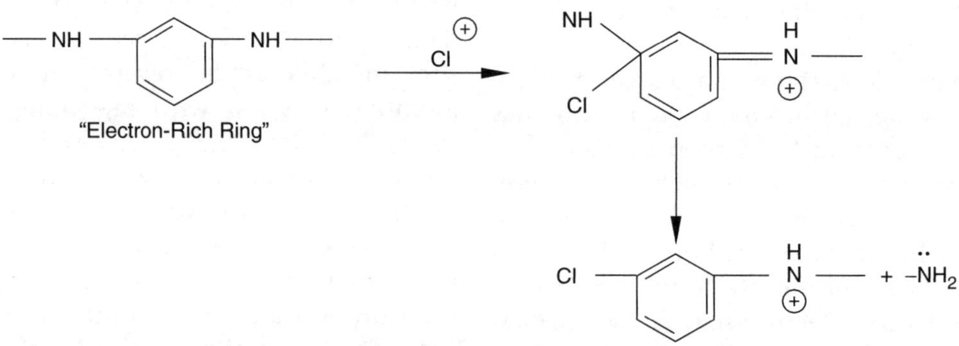

Figure 5.4 PA polymer chain cleavage reaction. Source: Glater et al.

to chlorination, i.e. lower molecular weight is reflected in lower viscosity measurements.

One way to suppress the ring chlorination reaction is to make the diamine rings less electron-rich. Nitro (—NO_3) groups are powerful electron acceptors and are known to drastically deactivate aromatic rings towards electrophilic substituents. Nitration of an aromatic ring is known to reduce its reactivity by 10 for further ring addition reactions such as chlorination. A reaction scheme where an NO_2 molecule attaches itself to an aromatic ring containing the amide groups is shown in Figure 5.5.

5.4 Fluid flow in microporous membranes

In the analysis of fluid flow in microporous membrane tubes and channels, it is tacitly assumed that the no-slip boundary condition that characterises flows with solid bounding

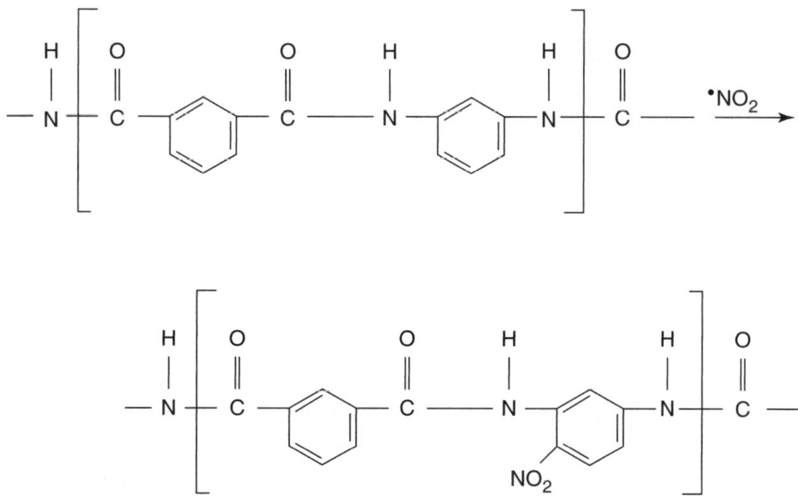

Figure 5.5 PA polymer nitration reaction. Source: Singh[7]

walls is applicable. This postulate is incorrect because the surface velocity at a porous wall is in fact not zero but proportional to the shear rate at the permeable boundary, i.e.

$$U_s = \phi.\, du/dy\big|_{y=0} \tag{4}$$

where U_s is slip velocity, ϕ is slip flow coefficient, and du/dy is shear rate at wall boundary. Slip velocity and slip coefficient for a UF membrane were measured with a dye-tracing photographic technique by Kohler.[11] Laminar flow velocity profiles in tubes and channels using slip coefficients measured by Kohler were subsequently investigated.[12,13] Profiles for channel flow (Figure 5.6) are shown in Figure 5.7, which is a plot of normalised velocity as a function of channel height [U $(=u/u_{avg})$ vs. λ $(=y/h)$] for an entrance Reynolds number of 1,000 $(\mathrm{Re} = 4u_{avg}h/\nu)$ and a longitudinal position, x/h equal to 500. The curves are plotted for slip coefficient equal to 0, 0.1, and 0.5, and a wall Reynolds number, $\mathrm{Re_w}$ of 0.1 [$(\mathrm{Re_w} = \nu_w h/\nu)$ where ν_w is the fluid velocity through the membrane].

The curves show that the velocity at the membrane surface $(\lambda = 1)$ is 0 when ϕ is 0 as expected. As the slip velocity increases with increasing ϕ, the wall shear decreases, and the velocity profiles become flatter, approaching those for plug flow.[12–14] The effect of slip coefficient on axial pressure gradient *(P)* is as follows: an increase in ϕ results in a decrease in wall shear stress so that P also decreases. In addition, the transition from laminar to turbulent flow at a porous surface occurs at a Re of less than 2,000 that is also the case with membrane systems.

The effect of slip coefficient on concentration polarisation (CP) was mathematically modeled for flat membrane and tubular membrane systems.[12,13] CP has been discussed in Chapter 1. Lowering of CP due to slip coefficient ϕ as a function of product water recovery (ξ) for different normalised diffusion coefficients (α) is shown in Figure 5.8. The data shows that CP decreases both with ϕ and α. Since α is a measure of particle diffusion from the membrane surface to the bulk solution, slip-flow possibly augments diffusive back transport of particles from the membrane surface to the bulk solution. Thus, the slip-flow velocity model possibly accounts for actual UF/MF flux that is under

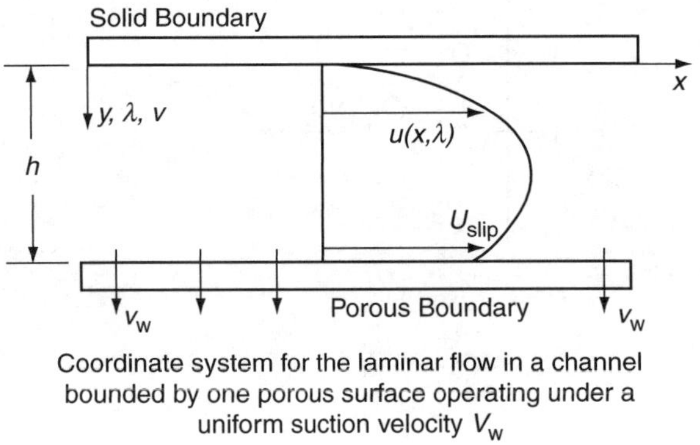

Figure 5.6 Schematic of a channel with a porous wall

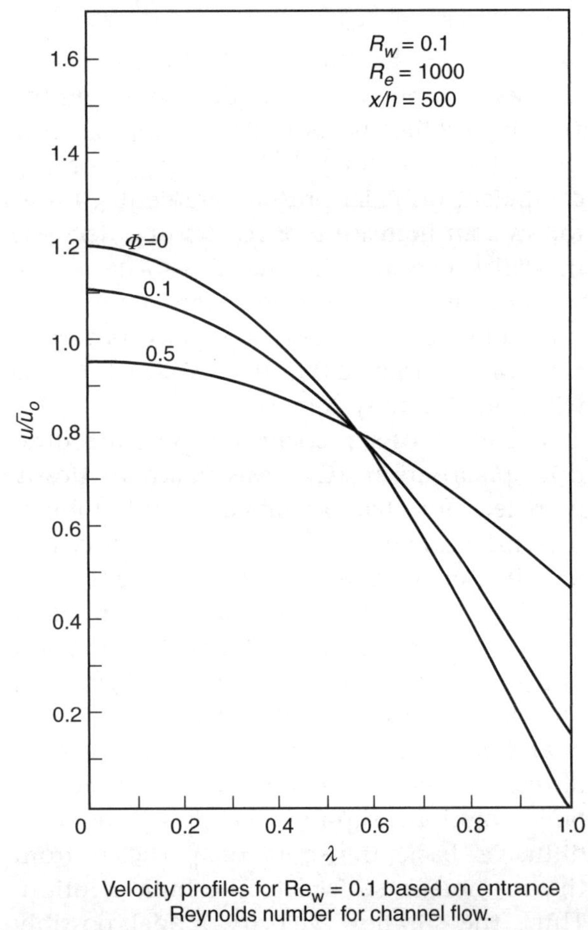

Figure 5.7 Slip-flow velocity profiles. Source: Singh and Laurence

predicted by the gel polarisation model discussed in Chapter 1.

5.5 Surfactant micelles size correlation

Surfactant-based ultrafiltration (SBUF) has been investigated since the 1980s for removing dissolved organic compounds especially hazardous wastewater, as discussed in Chapter 1. The basis of SBUF is that at concentrations above the critical micelle concentration (CMC), surfactant micelles attach to each other to form organised aggregates called micelles. These aggregates average 30–200 monomers and are easily rejected by UF membranes. Thus sodium dodecyl sulphate, an anionic surfactant with a molecular weight of 288 is too small to be rejected by a UF membrane.[15] However, the anionic surfactant micelles were easily rejected by a 10,000 MWCO membrane. The micelles form in such a way that the hydrophobic portions of the molecules align towards the centre, away from the surrounding hydrophilic aqueous solution.

Surfactants, both ionic and non-ionic, are commonly available. Non-ionic surfactants form larger micelles with aggregation numbers an order of magnitude higher than ionic surfactants but at a much lower CMC, e.g. the CMC value for sodium dodecyl sulphate is 2,360 mg/l (MW = 288) as compared to 16 mg/l in the case of Igepal RC-520 (MW = 526), a non-ionic surfactant. Since only micelles are rejected by UF membranes, a high CMC value results in a substantial leakage of surfactant monomers through the membrane. Hence, the size of micelles is an important parameter for selecting a UF membrane with the largest MWCO in order to achieve both high flux and high rejection.

The micelle diameter can be calculated by measuring the micelle diffusion coefficient

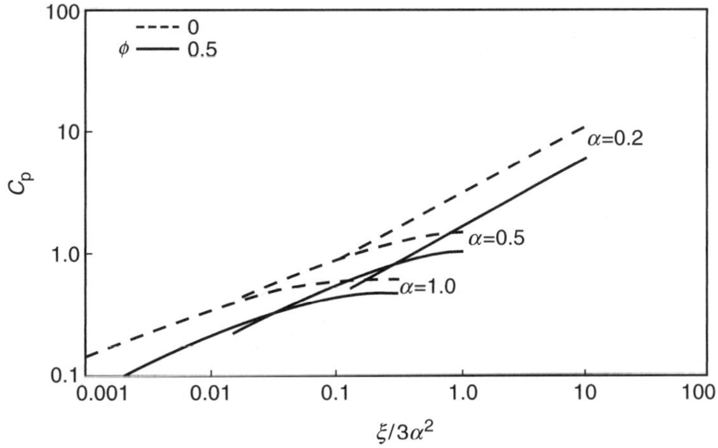

Concentration polarisation, C_p as a function of slip-coefficient, ϕ. ξ is fraction of water removed (1 – R_w/R_e (x/h) or channel length. α is normalised diffusion coefficient, ($D/v_w h$).

Figure 5.8 Concentration polarisation profiles in a membrane channel. Source: Singh and Laurence

using the technique of dynamic light scattering (DLS). If one assumes that all micelles are spherical in shape, the radius of a micelle in solution may be calculated by using the Stokes–Einstein relationship:

$$D = kT/(6\eta R) \quad (5)$$

where D is the diffusion coefficient, k is Boltzmann's constant, T the absolute temperature, η the solvent viscosity, and R the hydrodynamic radius of the micelle. Using the DLS technique, the hydrodynamic radius of micelles of ionic surfactants, Tween 80 and Igepal RC-520 was found to be 37–39 Å and 400–600 Å, respectively.[16] The diffusion coefficient of the two non-ionic surfactants calculated using equation (5) was 5.5×10^{-7} cm²/s and 4.5×10^{-8} cm²/s, respectively. The diameter of the anionic surfactant was considerably smaller, 24 Å, which was in good agreement with the value in the literature.

The surfactant diffusion data was compared with the molecular weight (MW) and diffusion data of selective proteins from the literature to estimate pseudo-molecular weights (PMW) of the surfactant micelle in Figure 5.9.[16] The correlation indicates that the PMW of Tween 80 and RC-520 micelles was 92,000 and 346,000, respectively. A rule of thumb is that the MWCO of a UF membrane should be at least one-half of the solute to be separated. Thus, a new tool was developed where the relationship between the micelle diameter, diffusion coefficient, and a PMW could be used to screen high MWCO membranes for SBUF applications.

5.6 Deioniser design

Deionisation (DI) requires replacement of all cations and anions in the aqueous solution, as discussed in Chapter 2. The simplest DI system is a dual-bed system consisting of a strong acid cation exchanger followed by a strong base anion exchanger operating in series. For higher purity water production, a mixed-bed deioniser is required downstream of the two-bed cation–anion exchanger system or an RO system. A general rule of thumb is that when the TDS exceeds 300 mg/l,

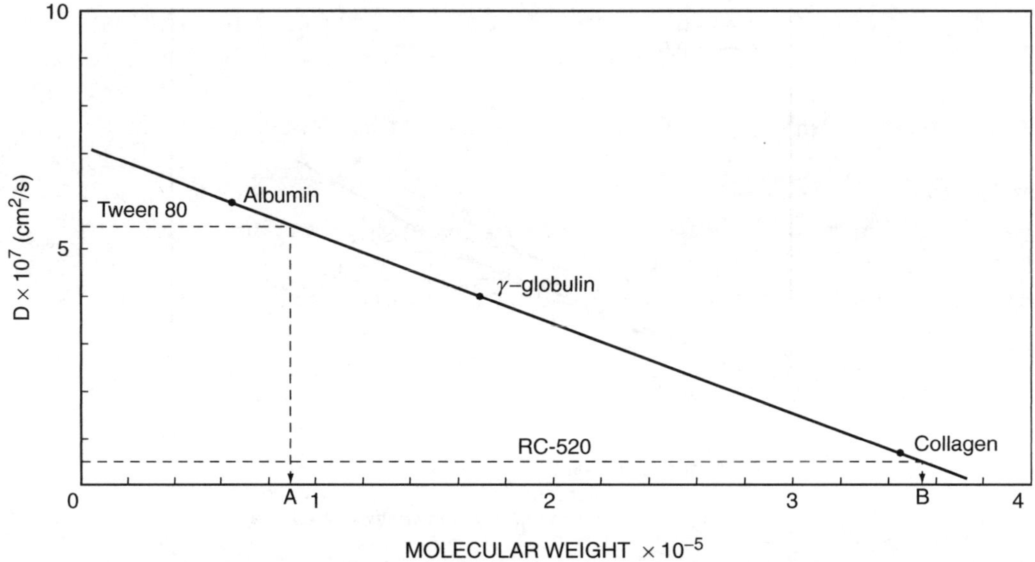

Diffusion coefficient vs molecular weight plot for macromolecules. Diffusion coefficients for Tween 80 and RC-520 Points A and B refer to pseudomolecular weights of Tween 80 and RC-520, respectively.

Figure 5.9 A correlation between the size of surfactant micelles and molecular weight of macromolecules. Source: Singh[16], Copyright 1996 from Separation Science and Technology by R. Singh. Reproduced by permission of Taylor & Francis Group, LLC, http://www.taylorandfrancis.com

RO becomes more economical than dual-bed IX systems. IX systems capabilities are listed below:[17]

- Sodium cycle softeners with co-flow regeneration: softened water hardness = 1–3 mg/l.
- Sodium cycle softeners with counter-flow regeneration: softened water hardness = 0.5 mg/l.
- Dual-bed IX systems with co-flow regeneration: DI water conductivity = 5–10 μS/cm, and silica = 50 μg/l.
- Dual-bed IX system with counter-flow regeneration: DI water conductivity = 1–3 μS/cm, and silica = 10–20 μg/l.
- A dual-bed IX system with co-flow regeneration followed by mixed-bed IX: DI water conductivity = 0.06–0.1 μS/cm, and silica = 5–10 μg/l.

5.6.1 DI system design basis

A dual-bed IX system is designed for producing DI water with 1.0 μS/cm conductivity for a medium pressure (25–60 bar g) boiler.[18,19] The feed water has the following components: Ca^{2+} = 75 mg/l; Na = 50 mg/l; Mg^{2+} = 25 mg/l; Cl^{-} = 30 mg/l; SO_4^{2-} = 80 mg/l; HCO_3 = 40 mg/l; and SiO_2 = 10 mg/l where concentrations are in ppm as $CaCO_3$ equivalent. The feed water flow rate is 57 m^3/h at 20°C. The design procedure is given in Sections 5.6.2 and 5.6.3 using the resin manufacturer's data given below:

Strong acid cation resin bed:

- Regeneration level of 6 lb H_2SO_4 per ft^3 (96 kg H_2SO_4 per m^3) for maximum sodium leakage of 2.0 ppm, and an operating capacity of 15.6 kgr/ft^3 (35 kg/m^3);

- Pressure drop per foot (per 0.305 m) of bed depth at a linear velocity of 8.6 gpm/ft^2 (21 m^3/h/m^2) is 0.75 psi (5.2 kPa);
- Backwash flow rate of 6.4 gpm/ft^2 (15.6 m^3/h/m^2) for a bed expansion of 60%; and
- Rinse water flow is 25–50 gallons/ft^3 (3.34–6.68 m^3/m^3) using DI water or RO permeate. The rinse water flow rates should be between 1 gpm/ft^3 (8 m^3/h/m^3) and 1.5 gpm/ft^3 (12 m^3/h/m^3).

Strong base anion resin bed:

- Regeneration level of 4 lb NaOH per ft^3 (64 kg NaOH per m^3) for maximum silica leakage of 0.05 ppm and an operating capacity of 15.3 kgr/ ft^3 (35 kg/ m^3);
- Pressure drop per foot (per 0.305 m) of bed depth at a linear velocity of 8.5 gpm/ft^2 (20.8 m^3/h/m^2) is 0.85 psi (5.9 kPa);
- Backwash flow rate of 2.6 gpm/ft^2 (6.34 m^3/h/m^2) for a bed expansion of 60%; and
- Rinse water flow is 40–90 gallons/ft^3 (5.34–12.0 m^3/m^3) using DI water or RO permeate. The rinse water flow rates should be between 1 gpm/ft^3 (8 m^3/h/m^3) and 1.5 gpm/ft^3 (12 m^3/h/m^3).

5.6.2 Cation ion-exchanger design

Quantity of water to be treated per service cycle. The amount of process water deionised per service run cycle is (250 gpm)(60 min/h) (12 h/cycle) = 180,000 gallons (682 m^3).

Cation load. Since the influent cation load is expressed as ppm equivalents as $CaCO_3$, it is necessary to convert to units consistent with resin manufacturer's capacity data expressed as kilograins (as $CaCO_3$) per ft^3 of resin. Total cation load in this case is 75 + 50 + 25 = 150 ppm. Convert to kgr: (150 ppm)(180,000 gallons/cycle)/(1,000 gr/kgr)(17.1 ppm/gr/gallon) = 1,580 kgr as $CaCO_3$ per cycle.

Quantity of cation-exchange resin. The cation load per cycle is 1,580 kgr from above, and the resin capacity is given as 15.6 kgr/ft^3 in Section 5.6.1. Therefore, the amount of resin needed is 1,580 kgr/(15.6 kgr/ft^3) = 101 ft^3 (2.86 m^3).

Bed velocity. For water treatment, the space velocity (SV) is usually between 1 and 5 gpm/ft^3 resin (8 and 40 m^3/h/m^3). Therefore the SV is 250 gpm/101 ft^3 = 2.5 gpm/ft^3. The normal linear velocity range is 4–10 gpm/ft^2 (9.75–24.4 m^3/h/m^2). Assuming a bed depth of 3.5 ft (1.07 m), the cross-sectional area is 101 ft^3/3.5 ft = 28.9 ft^2. The linear velocity therefore is: 250 gpm/ 28.9 ft^2 = 8.6 gpm/ft^2, which is within range. If either SV or linear velocity is considerably greater than the normal range, more resin would be required.

Vessel dimensions. The vessel diameter is (cross section area × 4/π)$^{1/2}$ = (28.9 × 4/π)$^{1/2}$ = 6.1 ft (1.86 m). In order to determine the vessel side shell height (SSH), assume 100% bed expansion; hence, the vessel SSH is 7 ft (2.13 m) or twice the bed height.

Resin bed pressure drop and backwash flow rate. A minimum head-space (freeboard) is required for bed expansion at the design backwash flow rate. Typically the bed expansion is 50–60%. The flow rate required for this expansion is specified as 6.4 gpm/ft^2. Thus the backwash flow rate is 6.4 gpm/ft^2 × 28.9 ft^2 = 185 gpm (42 m^3/h). The pressure drop (ΔP) per foot of bed depth is specified as 0.75 psi. Thus ΔP for the cation-resin bed is (0.75 psi/ft) × (3.5 ft) = 2.6 psi (17.9 kPa). This excludes the pressure drop due to valves, fittings, or liquid distributors or collectors. Overall the ΔP should be $<$10 psi (68.9 kPa).

Regeneration calculations. The acid regeneration level to hold Na^+ leakage to 2 ppm is 6 lb/ ft^3 (96 kg H_2SO_4 per m^3), as specified in Section 5.6.1. The total acid required per cycle, therefore, is (6 lb/ft^3) (101 ft^3) = 606 lb (275 kg) per cycle. To avoid precipitation of $CaSO_4$, regenerate in two steps. First use 50% of the regenerant as a 2% solution, and then the rest as a 4% solution. For the 2% step process: (1/2 × 606 lb H_2SO_4)/ (8.43 lb solution per gallon)(0.02 lb H_2SO_4 per lb solution) or about 1800 gallons (6.8 m^3) of 2% acid solution. The 4% step requires (1/2 × 606)/ (8.54 × 0.04), or about 890 gallons (3.37m^3) of 4% acid solution. Note 8.43 lb/gallon and 8.54 lb/gallon are densities of the 2% and 4% acid solutions, respectively.

Rinse water flow rates. The column must be rinsed with DI/RO water after regeneration to remove traces of acid. Bed rinse is done at two flow rates to remove all traces of acid. The rinse-water requirements are specified as 25–50 gallons/ft^3 resin For the resin volume of 101 ft^3, the rinse water requirement is about 2,500–5,000 gallons (9.5–19 m^3). For the slow rinse step, the flow rate is (one bed volume – 101 ft^3 × 7.48 gallons/ft^3, or ~750 gallons) at 50 gpm (11.4 m^3/h) for 15 min. For the fast rinse-step, the flow rate is based on the remainder volume, i.e. at 150 gpm (34.1 m^3/h) for ~50 min.

5.6.3 Anion ion-exchanger design

Quantity of water to be treated per service cycle. The amount of process water deionised per service run cycle is (250 gpm) (60 min/h) (12 h/cycle) = 180,000 gallons (682 m^3).

Anion load. Total anion load is 30 + 80 + 40 = 150 ppm. Convert to kgr as follows: (150 ppm)(180,000 gallons/cycle)/(1,000 gr/kgr) (17.1 ppm/gr/gallon) = 1,580 kgr as $CaCO_3$ per cycle.

Quantity of anion-exchange resin. The anion load per cycle from above is 1,580 kgr, and the resin capacity is 15.3 kgr/ft^3 from Section 5.6.1. So the amount of resin needed is 1580/15.3 = 103 ft^3 (2.91 m^3).

Bed velocity. The space velocity is 250 gpm/ 103 ft^3 = 2.4 gpm/ft^3 resin. The normal linear velocity range is 4–10 gpm/ft^2 (9.75–24.4 m^3/ h/m^2). Assume a bed depth of 3.5 ft (1.07 m). Then the bed cross–sectional area is 103 ft^3/ 3.5 ft = 29.4 ft^2, and the linear velocity is 250 gpm/29.4 ft^2 = 8.5 gpm/ft^2, which is within range. If either SV or linear velocity is considerably greater than the normal range, more resin would be required.

Vessel dimensions. The vessel diameter is $(\text{area} \times 4/\pi)^{1/2} = (29.4 \times 4/\pi)^{1/2} = 6.12$ ft (1.86 m). In order to determine the vessel side shell height (SSH), assume 100% bed expansion; hence, the vessel SSH is 7 ft (2.13 m) or twice the bed height.

Resin bed pressure drop, and backwash flow rate. A minimum head-space (freeboard) is required for bed expansion to permit the requisite backwash flow rate. Typically the design bed expansion is 50–60%. The flow rate required for this expansion is specified as 2.6 gpm/ft^2 (lower linear velocity because the anion resin is lighter than the cation resin). Hence, the backwash flow rate is (2.6 gpm/ft^2) × (29.4 ft^2) = 76 gpm (17.3 m^3/h).

The pressure drop (ΔP) per foot of bed depth is specified as 0.85 psi. Thus, ΔP for the anion-exchange bed is (0.85 psi/ft) × (3.5 ft) = 3 psi (2.07 kPa). This excludes the pressure drop due to valves, fittings, or liquid distributors, or collectors. Overall the ΔP should be <10 psi (68.9 kPa).

Regeneration calculations. The caustic soda regeneration level is 4 lb NaOH per ft^3 as specified. Total caustic soda requirement, therefore, is (4 lb/ft^3)(103 ft^3) = 412 lb (187 kg) NaOH per cycle. Regenerant concentration is typically 4% NaOH solution having a density of 8.68 lb/gallon. Total regenerant solution requirement per cycle is (412 lb NaOH)/0.04 lb NaOH per lb of solution) (8.68 lb solution per gallon) or 1190 gallons per cycle (4.5 m^3 per cycle). The regenerant flow rate is 0.5 gpm/ft^3 (4 m^3/h/m^3).

Rinse water flow rates. The column is rinsed with DI/RO water after regeneration to remove caustic remaining in the bed. Bed rinse is done at two flow rates to remove all traces of caustic. The rinse water requirements are specified as 40–90 gallons/ft^3 resin, i.e. for a resin volume of 103 ft^3, the requirement is 4,000–9,000 gallons (15.1–34.1 m^3). For the slow-rinse step, the flow rate is (one bed volume – 103 $ft^3 \times 7.48$ gallons/ft^3, or 750 gallons) at 50 gpm (11.4 m^3/h) for 15 min. For the fast-rinse step, the flow rate is based on the remainder volume, i.e. at 150 gpm (34.1 m^3/h) for ~50 min.

5.7 Process controls

Microprocessor-based control systems generally consist of a programmable logic controller (PLC) along with sensors and alarms. The PLC controls the parameters and devices that are automated. The PLC is programmed in ladder logic and provides logic control in a step-wise fashion. PLCs are rugged and inexpensive, and offer many options for interfacing with process equipment. This flexibility allows the PLC to control all the processes within the plant as well as provide data to external systems via analogue signals or serial data. The PLC itself, however, is not suited to memory intensive requirements such as data logging or sophisticated operator interface functions.

An operator interface system (OIS) is used to indicate and record data as gathered by the PLC. The OIS may be a specialised machine with a built-in display or a personal computer (PC) with OIS application software and a PLC interface card. A PC–based system offers greater flexibility and storage capacity but is generally more expensive. The PLC system can provide PID functionality but without an interface system that allows the operator to directly adjust parameters. The PLC, however, can be connected via a digital communication link to an OIS computer. The computer can be programmed with graphical-equipment/process screens to enable the operator to view the system from a primary location and adjust the alarm setpoints. This arrangement is capable of encompassing a whole plant. The operator and the plant, though, are completely dependent upon a functioning PLC and OIS computer. The operator must start and stop each equipment from the control panel. Once the system is running, it functions automatically. Systems are switched 'on' and 'off' by the PLC automatically fashion based upon process feedback signals – unless there is a fault in the system, which then requires manual intervention to clear the fault and restart the equipment.

Most industrial sensors, including temperature, pressure and flow transmitters provide an analogue output in the form of a voltage or current signal. For example, a temperature sensor might generate 1 V at 10°C and 5 V at 50°C. Analogue signals in a current form (4–20 mA) are especially common because they are more impervious to noise and can conveniently provide a power source for low demand instruments. Another advantage is that using 4 mA to represent zero for whatever is being measured allows the system to differentiate a faulty instrument from one that is simply reading zero.

An analogue signal cannot be used directly by the PLC; so it is transformed into a digital signal with the aid of an analogue-to-digital (A/D) converter. The accuracy of an A/D converter depends on the number of bits (the unit for binary information) it sends to the computer. A 10-bit converter will work in the range of 0–1,024 and is, therefore, able to divide an analogue signal voltage into 1,024 steps, This represents an accuracy of 1/1,024, which is approximately 0.1%. Many A/D converters are capable of 12-bit resolution that represents an accuracy of 0.02%. The PLC generates analogue output signals for controlling the process such as valve position or motor speed in a similar manner using digital to analogue conversion. Most PLC systems also resolve the output signal into 12 bits.

More sophisticated instruments of many different types, provide data in the form of a serial string. In such cases, A/D conversion is not necessary, but the PLC must be programmed to accept the format of the serial data. This method can provide much more data than just a single process value and eliminates the inherent inaccuracy of digital-to-analogue D/A and A/D converters. Addressable instruments utilise widely accepted protocols. These devices have the advantages of multi-parameter data transfer of serial instruments but afford the ease of wiring and use of analogue instruments. In many situations a device that switches between only two states may be adequate to control a process. In such cases, it is more cost effective to use a digital sensor that changes from one state to another based on a change in the process, for instance, when the pump discharge pressure exceeds a preset value, a pressure switch may change its state to close a circuit and notify the PLC. A two-state sensor may also provide backup for an analogue instrument in a critical process.

All instruments and control systems must provide a way to implement process control inter-locks, alarms, and algorithms to prevent a dangerous upset from occurring, and to safely shutdown the process equipment if one does occur. Alarms let an operator know when a process deviates far enough from normal conditions, and immediate action is warranted to prevent safety or quality incidents. The tendency to over-alarm the system should be avoided. Individual warning lights or annunciator panels are commonly used to indicate specific system faults, which also may be indicated in the OIS system. The latter gives the operator the ability to see the time of occurrence and associate each alarm with other conditions in the system for troubleshooting purposes. System-wide alarm conditions are indicated with a visual beacon and an audible alarm. Alarms and graphical indicators are usually driven by the PLC.

5.8 Centrifugal pumps

Centrifugal pumps are the prime fluid movers in membrane plants for liquid separations. In general, centrifugal pumps are used in moderate-to-high flow under low-head conditions, and operate within the rated ranges of head and velocity. Pump specifications depend on process requirements, for example, fluid flow rate determines the capacity of the pump whereas pump head depends on fluid density and viscosity. Pump head is commonly used as a measure of energy. The total dynamic head (TDH) is a reflection of the energy a pump imparts to a fluid, and takes into account differences in pressure, fluid elevation, and velocity between source and destination. In addition, TDH accounts for pipe friction losses and the pressure drop through in-line items such as instruments, valves, strainers, and static mixers.

5.8.1 Pump selection criteria

Material and energy balances are used to calculate the required flow rate while the maximum pump pressure or 'shutoff' pressure is the maximum pressure a pump will develop under no-flow conditions, e.g. when the pump outlet line is fully blocked. The amount of work needed to drive the pump increases with increase in fluid density and viscosity. The maximum values of these properties usually occur during start-up, shut-down, and process upsets. Similarly, fluid temperature and the presence of suspended solids are taken into consideration specifying a pump.

Pump brake horsepower (BHP) is related to the flow rate (Q (gpm)), total dynamic head (TDH or H (ft)), specific gravity (S.G.) and efficiency (η) by the relationship, $\text{BHP} = (H \times Q \times \text{S.G.})/(3960 \times \eta)$. The efficiency is set by the pump manufacturer when the final pump selection is made. It is usually based on their shop tests for the same model and size pumps. The pump efficiency can vary between 10 and 80%. Pump electric motor specifications require mechanical and electrical requirements. Motors can vary in size depending upon power, speed (rpm), frame size, area classification, orientation, service factor, and type of enclosure (e.g. totally enclosed fan closure).

Pumps are specified to avoid cavitation. When the suction pressure of the pump fluid drops below its vapour pressure, vapour bubbles form resulting in cavitation. When the fluid gets re-pressurised, these bubbles implode which results in pitting of the impeller and other pump components. Further, since vapour has a lower density than liquid, cavitation results in lowering pump capacity and efficiency.

Pump manufacturers have established guidelines to ensure each pump they supply is not exposed to conditions that result in cavitation. The design standard is called NPSHR or net positive suction head required. The NPSHR takes into account any potential head losses that might occur between the pump's suction nozzle and impeller thereby ensuring that the liquid does not drop below its vapour pressure (bubble point). The NPSH is a measure of the proximity of a liquid to its vapour pressure, and must exceed the pump manufacturer's pump NPSHR. There are two process variables that can be adjusted, in case the available NPSH is less than the NPSHR; raise the static head and lowering friction losses. Conversely, the NPSHR can be reduced by using a larger, slower speed pump, a double-suction impeller, a larger impeller inlet area, an oversised pump, and a secondary impeller placed ahead of the primary impeller.

Every pump manufacturer provides a specific curve for a pump, as shown in Figure 5.10. The curve relates head, flow rate, power, NPSHR, and efficiency for specific impeller diameters for the particular pump. The pump should not be operated below the manufacturer's minimum continuous flow rate rating. The pump impeller size is indicated in horizontal lines. The pump curve shows the pump has a specific NPSHR that varies depending on the head and flow. The goal is to select a pump with rated (or design) point as close as possible to the best efficiency point as determined by the manufacturer. Since pumps are rarely operated at their exact design point, flow rate and head need to be changed to increase pump efficiency or to accommodate changes in process requirements. Hence, the impeller selected should be one size smaller than the maximum size for the given pump casing so that it can be replaced with a larger one without replacing the casing.

The materials of construction of a pump depend on fluid properties such as corrosiveness, presence of dissolved gases, and product application. Pump seals around the pump shaft are very important for preventing process fluid leaks. The seals

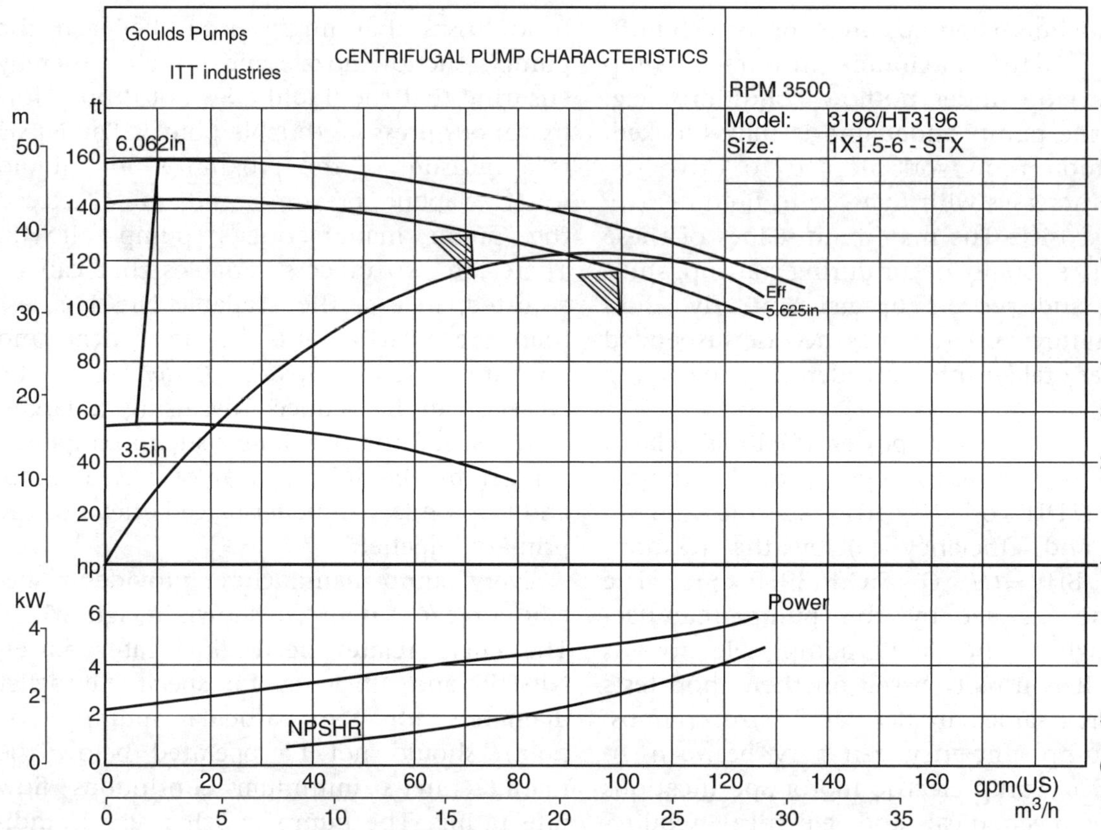

Figure 5.10 Typical centrifugal pump curves

are of two types: packing and mechanical. Mechanical seals are the most widely used. Another class of pumps, the magnetically coupled pumps, are seal-less with no rotating shaft seals. These pumps, however, may not be suitable for high-purity water systems because of the potential for bacteria growth in the gap between the inner magnet and the containment can.[20]

5.8.2 Pump controls

The purpose of pump controls is to maintain a set of process conditions and operate it safely. These include adjusting (a) flow rate by throttling pump discharge, (b) controlling pump speed, and (c) switching the pump on and off. Throttling is achieved with a flow controller and a control valve in the discharge line. Controlled flows are most often accomplished with a self-regulating (modulating) pressure-control valve on the pump discharge (as in RO high-pressure pump discharge line), or a flow-control valve and controller with a flow configured at the set point (as in RO reject flow control).

The pump speed can be controlled by using a variable-frequency driver (VFD) especially, when the capacity of the pump varies greatly. VFDs include electric, electromechanical, mechanical, and hydraulic power recovery turbines.

Pump on–off control is achieved with level switches or level transmitters that have built-in dead bands on the suction vessel (storage tank) to start and stop the pump. Frequent starts and stops can damage the pump and the motor; hence, the storage tank is sised to accommodate changes in level by providing sufficient storage (residence time). Low suction pressure switches are used as safety inter-locks for shutting off the pump when the suction pressure is below the set-point. Conversely, high-pressure switches are installed on the pump discharge so as not to exceed the maximum allowable output pressure resulting in dead-heading the pump.

5.9 Control valves

A control valve is a variable orifice used to regulate the flow of a process fluid. Control valves are a key component in any process plant. Proper selection of these valves can have a significant impact on how well the process is controlled and the overall cost of the plant. Characteristics of commonly used control valves in membrane plants are given in Table 5.2. In order to select a control valve, the following system operating conditions and fluid properties must be known: fluid state (vapour, liquid, or two-phase), vapour pressure, flow rate, inlet and outlet pressure, inlet temperature, fluid density, molecular weight and viscosity, specific heat ratio, critical temperature, and critical pressure. In about 80% of the applications, the valves are selected to be fail-safe closed under spring actuator action when there is a loss of power or signal. Hence, information about the valve's fail-safe position and maximum shut-off pressure is also required.

A valve should satisfy the following conditions to be effective: operate over a wide range of flows (wide range-ability); accurately respond to any signal across its operating range; exhibit little dead time (hysteresis); react to incremental adjustments from the controller (resolution); and respond with the required speed (stroking speed).[21] The quality of a control valve is quantified in terms of its (a) gain, (b) time constant, and (c) dead-time lag. Gain is the ratio of the percentage change in a process variable to the percentage change of valve travel, and depends on the valve characteristics and process conditions. The gain should never be less than 0.5.

The ability of a valve to control flow depends upon the quality of its actuator. A positioner is added when tighter control is required for the modulating valve, as shown in Figure 5.11. Positioners improve performance by amplifying the controller's signal, thereby achieving a more precise response. Control valves are either motorised (electrically operated) or pneumatic (air operated). A control valve should react instantaneously to any change in the signal. A fast response, however, may not be suitable for all applications. The signal is generated by a control loop, which consists of a sensing element, a controller, and the final control element, as shown in Figure 5.12. The sensing element transmits a signal to a local controller or a remote controller such as a PLC. The controller compares the signal with the set-point, and makes the required corrections by sending a signal to the control valve. The correction is measured and verified by the sensing element. Also shown in the figure is the I/P transducer converting the electronic signal to a pneumatic signal.

There are two main features of control valves, valve capacity and valve characteristics. The most important valve parameter for calculating the size of a control valve is the flow coefficient, C_v that gives the flow rate of water through a valve for a unit pressure drop at standard conditions. It varies with size and style of valve. For

Table 5.2 Characteristics of process valves

Ball	Globe	Butterfly	Diaphragm
• On/off (Rotary)	• Throttling	• Throttling	• Throttling
• Virtually no pressure loss across the valve face	• Good sealing characteristics	• Lightweight, compact design	• Almost no leakage; process fluid is isolated from valve stem
• Lightweight, compact design	• Can be used in frequent open/close service	• Minimal pressure loss across valve face	• Self-cleaning
• High capacity	• Quick change of trim without removing valve from line	• Low cost	
• Good rangeability	• High capacity	• High throughput capacity	
• Tight shut-off	• Good rangeability	• Smaller shaft and actuator	
	• Low-noise trim available		
	• Smooth control		
• Sealability poor withmetal seats used at high temperatures	• High-pressure losses due to contorted path through the valve	• Poor sealing characteristics	• Limited operating pressure
• Limited temperature range with resilient seats	• Low-noise trim reduces capacity	• Good control limited to 60° opening	• Limited temperature
• Choke flow problems			
• Cavitation problems	• For flow regulation		
	• When tight shut-off is required.	• Tight shut-off requires special lining: plus over-sised shaft and actuators	• High wear and tear
			• Poor control over 50% opening
• Requires removal for maintenance		• Lining imposes temperature limitations	
		• Low-pressure applications	• Water treatment service
			• Chemical and abrasive service
• Not for service with highly corrosive fluids			
• Most suitable for slurries			

Source: Bishop et al., Reproduced with permission from Chemical Engineering Progress, Nov. 2002. Copyright 2002 AIChE

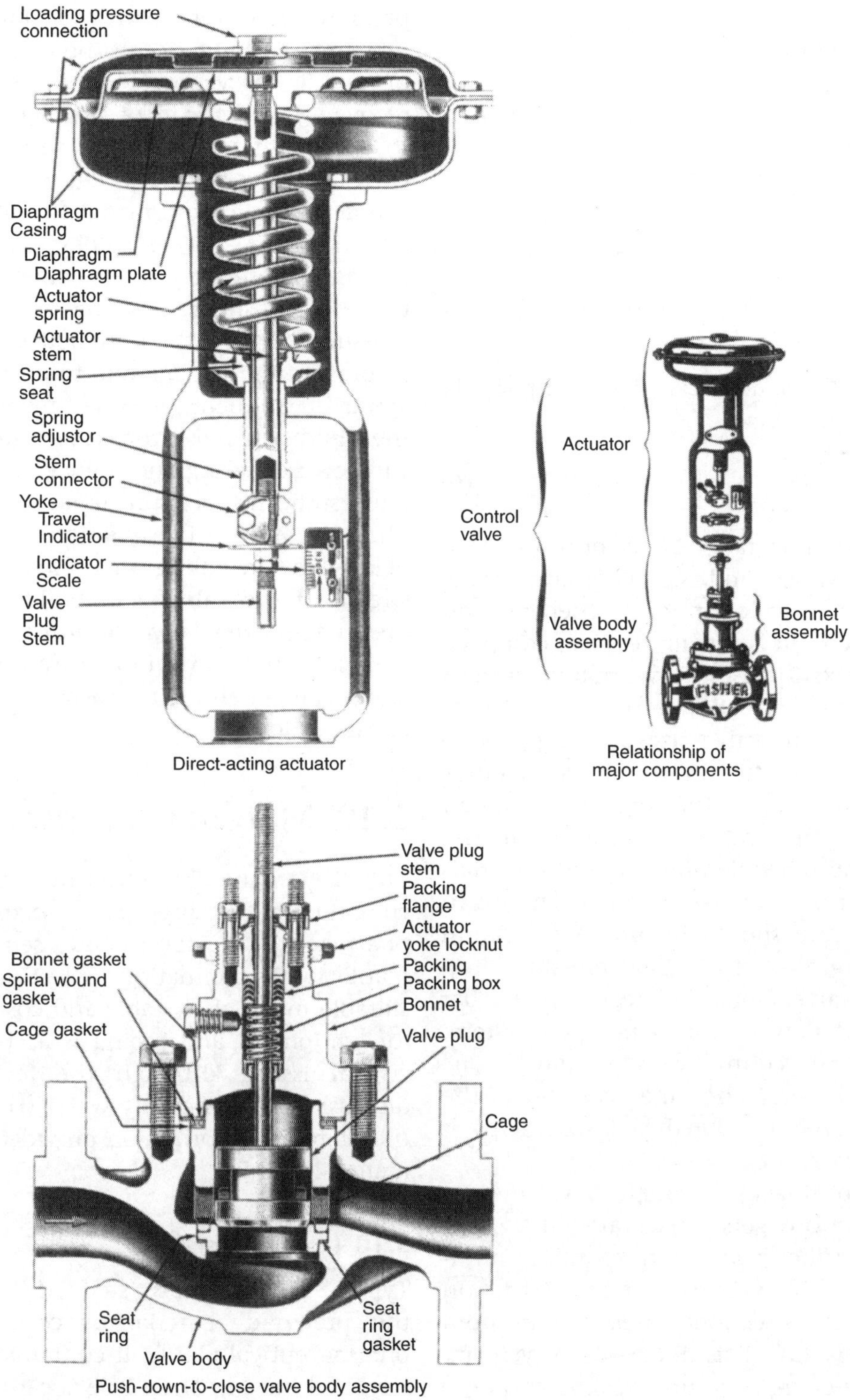

Figure 5.11 A typical Fisher control valve assembly with actuator

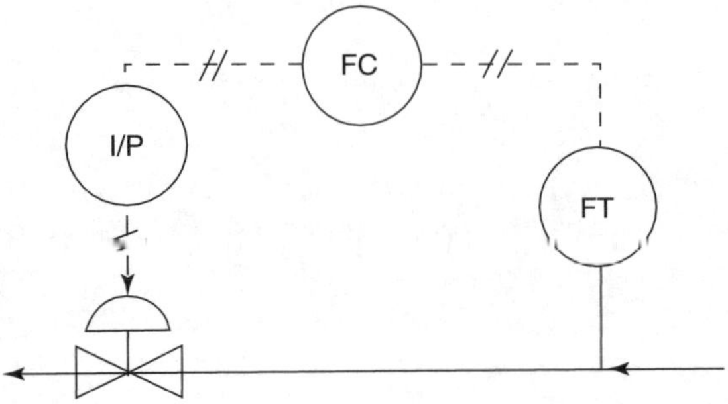

Figure 5.12 Control valve controller action

liquids, the C_v value is calculated empirically using the equation:

$$C_v = Q/(s.g./\Delta P)^{1/2} \quad (6)$$

where, Q is flow rate, ΔP is pressure drop across the valve, and s.g. is fluid specific gravity. The calculation of C_v depends on whether the fluid is incompressible, compressible, or mixed-phase. Valve manufacturer's provide actual valve flow capacities in terms of C_v for various valve sizes and types. The valve should operate in the 10–80% range of its stroke range; otherwise it is difficult to control the valve. System geometry, pipe size and schedule number, and materials of construction must be specified. The maximum flow rate should include a minimum design margin of 10%. The pressure drop through a valve should be greater than 1 bar at the rated flow rate in a pumped system. The size of the control valve should be one size smaller than the line size since the pressure drop due to reduced size helps it attain the correct C_v.

The fluid flowing through a valve or a flow restriction gets accelerated to some maximum velocity with corresponding minimum pressure downstream of the restriction where the cross-sectional area is minimum (vena contracta). This increase in velocity or kinetic energy occurs at the expense of the pressure or potential energy. If the pressure is lower than the liquid vapour pressure, flashing occurs producing bubbles or cavities of vapour. The pressure rises or recovers downstream of the vena contracta. If the pressure exceeds the vapour pressure, the bubbles or cavities collapse causing noise similar to gravel flowing through a line, vibration, loss of efficiency, and physical damage. These gas pockets collapse violently; the resulting shock waves can generate impact pressure greater than 10,000 bar. The key to preventing damage due to cavitation is to avoid a sudden pressure recovery when the fluid emerges from the throat of the valve. Accelerating the fluid gradually maintains sufficient pressure to avoid cavitation. The initiation of cavitation can also be delayed by installing an orifice plate downstream of the valve. This builds up a backpressure on the valve resulting in ΔP that is below the cavitation threshold. Orifice plates are useful in processes with fairly constant flow.

5.10 Materials properties

The key issue in selecting materials of construction for high-purity plants such as pharmaceutical, food and semiconductor plants is contamination. Selecting the most suitable materials – stable and cost effective – for equipment and piping is a complex and critical issue. Although the focus of this discussion is high-purity water (HPW) plants, the materials information provided is a useful materials guide.

5.10.1 Stainless steel

Type 316L stainless steel (SS) has been the preferred material of construction in pharmaceutical plants. It contains about 70% iron and has less than 0.03% carbon (L is for low carbon). In HPW systems, rust deposits formed on SS surfaces are thin called 'rouge' due to their reddish brown hue. Rouging is the result of dissolved ferrous ion (Fe^{2+}) oxidising to the insoluble ferric iron (Fe^{3+}) in the presence of oxygen and precipitating as either Fe_2O_3 or $Fe(OH)_3$. One gram of iron will cover an area of 390 m^2, three atoms deep.[22] Oxygen can enter a system through pump seals or by diffusion–sorption through plastics.

There are three types of rouge: (a) Type I is due to the dissolution of steel such as in the case of pump impellers; (b) Type II results from active corrosion, e.g. on non-stainless steel components; (c) Type III is due to high temperature. Passivation is used to remove rogue, but it does not prevent Type I and III rogues from reoccurring if oxygen permeates the system. Type II rogues is usually the result of improper welding that leaves a heat-affected zone near the weld. Type II attacks the heat-affected zone only. Rogue can be avoided by using thermoplastics as will be discussed in the Section 5.10.2. However, plastics are permeable to gases such as O_2, N_2, He, H_2, CO_2 as well as many solvents. This permeability results in 'blistering' behind plastic-lined vessels and pipes over time.

The 300 series SS are also sensitive to chlorides at a pH of 6.5–8 and at temperatures below 60°C. Type 316L SS tolerates chlorides up to 1,000 ppm, but concentrations can reach 26,000 ppm in dry zones (for $MgCl_2$) resulting in failure. Stainless steel is impermeable to oxygen unlike plastics, it has 10 times the thermal conductivity of plastics and does not creep at temperatures below 425°C (at room temperature for plastics). SS pipes can withstand high pressures (up to 110 bar for a 25 mm O.D. 300-series SS vs. 5 bar for plastic pipe). Its yield and tensile stresses are an order of magnitude greater than those of thermoplastics.

5.10.2 Plastics

Unlike stainless steel, the properties of plastics vary depending on the type of polymer. Polymers are substances formed by linking one or more species of atoms or groupings of atoms by covalent bonds. The method and type of linking converts simple molecules into macromolecular structures that determine the chemical and physical properties of the synthesised polymer. Polymer properties depend on many factors such as polymer chain length and conformation, cross-linking of chains, intense polar interactions, and the type and size of attached side groups, to name a few controlling factors. For example, branched polyethylene (PE) has a lower melting point than linear PE even though they are chemically identical. For additional information, refer to 'Glass transition temperature' and 'Thermoplastics' in the Glossary section.

Commonly used thermoplastics such as PE, polyvinyl chloride (PVC), and polypropylene (PP) are less costly than fluoropolymers, e.g. polyvinylidene fluoride (PVDF) but do not perform as well in high-temperature conditions. They also include plasticisers, heat stabilisers, and fire retardants that can leach out. PVDF in its virgin form is highly pure and does not contain additives.

Fluropolymers such as PVDF and perfluoroalkoxy (PFA) have excellent chemical resistance to deionised water, high thermal stability and are resistant to degradation by sunlight. They have a low coefficient of friction unlike metal surfaces, thereby preventing microorgansims (fungi and bacteria) from growing. This is critical since unlike chemical plants, biocides cannot be added to HPW to prevent microbial growth. PFA polymer can be used at temperatures up to 260°C. However, it is highly expensive, and PVDF is the preferred material. PVDF is

316L SS's main competitor in HPW systems, and is widely used in high-purity fluid loops in semiconductor plants. PVDF is among the hardest fluoropolymers and has the highest tensile strength of all plastics. It also has better abrasion strength than SS.

Since PVC and PP have low melting temperatures (<90°C), components using these materials can only be sterilised by chemicals such as hydrogen peroxide and chlorine, both of which require rinsing after sterilisation. The melt temperature of PVDF, on the other hand, is 178°C that makes it amenable to steam sterilisation. PVDF can also be sterilised by ozone. Since ozone has a short half-life (minutes), PVDF systems do not need rinsing after sterilisation.

5.11 Process validation

Membrane plants used for producing high-purity water (HPW) in the manufacture of pharmaceuticals require validation like other equipment and processes. The FDA has defined process validation as 'establishing documented evidence which provides a high degree of assurance that a specific process will consistently produce a product meeting its predetermined specification and quality attributes.' In other words, validation involves detailed testing and documentation of system equipment and components, and system functional operation to ensure that design and regulatory specifications are met, and that system performance will be consistently acceptable during service.

Process validation is part of current good manufacturing practices (cGMP), and is required in the US and the EU for a manufacturing licence. It involves the identification, monitoring, and control of sources of variation that can contribute to changes in the product. Process development normally involves identifying critical variables, defining set-points for each unit operation, and establishing operating ranges. Cleaning validation shows the ability of cleaning procedures to allow reuse of processing components and equipment without deterioration of product quality.

In the case of pharmaceutical plants, a HPW system must comply with the rules of the US Food and Drug Administration (FDA) as stated in the 21 CFR (Code of Federal Regulations), Parts 210, 211, and 820. According to the FDA regulatory code, validation implies 'the conformation by documented examination and provision of objective evidence that the particular requirements for a specific intended use can be consistently fulfiled.' These regulations specify minimum current good manufacturing practice. Compliance, which must be validated, ensures that drugs meet the requirements of safety, identity, strength, quality, and purity.

The system validation documentation includes but is not limited to Quality Assurance Plan (QAP), Operation and Maintenance Manual (O&M), and Factory Acceptance Test (FAT) reports. A properly developed QAP and its supporting documentation help to reduce cost, boost product quality, ensure regulatory compliance, and increase profitability. The FAT validation is the first verification step of the customer's validation process, which may include design qualification (DQ), installation qualification (IQ), operational qualification (OQ), and performance qualification (PQ).

5.12 RO/NF feed water analysis

Feed water constituents must be analysed prior to designing a membrane system. The ASTM Designation D4195-88 'Standard

Guide for Water Analysis for Reverse-Osmosis Applications' components and properties that require analysis is given below. Water treatment techniques are discussed in Chapter 2.

Carbon dioxide (CO_2) is not rejected by RO/NF membranes. If it is the major dissolved constituent in the permeate, it will increase the ionic load on downstream ion-exchange polisher anion resins.

Free chlorine (Cl_2) is an oxidant. Polyamide RO and NF membranes are sensitive to chlorine, as discussed in Chapter 5. Hence, water must be dechlorinated by passing it through an activated carbon filter or by the addition of a reducing chemical such as sodium sulphite, sodium bisulphite, or sodium metabisulphite to feed water. However, it is necessary for CA membranes to protect them from bacterial attack.

Hydrogen sulphide (H_2S) is not present in chlorinated water supplies because it is quickly converted to sulphate. It is, however, associated with some sub-surface water supplies that have been depleted of oxygen. H_2S forms sulphur particles in the water and sulphide ion may form insoluble metal sulphides. Both contribute to membrane fouling.

Langelier Saturation Index. See Glossary.

Oxygen (O_2) is indicative of the character and source of the water supply. Water with depleted oxygen/air is likely to contain soluble iron, manganese, and hydrogen sulphide, often in appreciable quantities. When oxygen is added to such a water supply, precipitates (e.g. ferric compounds) are formed.

Silt Density Index. See Glossary.

Silica. The solubility of silica (SiO_2) is a function of pH and temperature (Figure 5.13). Exceeding the saturation level in cold water (<10°C) is not a serious problem because the precipitation is very slow. However, exceeding 120 ppm in any temperature presents a potential problem. Since water supplies often contain >20 ppm silica, silica very often limits recovery. Precipitated silica is virtually impossible to remove from the elements, and must be avoided. Metal hydroxides (e.g. iron, aluminium, magnesium) absorb or complex silica and catalyse the precipitation.

Turbidity. See Glossary.

Bicarbonate (HCO_3^-) results from the chemical decomposition of dissolved CO_2 with the hydroxide ion in water. A portion of the bicarbonate present in the feed water can be converted to carbonate as a result of pH changes owing to concentration of salts on the concentrate side of the membrane. Such cases require the addition of acid or a crystal growth inhibitor to avoid calcium carbonate scaling.

Carbonate (CO_3^{2-}) forms a sparingly soluble salt with calcium, which readily fouls RO membranes. Calcium carbonate can easily be removed by lowering the pH below 5.0 or raising the pH above 8.0.

Chloride (Cl^-) does not generate insoluble salts. Because it passes through RO membranes more easily than most anions, it is one of the predominant ions in the permeate.

Fluoride (F^-) forms insoluble salts with divalent metal ions such as calcium, magnesium, barium, and strontium.

Nitrate (NO_3^-) is similar to chloride in that it is not aggressive to the membrane nor does it tend to form insoluble salts.

Phosphate (PO_4^{3-}) forms a low solubility salt with calcium. Water containing phosphate generally require the addition of an antiscalant.

Sulphate (SO_4^{2-}) forms a sparingly soluble salt with calcium, strontium, or barium. Sulphate does not usually limit the cycles of concentration, unless one or more of these three cations are present in high concentrations.

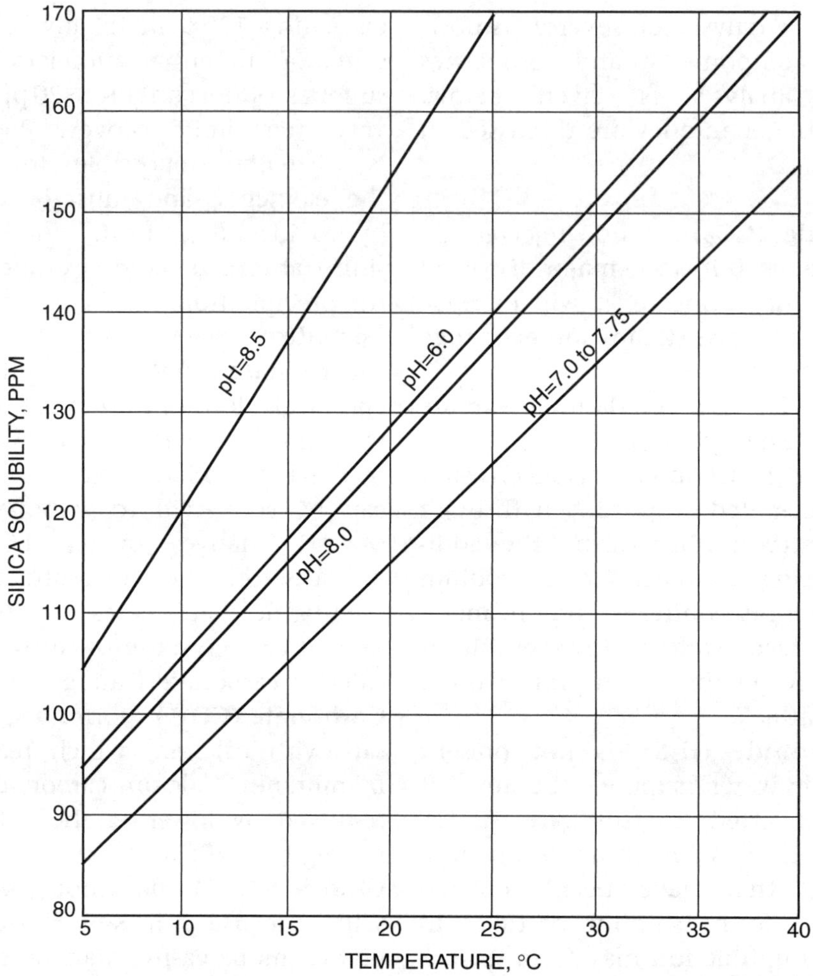

Figure 5.13 Silica solubility plot. Source: Amjad, Reverse Osmosis, 1993

Aluminium (Al^{+++}) is generally found as the hydroxide, which causes particulate fouling of the RO membrane. Aluminium based compounds such as sodium aluminate and aluminium sulphate are common coagulants used in potable water treatment. Under proper circumstances, the aluminium precipitate formed in the water treatment plant is completely filtered. Occasional fluctuations in pH result in excessive amounts of dissolved Al^{+++} which pass into the distribution system.

Barium (Ba^{++}) forms a low solubility compound with sulphate ion. Strontium and barium tend to occur together (see Strontium).

Calcium (Ca^{++}) forms insoluble or slightly soluble salts with common ions such as carbonate and sulphate. Both can be adequately controlled with crystal growth inhibitors or by softening RO feed water. The precipitate can readily be removed by cleaning the RO membrane. Calcium concentration, in conjunction with alkalinity and sulphate concentration, often establishes the upper limit to the water recovery of RO units.

Iron is generally present in the dissolved ferrous (Fe^{++}) form, but can oxidise to the ferric (Fe^{+++}) state and precipitate as the hydroxide. Municipal potable water supplies contain <0.3 ppm of iron, but sub-surface supplies may contain levels of iron in excess of 10–20 ppm. Prior to use in an RO system, a water supply containing iron should be treated to remove the iron, or steps should be taken to avoid contact of the water with air or oxidising substances such as chlorine.

Magnesium (Mg^{++}) forms sparingly soluble salts such as magnesium silicate and, under high pH conditions, magnesium hydroxide. Both are uncommon in RO systems.

Manganese (Mn^{++}) like iron is usually at levels of less than 0.3 ppm in public water supplies. Private well water supplies often contain higher manganese levels, often in a dissolved form, which precipitates on exposure to air. Well water supply containing manganese should be pretreated, or steps should be taken to eliminate contact with air or oxidants to assure that the manganese remains soluble.

Potassium (K^+) is present in small quantities. It is soluble like sodium, and does not cause fouling or scaling.

Sodium (Na^+) forms relatively soluble salts with most anions, including bicarbonate, sulphate, and chloride, and thus seldom presents a fouling issue. However, since it passes most readily through the RO membrane, it is present at the highest concentrations in the permeate. It generally controls overall salt rejection.

Strontium (Sr^{++}) is found in less than 20% of water supplies. It forms a salt with sulphate ion that is soluble up to one ppm in concentration. It is kept in suspension with a crystal growth inhibitor. When the feed water is secondary effluent from a municipal plant, analysis of total oil, grease and surfactants are also required.

5.13 Conversion factors

Density:

1 lb(mass)/ft^3 = 16.02 g/l

Flow:

1 m^3/h = 4.4 US gpm

1 million gallons/day = 3.8×10^3 m^3/day

Flux:

1 gallon/ft^2/day = 1.7 l/m^2/h

Force and pressure:

1 dyne = 1 g cm/s^2

1 kg · m/s^2 = 1 Newton (N)

1 bar = 10^5 Pascal (Pa) = 10^5 N/m^2 = 14.5 psi (lb/in^2)

1 psi (lb$_f$/in^2) = 6.8947 kPa = 6.8947×10^3 N/m^2

1 psi = 2.036″ Hg = 51.715 mm Hg

1 atmosphere (atm) = 14.696 psi = 1.01 bar = 101.325 kPa = 760 mm Hg = 33.9 ft water

1 kPa = 0.145 psi

1 kg$_f$ /cm^2 = 14.223 psi

Length:

1 Angstrom (Å) = 10^{-10} m = 10^{-4} micrometre (micron) = 0.1 nm

1 inch = 2.54 centimetres

1 metre = 39.37 inches

1 μm = 10^{-6} m = 10^{-4} cm

1 mil = 0.001 inch = 25.4 μm

1 nanometre (nm) = 10 Å = 10^{-9} m

Mass:

1 grain (gr) = 17.1 ppm per gallon

1 pound (lb) = 16 ounces (oz) = 453.59 grammes (g)

1 kilogramme (kg) = 2.205 lb

1 ton, short (US) = 2,000 lb

1 ton, long = 2,240 lb
1 ton, metric = 1,000 kg

Power, work, and energy:
1 horsepower (hp) = 0.746 kilowatts (kW)
1 watt (W) = 14.3 cal/min
1 joule/s (J/s) = 1 Watt
1 Btu = 1,055 J
1 kilocalorie (kcal) = 4.184 kJ
1 hp h = 0.7457 kWh = 2,544 Btu

Temperature:
T (°F) = [T (°C) × 1.8] + 32
T (°C) = [T (°F) – 32]/1.8

Velocity:
1 ft/s = 0.305 m/s
1 m/s = 3.281 ft/s

Viscosity:
1 centipoise (cP) = 0.01 poise = 10^{-3}
Pa.s = 10^{-3} N.s/m^2
1 lb/ft/s = 1488 cP

Volume:
1 gallon (US) = 4 quarts (qt) = 3.785 L
1 gallon (Imperial) = 1.2 US gallons
1 L = 1,000 cm^3
1 m^3 = 1,000 L = 264 US gallons
1 in^3 = 16.387 cm^3
1 ft^3 = 28.317 L

5.14 Tables

Table 5.3 Viscosity and density data

Fluid	Viscosity[1]	Density[2]
Air	0.00002	1.3
Water	0.001	1,000
Glycerine	1.8	1,000
Golden syrup	120	1,000

[1] μ (kg m^{-1} s^{-1})
[2] ρ (kg m^{-3})

Table 5.4 Typical cleaning solutions for water treatment heat exchangers

Foulant type	Cleaning solution
Encrustation – Scaling	
Calcium carbonate; calcium sulphate, silicates.	Nitric acid, sulphamic acid, citric acid, phosphoric acid, complexing agents (EDTA, NTA), or sodium polyphosphates (< 4% by wt.; < 60°C).
Sediment	
Corrosion products; metal oxides; silt; alumina; diatomic organisms and their excrement of various colours.	Nitric acid, sulphamic acid, citric acid, phosphoric acid, complexing agents (EDTA, NTA), or sodium polyphosphates (< 4% by wt.; < 60°C).
Biological growth	
Bacteria; nematodes; protozoa.	Sodium hydroxide, or sodium carbonate plus surfactants (< 4% by wt.; < 80°C).
Hydrocarbons	
Oil residue; asphalt; fats.	Paraffin- or naphtha-based solvent (e.g. kerosene)*.

* **Note:** Natural, butyl and EPDM gaskets swell in these solvents.

Table 5.5 Common chemicals used to treat water

Chemical	Common name	Typical specifications	Equivalent weight	Bulk density (lb/cu ft or lb/gal)	Approx. pH (1% solution)	Solubility
Aluminium sulphate $Al_2(SO_4)_3 \cdot 14H_2O$	Alum	Lump – 17% Al_2O_3	100[1]	60	3.4	4.2 lb/gallon at 60°F
		Liquid – 8.5% Al_2O_3		11		
Bentonitic clay	Bentonite	–	–	60	–	Insoluble
Calcium carbonate $CaCO_3$	Limestone	96% $CaCO_3$	50	80	9	Insoluble
Calcium hydroxide $Ca(OH)_2$	Hydrated lime Slaked lime	96% $Ca(OH)_2$	40[1]	40	12	Insoluble
Calcium hypochlorite $Ca(OCl)_2 \cdot 4H_2O$	HTH	70% Cl_2	103	55	6–8	3% at 60°F
Calcium oxide CaO	Burned lime Quicklime	96% CaO	30[1]	60	12	Slake at 10–20%
Calcium sulphate $CaSO_4 \cdot 2H_2O$	Gypsum	98% Gypsum	86[1]	55	5–6	Insoluble
Chlorine(Cl_2)	Chlorine	Gas – 99.8% Cl_2	35.5	gas	–	0.07 lb/gallon at 60°F
Copper sulphate $CuSO_4 \cdot 5H_2O$	Blue vitriol	98% Pure	121[1]	75	5–6	2 lb/gallon at 60°F
Dolomitic lime $Ca(OH)_2 \cdot MgO$	Dolomitic lime	36–40% MgO	67[2]	40	12.4	Insoluble
Ferric chloride $FeCl_3 \cdot 6H_2O$	Iron chloride	Lump – 20% Fe	91[1]	70	3–4	45% at 60°F
		Liquid – 20% Fe		13		
Ferric sulphate $Fe_2(SO_4)_3 \cdot 3H_2O$	Iron sulphate	18.5% Fe	51.5[1]	70	3–4	30% at 60°F
Ferrous sulphate $FeSO_4 \cdot 7H_2O$	Copperas	20% Fe	139[1]	70	3–4	1 lb/ gallon at 60°F
Hydrochloric acid HCl	Muriatic acid	30% HCl 20°Baume	120[1]	9.6	1–2	35% at 60°F
Sodium aluminate $NaAlO_2$	Aluminate	Flake – 46% Al_2O_3	100[1]	50	11–12	40% at 60°F

(Continued)

Table 5.5 Continued

Chemical	Common name	Typical specifications	Equivalent weight	Bulk density (lb/cu ft or lb/gal)	Approx. pH (1% solution)	Solubility
		Liquid – 26% Al_2O_3	13			
Sodium chloride NaCl	Rock salt, salt	98% pure	58.5	60	6–8	2.6 lb/gallon at 60°F
Sodium carbonate Na_2CO_3	Soda ash	98% pure				
		58% Na_2O	53	60	11	1.5 lb/gallon at 60°F
Sodium hydroxide NaOH	Caustic, Lye	Flake – 99% NaOH		65	12.8	70% at 60°F
		Liquid – 50–70%	40	12		
Sodium phosphate Na_2HPO_4	Disodium phosphate	49% P_2O_5	47.3	55	9	20% at 60°F
Sodium metaphosphate $NaPO_3$	Hexameta phosphate	66% P_2O_5	34	47	5–6	1 lb/gallon at 60°F
Sulphuric acid H_2SO_4	Oil of vitriol	94–96%				
		66° Baume	50[1]	15	1–2	Infinite

[1] Effective equivalent weight of commercial product
[2] Effective equivalent weight based on $Ca(OH)_2$ content
Source: Nalco

Table 5.6 Calcium carbonate equivalent of common substances

	Formula	Molecular weight	Equivalent weight	Substance to $CaCO_3$ equivalent (multiply by)	$CaCO_3$ equivalent to substance (multiply by)
Aluminium	Al^{3+}	27.0	9.0	5.56	0.18
Ammonium	NH_4^+	18.0	18.0	2.78	0.36
Barium	Ba^{2+}	137.4	68.7	0.73	1.37
Calcium	Ca^{2+}	40.1	20.0	2.50	0.40
Copper	Cu^{2+}	63.6	31.8	1.57	0.64
Hydrogen	H^+	1.0	1.0	50.0	0.02
Ferric iron	Fe^{3+}	55.8	18.6	2.69	0.37
Ferrous iron	Fe^{2+}	55.8	27.9	1.79	0.56
Magnesium	Mg^{2+}	24.3	12.2	4.10	0.24
Manganese	Mn^{2+}	54.9	27.5	1.82	0.55
Potassium	K^+	39.1	39.1	1.28	0.78
Sodium	Na^+	23.0	23.0	2.18	0.46
Strontium	Sr^{2+}	87.6	43.8	1.14	0.88
Zinc	Zn^{2+}	65.4	32.7	1.53	0.65
Bicarbonate	HCO_3^-	61.0	61.0	0.82	1.22
Carbonate	CO_3^{2-}	60.0	30.0	1.67	0.60
Chloride	Cl^-	35.5	35.5	1.41	0.71
Chromate	CrO_4^{2-}	116.0	58.0	0.86	1.16
Fluroide	F^-	19.0	19.0	2.63	0.38
Iodide	I^-	126.9	126.9	0.39	2.54
Hydroxyl	OH^-	17.0	17.0	2.94	0.34
Nitrate	NO_3^-	62.0	62.0	0.81	1.24
Phosphate (tribasic)	PO_4^{3-}	95.0	31.7	1.58	0.63
Phosphate (dibasic)	HPO_4^{2-}	96.0	48.0	1.04	0.96
Phosphate (monobasic)	$H_2PO_4^-$	97.0	97.0	0.52	1.94
Sulphate	SO_4^{2-}	96.1	48.0	1.04	0.96
Bisulphate	HSO_4^-	97.1	97.1	0.52	1.94
Sulphite	SO_3^{2-}	80.1	40.0	1.25	0.80
Bisulphate	HSO_3^-	81.1	81.1	0.62	1.62
Sulphide	S^{2-}	32.1	16.0	3.13	0.32

Source: USFilter

Table 5.7 Solubility product of sparingly soluble salts

Substance	Formula	Temperature (°C)	Solublity product (K_{sp})
Aluminium hydroxide	$Al(OH)_3$	20	1.9×10^{-33}
Barium carbonate	$BaCO_3$	16	7×10^{-9}
Barium sulphate	$BaSO_4$	25	1.08×10^{-10}
Calcium carbonate	$CaCO_3$	25	8.7×10^{-9}
Calcium fluoride	CaF_2	26	3.95×10^{-11}
Calcium sulphate	$CaSO_4$	10	6.1×10^{-5}
Cupric sulphide	CuS	18	3.5×10^{-45}
Ferric hydroxide	$Fe(OH)_3$	18	1.1×10^{-36}
Ferrous hydroxide	$Fe(OH)_2$	18	1.64×10^{-14}
Magnesium ammonium phosphate	$MgNH_4PO_4$	25	2.5×10^{-13}
Magnesium carbonate	$MgCO_3$	12	2.6×10^{-5}
Magnesium hydroxide	$Mg(OH)_2$	18	1.2×10^{-11}
Manganese hydroxide	$Mn(OH)_2$	18	4×10^{-14}
Strontium carbonate	$SrCO_3$	25	1.6×10^{-9}
Strontium sulphate	$SrCO_4$	17.4	2.81×10^{-7}
Zinc hydroxide	$Zn(OH)_2$	20	1.8×10^{-14}

Source: USFilter

References

1 R. Singh, Characteristics of a chlorine resistant reverse osmosis membrane, Desalination, 95 (1994), 27–37.
2 H.K. Lonsdale, U. Merten, R.L. Riley, Transport properties of cellulose acetate osmotic membranes, J. App. Poly. Sci. 9 (1965), 1341–1362.
3 R.E. Kesting, Synthetic Polymeric Membranes, John Wiley & Sons: New York, NY (1985).
4 V.L. Punzi, K.B. Hunt, G.P. Muldowney, Ind. & Eng. Chem. Res. 29 (1990), 259.
5 K. Ghosal, R.T. Chern, B.D. Freeman, AIChE Reprints (1992).
6 S. Chowdhury, P. Kumar, P.K. Bhattacharya, A. Kumar, Separation characteristics of modified polysulphone ultrafiltration membranes using NO_x, Sep. & Pur, Tech. 24 (2001), 271–282.
7 R. Singh, Polyamide polymer solution behaviour under chlorination conditions, J. Mem. Sci. 88 (1994), 285–287.
8 J. Glater, M.R. Zachariah, S.B. McCray, J.W. McCutchen, Reverse osmosis membrane sensitivity to ozone and halogen disinfectants, Desalination 48 (1983), 1.
9 D.B. Guy, R. Singh, On-site regeneration of reverse osmosis membranes, WSIA J. 9 (1) (1982), 35–43.
10 T. Matsuura, Future trends in reverse osmosis membrane research and technology, in Reverse Osmosis (ed. Z. Amjad), Van Nostrand Reinhold: New York, NY (1993), 37–75.
11 J.P. Kohler, An investigation of laminar flow through a porous-walled channel, Ph.D. Dissertation, University of Massachusetts, Amherst, 1973.
12 R. Singh, R.L. Laurence, Influence of slip velocity at a membrane surface on ultrafiltration performance – I. Channel flow system, Int. J. Heat and Mass Transfer 22 (1979), 721–729.
13 R. Singh, R.L. Laurence, Influence of slip velocity at a membrane surface on ultrafiltration performance – II. Tube flow

system. Int. J. Heat and Mass Transfer 22 (1979), 731–737.
14 S. Chellam, M.R. Wiesner, C. Dawson, Laminar flow in porous ducts. Rev. Chem. Eng. 11 (1) (1995), 53–98.
15 R. Singh, Removal of volatile organic compounds by ultrafiltration, Desalination 104 (1996), 203–214.
16 R. Singh, Investigation of ultrafiltration rejection of surfactant micelles by dynamic light scattering, Sep. Sci. and Tech. 31 (9) (1996), 1351–1356.
17 S. Siverns, C. Wilson, J. d'Ailly, Process water treatment: navigating the options, Chem. Eng. 107 (6) (2000), 64–71.
18 S.B. Applebaum, Demineralisation by Ion Exchange, Academic Press: New York, NY (1968).
19 N.P. Chopey, T.G. Hicks, Section 14, in Handbook of Chemical Engineering Calculations, McGraw-Hill: New York, NY (1984).
20 K. Fernandez, B. Pyzdrowski, D.W. Schiller, M.B. Smith, Understanding the basics of centrifugal pumps, CEP 98 (6) (2002), 52–56.
21 T. Bishop, M. Chapeaux, L. Jaffer, K. Nair, S. Patel, Ease control valve selectionl, CEP 98 (11) (2002), 52–56.
22 R. Greene, Choose your material, CEP 98 (6) (2002), 15–17.

Index

N

P